Holomorphic Function Theory in Several Variables

Universitext

For other titles in this series, go to
www.springer.com/series/223

Christine Laurent-Thiébaut

Holomorphic Function Theory in Several Variables

An Introduction

Christine Laurent-Thiébaut
Université Joseph Fourier
Institut Fourier
B.P. 74
38402 Saint-Martin d'Hères Cedex
France
Christine.Laurent@ujf-grenoble.fr

EDP Sciences ISBN 978-2-7598-0364-4
Translation from the French language edition:
Théorie des fonctions holomorphes de plusieurs variables by Christine Laurent-Thiébaut

http://www.edpsciences.org/
http://www.cnrseditions.fr/

ISBN 978-0-85729-029-8 e-ISBN 978-0-85729-030-4
DOI 10.1007/978-0-85729-030-4
Springer London Dordrecht Heidelberg New York

British Library Cataloguing in Publication Data
A catalogue record for this book is available from the British Library

Library of Congress Control Number: 2010934643

Mathematics Subject Classification (2000): 32A26, 32D05, 32D20

Cover design: SPI Publisher Services

Printed on acid-free paper

Springer is part of Springer Science+Business Media (www.springer.com)

Contents

Appendix

Preface

This book started life as a basic post-graduate course taught at the Institut Fourier during the academic year 1994–1995. It is intended as an introduction to the theory of holomorphic functions in several variables on both $\mathbb{C}^n$ and complex analytic manifolds, aimed mainly at advanced Masters students or beginning thesis students. We assume that the theory of holomorphic functions in one complex variable is known, but the basics of differential geometry and current theory needed in multivariable complex analysis are summarised in Appendix A and Chapter II.

We use integral representations together with Grauert's bumping method. The advantage of this point of view is that it offers a natural extension of single variable techniques to several variables analysis and leads rapidly to important global results whilst avoiding the excessive introduction of new tools. Once these techniques (presented here in the pseudoconvex setting) are mastered, it will be fairly easy for the reader to tackle Andreotti–Grauert theory for both complex analytic manifolds and CR manifolds (cf. [He/Le2] and [L-T/Le]). Our choice of applications focuses on global extension problems for CR functions, such as the Hartogs–Bochner phenomenon and removable singularities for CR functions.

Most of the subjects discussed in this book are classical, since they are part of the foundations of Complex Analysis, so it is difficult to be original. This book is therefore heavily influenced by previous work: the source material is quoted at the end of each chapter, along with some historical notes. The bibliography does not claim to be in any way encyclopædic, so many important works on the subject are not included. The reader looking for precise historical notes and a more exhaustive bibliography might do well to consult the end of chapter notes and the bibliography in R.M. Range's book [Ra].

Parts of this book (Sections 5 and 6 of Chapter IV, Section 5 of Chapter V and Chapter VIII) owe a great deal to the work of Guido Lupacciolu, who died before his time in December 1996.

And finally, I would like to thank all those who helped writing this book, particularly Alain Dufresnoy and Jürgen Leiterer. If this book has reached its final form, it is largely thanks to their comments on both the form and the content.

Many thanks also to Myriam Charles for having typed a text which is particularly rich in mathematical formulae and Arlette Guttin-Lombard for her TEX-nical advice.

Introduction

At the start of this century, F. Hartogs discovered special extension properties of holomorphic functions of several variables by finding a domain in $\mathbb{C}^2$ which is not the domain of definition of a holomorphic function. (No such open set exists in $\mathbb{C}$.) Understanding this phenomenon then became one of the main problems in multivariable complex analysis, the objective being that of finding good characterisations of domains of definition of holomorphic functions, called domains of holomorphy.

The first work on this subject by F. Hartogs [Har] in 1906 and E.E. Levi [Lev] in 1910 showed that domains of holomorphy have certain convexity properties, which we will generically call pseudoconvexity. The equivalence of the various definitions of pseudoconvexity which appeared as time went on was proved by K. Oka [Ok] in the 1940s. The appropriate tools for studying pseudoconvexity are the plurisubharmonic functions introduced independently by P. Lelong [Lel1] and K. Oka [Ok]. In the 1930s, H. Cartan and P. Thullen [Ca/Th] found an intrinsic global characterisation of domains of holomorphy in terms of convexity with respect to the algebra of holomorphic functions on the domain: this notion of "holomorphic convexity" is a fundamental concept in complex analysis. The characterisation of domains of holomorphy in terms of pseudoconvexity (also called the solution to the Levi problem) for domains in $\mathbb{C}^n$ was given independently at the start of the 1950s by K. Oka [Ok], H. Bremermann [Br1] and F. Norguet [No]: the problem was solved for complex analytic manifolds by H. Grauert [Gr] in 1958. This work relied heavily on the theory of coherent analytic sheaves, which has proved to be a powerful tool for the study of analytic spaces. The solution of the Levi problem presented here follows Grauert's ideas, but relies on integral representation methods to solve certain technical problems.

The theory of integral representations in complex analysis originates in the work of H. Grauert, G.M. Henkin, I. Lieb and E. Ramirez [Gr/Li, He1, He2, Ram] at the beginning of the 1970s, and has not stopped developing since. It has enabled us to solve previously inaccessible problems and can be used to reprove more precise versions of fundamental results in several variable holomorphic function theory initially obtained using other methods. The main aim is to construct good integral operators for solving the Cauchy–Riemann equation. Solving this equation is at the heart of most problems in complex analysis, and this is how we construct the non-extending holomorphic function in the solution to the Levi problem.

Hartogs' result naturally leads us to consider the extension problem for holomorphic functions defined in a neighbourhood of all or part of the boundary of a domain. More generally, it leads us to consider the extension problem for CR functions (i.e. restrictions of holomorphic functions) defined on all or part of the boundary of a domain in a complex analytic manifold. A rigorous proof of Hartogs' result was given around 1940 by S. Bochner [Bo] and E. Martinelli [Ma2] independently using an integral formula, known today as the "Bochner–Martinelli formula". This formula has since played an important role in the study of the extension problem for CR functions in $\mathbb{C}^n$, but unfortunately cannot be used to solve global extension problems in complex analytic manifolds. In 1961, L. Ehrenpreis [Eh] noted the link between Hartogs–Bochner extensions and solving the Cauchy–Riemann equation with support conditions. This is a crucial element in the study of extensions of CR functions in manifolds. In the mid-1980s, G. Lupacciolu and G. Tomassini [Lu/To] studied a special case of the extension problem for CR functions defined on part of the boundary of a domain. Many mathematicians have contributed to the solution of this extension problem over the last ten years. The results presented here are mostly due to G. Lupacciolu [Lu1, Lu2].

These global extension problems are of course linked to the problem of finding an envelope of holomorphy, but they are also linked to a more geometric problem, namely the problem of constructing holomorphic chains with prescribed boundary. If we consider the Hartogs phenomenon in terms of graphs then the graph of the holomorphic extension is a holomorphic chain whose boundary is the maximally complex CR manifold defined by the graph of the given CR function.

The book is organised as follows.

In Chapter I we discuss the elementary local properties of multivariable holomorphic functions which can be deduced from single-variable holomorphic function theory.

The first part of Chapter II deals with currents. The introduction of the Kronecker index of two currents enables us to obtain a Stokes-type formula in a fairly general setting. In the second part we discuss complex analytic manifolds and define various objects linked to their complex structure, such as (p, q) differential forms, the operator $\overline{\partial}$ and Dolbeault cohomology.

In Chapter III we prove our first integral representation formula, the Bochner–Martinelli–Koppelman formula. The proof given here is based on Stokes' formula for the Kronecker index. We begin studying the Cauchy–Riemann equation using this Bochner–Martinelli–Koppelman formula.

In Chapter IV we consider the extension problem for CR functions defined on the boundary of a bounded domain in $\mathbb{C}^n$. Bochner's extension theorem is proved and we also study a special case of extensions where the function is only defined on part of the boundary.

Chapter V deals with the extension problem for CR functions defined on all or part of the boundary of a relatively compact domain in a complex analytic manifold. We study the relationship between these extension phenomena and the vanishing of certain Dolbeault cohomology groups.

In Chapter VI we define domains of holomorphy, holomorphic convexity and pseudoconvexity for open sets in $\mathbb{C}^n$. We prove that domains of holomorphy and holomorphically convex domains are the same thing and we show that every domain of holomorphy is pseudoconvex. The converse, known as the Levi problem, is considered in Chapter VII.

Chapter VII deals with the solution of the Levi problem. Our method is based on local resolutions of $\overline{\partial}$ with Hölder estimates and results on the invariance of Dolbeault cohomology using Grauert's bumping method. There is one novelty in the proof presented here: it uses a result of Laufer's [Lau] which enables us to deduce the vanishing of Dolbeault cohomology groups from finiteness theorems obtained via local resolutions of $\overline{\partial}$. We end the chapter by solving the Levi problem for complex analytic manifolds and stating several vanishing theorems for Dolbeault cohomology which enable us to give sufficient geometric conditions for the extension of CR functions studied in Chapter V.

Chapter VIII gives necessary and sufficient conditions for the extension of CR functions defined on part of the boundary of a strictly pseudoconvex domain.

Here are some remarks on the organisation of the book which should be helpful to the reader:

- All the theory needed to solve the Levi problem – namely, the facts that a domain of holomorphy and a pseudoconvex open set in $\mathbb{C}^n$ are the same thing and that a Stein manifold and a manifold with a strictly plurisubharmonic exhaustion function are the same thing – is worked out in Chapter III, Sections 3 and 4 of Chapter V and Chapters VI and VII.
- Chapter IV, Sections 1, 2 and 5 of Chapter V, Section 8.3 of Chapter VII and Chapter VIII deal with global extension phenomena for CR functions.

I

Elementary local properties of holomorphic functions of several complex variables

In this chapter we study the local properties of holomorphic functions of several complex variables which can be deduced directly from the classical theory of holomorphic functions in one complex variable. The basis for our work is a Cauchy formula for polydiscs which generalises the classical Cauchy formula. Most of the theorems proved in this chapter extend well-known theorems for holomorphic functions in dimension 1 (such as the open mapping theorem, the maximum principle, Montel's theorem and the local inversion theorem) to multivariable analysis. However, when we try to extend holomorphic function a phenomenon which is specific to n-dimensional space with $n \geqslant 2$ appears, namely Hartog's phenomenon. A special case of this phenomenon, which is studied in detail in Chapter III, is discussed at the end of this chapter.

1 Notation and definitions

We write $\mathbb{N}$ for the set of natural numbers, $\mathbb{R}$ for the field of real numbers and $\mathbb{C}$ for the field of complex numbers. For any positive integer $n \in \mathbb{N}$ the set $\mathbb{C}^n$ is equipped with the usual vector space structure and for any $z = (z_1, \dots, z_n) \in \mathbb{C}^n$ the norm of z is given by $|z| = (|z_1|^2 + \cdots + |z_n|^2)^{1/2}$. We define an isomorphism of $\mathbb{R}$-vector spaces between $\mathbb{C}^n$ and $\mathbb{R}^{2n}$ by setting $z_j = x_j + iy_j$ for any $z = (z_1, \dots, z_n) \in \mathbb{C}^n$ and $j = 1, \dots, n$.

The holomorphic and anti-holomorphic differential operations are given by

$$(1.1) \qquad \begin{cases} \dfrac{\partial}{\partial z_j} = \dfrac{1}{2}\Big(\dfrac{\partial}{\partial x_j} + \dfrac{1}{i}\dfrac{\partial}{\partial y_j}\Big) = \dfrac{1}{2}\Big(\dfrac{\partial}{\partial x_j} - i\dfrac{\partial}{\partial y_j}\Big), & j = 1, \dots, n \\ \dfrac{\partial}{\partial \overline{z}_j} = \dfrac{1}{2}\Big(\dfrac{\partial}{\partial x_j} - \dfrac{1}{i}\dfrac{\partial}{\partial y_j}\Big) = \dfrac{1}{2}\Big(\dfrac{\partial}{\partial x_j} + i\dfrac{\partial}{\partial y_j}\Big), & j = 1, \dots, n. \end{cases}$$

C. Laurent-Thiébaut, *Holomorphic Function Theory in Several Variables: An Introduction*, Universitext, DOI 10.1007/978-0-85729-030-4_1, © Springer-Verlag London Limited 2011

If $\alpha = (\alpha_1, \ldots, \alpha_n) \in \mathbb{N}^n$ and $\beta = (\beta_1, \ldots, \beta_n) \in \mathbb{N}^n$ are multi-indices and $x = (x_1, \ldots, x_n)$ is a point in $\mathbb{R}^n$ then we set

$$|\alpha| = \alpha_1 + \cdots + \alpha_n, \quad \alpha! = \alpha_1! \cdots \alpha_n!, \quad x^\alpha = x_1^{\alpha_1} \cdots x_n^{\alpha_n},$$

$$D^\alpha = \frac{\partial^{|\alpha|}}{\partial x_1^{\alpha_1} \cdots \partial x_n^{\alpha_n}} \quad \text{and} \quad D^{\alpha\overline{\beta}} = \frac{\partial^{|\alpha|+|\beta|}}{\partial z_1^{\alpha_1} \cdots \partial z_n^{\alpha_n} \partial \overline{z}_1^{\beta_1} \cdots \partial \overline{z}_n^{\beta_n}}. \tag{1.2}$$

We write D^α instead of $D^{\alpha\overline{0}}$ and $D^{\overline{\beta}}$ instead of $D^{0\overline{\beta}}$ whenever there is no risk of confusion.

If D is an open set in $\mathbb{R}^n$ then we denote by $\mathcal{C}^0(D)$ or $\mathcal{C}(D)$ the vector space of complex-valued continuous functions on D and we denote by $\mathcal{C}^k(D)$ the set of k times continuously differentiable functions for any $k \in \mathbb{N}$, $k > 0$. The intersection of the spaces $\mathcal{C}^k(D)$ for all $k \in \mathbb{N}$ is the space $\mathcal{C}^\infty(D)$ of functions on D which are differentiable to all orders. It is easy to check that $f \in \mathcal{C}^k(D)$ if and only if $D^{\alpha\overline{\beta}} f \in \mathcal{C}(D)$ for any pair $(\alpha, \beta) \in \mathbb{N}^n \times \mathbb{N}^n$ such that $|\alpha| + |\beta| \leqslant k$. If $k \in \mathbb{N}$ then the vector space of functions f contained in $\mathcal{C}^k(D)$ whose derivatives $D^\alpha f$, $|\alpha| \leqslant k$, are continuous on $\overline{D}$ is denoted by $\mathcal{C}^k(\overline{D})$ and we denote by $\mathcal{C}^\infty(\overline{D})$ space of infinitely differentiable functions on D all of whose derivatives are continuous on $\overline{D}$.

If $D \Subset \mathbb{R}^n$ and $f \in \mathcal{C}^k(\overline{D})$, $k \in \mathbb{N}$, then we define the $\mathcal{C}^k$ norm of f on D by

$$\|f\|_{k,D} = \sum_{\substack{\alpha \in \mathbb{N}^n \\ |\alpha| \leqslant k}} \sup_{x \in D} |D^\alpha f(x)|;$$

we write $\|f\|_D$ for $\|f\|_{0,D}$.

Definition 1.1. Let D be an open set in $\mathbb{C}^n$. A complex-valued function f defined on D is said to be *holomorphic* on D if $f \in \mathcal{C}^1(D)$ and

$$\frac{\partial f}{\partial \overline{z}_j}(z) = 0 \quad \text{for every } z \in D \text{ and } j = 1, \ldots, n. \tag{1.3}$$

The system of partial differential equations (1.3) is called the *homogeneous Cauchy–Riemann system.*

Remark. It is clear that if f is holomorphic then f is holomorphic with respect to each variable individually. More precisely, for any $z = (z_1, \ldots, z_n) \in D$ it is the case that on setting

$$D_j = \{t \in \mathbb{C} \mid (z_1, \ldots, z_{j-1}, t, z_{j+1}, \ldots, z_n) \in D\},$$

the function $f_j : D_j \to \mathbb{C}$ defined by $t \mapsto f(z_1, \ldots, z_{j-1}, t, z_{j+1}, \ldots, z_n)$ is holomorphic. We denote by $\mathcal{O}(D)$ the set of holomorphic functions on the open set D in $\mathbb{C}^n$.

The following theorem follows directly from Definition 1.1.

Theorem 1.2. *Let D be an open set in $\mathbb{C}^n$. The set $\mathcal{O}(D)$ is then a $\mathbb{C}$-algebra and if $f \in \mathcal{O}(D)$ has the property that $f(z) \neq 0$ for every $z \in D$ then $1/f \in \mathcal{O}(D)$.*

If D is an open set in $\mathbb{C}^n = \mathbb{R}^{2n}$ and f is a $\mathcal{C}^1$ function on D, then we denote its differential at $a \in D$ by $df(a)$: this is the unique $\mathbb{R}$-linear map $\mathbb{R}^{2n} \to \mathbb{R}^2$ such that $f(z+a) = f(a) + df(a)(z) + o(|z|)$ when z tends to 0 in $\mathbb{C}^n$.

A simple calculation shows that

$$df(a) = \sum_{j=1}^{n} \frac{\partial f}{\partial z_j}(a)dz_j(a) + \frac{\partial f}{\partial \overline{z}_j}(a)d\overline{z}_j(a). \tag{1.4}$$

A function $f \in \mathcal{C}^1(D)$ is therefore holomorphic on D if and only if

$$df(a) = \sum_{j=1}^{n} \frac{\partial f}{\partial z_j}(a)dz_j(a)$$

for every $a \in D$.

Theorem 1.3. *A function $f \in \mathcal{C}^1(D)$ satisfies the homogeneous Cauchy–Riemann system at a point $a \in D$ if and only if its differential $df(a)$ at a is $\mathbb{C}$-linear. In particular, $f \in \mathcal{O}(D)$ if and only if its differential is $\mathbb{C}$-linear at every point in D.*

Proof. Consider $f \in \mathcal{C}^1(D)$ and $a \in D$. By (1.4), we can decompose the $\mathbb{R}$-linear map $df(a)$ from $\mathbb{C}^n$ to $\mathbb{C}$ as

$$df(a) = S_a + \overline{T}_a,$$

where

$$S_a = \sum_{j=1}^{n} \frac{\partial f}{\partial z_j}(a)dz_j(a) \quad \text{and} \quad \overline{T}_a = \sum_{j=1}^{n} \frac{\partial f}{\partial \overline{z}_j}(a)d\overline{z}_j(a).$$

The map S_a is clearly $\mathbb{C}$-linear and $\overline{T}_a$ is the conjugate of the $\mathbb{C}$-linear map T_a defined by $T_a = \sum_{j=1}^{n} \frac{\partial \overline{f}}{\partial z_j}(a)dz_j(a)$. As every $\mathbb{R}$-linear map from $\mathbb{C}^n$ to $\mathbb{C}$ has a unique decomposition of this form, $df(a)$ is $\mathbb{C}$-linear if and only if $T_a \equiv 0$, which is another way of saying that f satisfies the homogeneous Cauchy–Riemann system. □

2 The Cauchy formula for polydiscs

A subset P in $\mathbb{C}^n$ is an open (respectively closed) *polydisc* if there are open (respectively closed) discs $P_1, \dots, P_n$ in $\mathbb{C}$ such that $P = P_1 \times \cdots \times P_n$. If ξ_j is the centre of P_j then the point $\xi = (\xi_1, \dots, \xi_n)$ is called the *centre* of P and if r_j is the radius of P_j then $r = (r_1, \ldots, r_n)$ is called the *multiradius* of P. The set $\partial_0 P = \partial P_1 \times \cdots \times \partial P_n$ is the *distinguished boundary* of P. We note that $\partial_0 P$ is not the boundary of P for $n > 1$. We denote by $P = P(\xi, r)$ the polydisc of centre ξ and multiradius r.

Let $P = P(\xi, r)$ be a polydisc in $\mathbb{C}^n$ and let $g \in \mathcal{C}(\partial_0 P)$ be a continuous function on the distinguished boundary of P. The integral of g on $\partial_0 P = \partial P_1 \times \cdots \times \partial P_n$ is defined by

$$\int_{\partial_0 P} g(\zeta) d\zeta_1 \cdots d\zeta_n = i^n r_1 \cdots r_n \int_{[0,2\pi]^n} g(\zeta(\theta)) e^{i\theta_1} \cdots e^{i\theta_n} d\theta_1 \cdots d\theta_n,$$

where $\zeta(\theta) = (\zeta_1(\theta), ..., \zeta_n(\theta))$ and $\zeta_j(\theta) = \xi_j + r_j e^{i\theta_j}$ for $j = 1, \ldots, n$.

If r and $r' \in \mathbb{R}^n$ then we say that $r < r'$ if and only if $r_j < r'_j$ for all $j = 1, \ldots, n$.

The following theorem gives us a Cauchy formula for holomorphic functions in several complex variables which generalises the classical Cauchy formula. It is a fundamental tool allowing us to generalise the elementary local properties of holomorphic functions of a single complex variable to the multivariable case.

Theorem 2.1. *Let $P = P(a, r)$ be a polydisc in $\mathbb{C}^n$ and let $f \in \mathcal{C}(\overline{P})$ be a function which is holomorphic with respect to each variable separately in P. Then*

$$f(z) = \frac{1}{(2i\pi)^n} \int_{\partial_0 P} \frac{f(\zeta) d\zeta_1 \cdots d\zeta_n}{(\zeta_1 - z_1) \cdots (\zeta_n - z_n)} \quad \textit{for every } z \in P. \tag{2.1}$$

Proof. We start by proving that formula (2.1) holds for any polydisc $\widetilde{P} = P(a, \widetilde{r})$, $0 < \widetilde{r} < r$ contained in P. We proceed by induction on the number of variables. Consider the statement

(C_n) Let P be a polydisc in $\mathbb{C}^n$ and let f be a continuous function which is holomorphic with respect to each of the variables separately in a neighbourhood of $\overline{P}$. Then formula (2.1) holds.

The proposition (C_1) is the classical Cauchy formula, which we assume known. Suppose that (C_{n-1}) holds for some $n > 1$. Fix $z = (z_1, \ldots, z_n) \in P$ and apply (C_{n-1}) relative to the $(n-1)$ last variables on the polydisc $P' = P_2 \times \cdots \times P_n$ in $\mathbb{C}^{n-1}$. We then get

$$f(z_1, \ldots, z_n) = \frac{1}{(2i\pi)^{n-1}} \int_{\partial_0 P'} \frac{f(z_1, \zeta_2, \ldots, \zeta_n)}{(\zeta_2 - z_2) \cdots (\zeta_n - z_n)} d\zeta_2 \cdots d\zeta_n. \tag{2.2}$$

For given $(\zeta_2, \ldots, \zeta_n)$ in $\partial_0 P'$, the function g defined by $g(\xi) = f(\xi, \zeta_2, \ldots, \zeta_n)$ is holomorphic in a neighbourhood of $\overline{P}_1$, so we can apply Cauchy's formula. We get

$$f(z_1, \zeta_2, \ldots, \zeta_n) = g(z_1) = \frac{1}{2i\pi} \int_{\partial P_1} \frac{f(\zeta_1, \ldots, \zeta_n)}{\zeta_1 - z_1} d\zeta_1. \tag{2.3}$$

Proposition (C_n) can now be proved by substituting (2.3) in (2.2) and applying Fubini's theorem to some choice of parameterisations of ∂P_1 and $\partial_0 P'$.

If f satisfies the hypotheses of Theorem 2.1 for the polydisc $P = P(a,r)$, then it satisfies the hypotheses of proposition (C_n) for any polydisc $\widetilde{P} = P(a,\widetilde{r})$ such that $0 < \widetilde{r} < r$ and formula (2.1) then holds for $\widetilde{P}$. In other words

$$f(z) = \frac{1}{(2i\pi)^n} \int_{\partial_0 \widetilde{P}} \frac{f(\zeta)d\zeta_1 \cdots d\zeta_n}{(\zeta_1 - z_1)\cdots(\zeta_n - z_n)}.$$

As the function $\zeta \mapsto f(\zeta)/(\zeta_1 - z_1)\cdots(\zeta_n - z_n)$ is continuous on $\overline{P} \setminus \{z\}$ we can apply Lebesgue's Theorem to some choice of parameterisation of $\partial_0 \widetilde{P}$ and letting $\widetilde{r}$ tend to r to obtain (2.1). □

Corollary 2.2. *Let D be an open set in $\mathbb{C}^n$ and let $f \in \mathcal{C}(D)$ be a function which is holomorphic with respect to each variable separately. The function f is then $\mathcal{C}^\infty$ on D and hence $f \in \mathcal{O}(D)$. Moreover, for every $\alpha \in \mathbb{N}^n$, $D^{\alpha\overline{0}} f \in \mathcal{O}(D)$ and $D^{\alpha\overline{\beta}} f \equiv 0$ whenever $|\beta| \neq 0$.*

Proof. For any $a \in D$, let P_a be a polydisc such that $P_a \Subset D$. We can then apply Theorem 2.1 to f and the polydisc P_a. As the function $(\zeta, z) \mapsto f(\zeta)/(\zeta_1 - z_1)\cdots(\zeta_n - z_n)$ is continuous on $\partial_0 P_a \times P_a$ and is $\mathcal{C}^\infty$ with respect to z, we can simply differentiate under the integral sign in formula (2.1) (which is possible) as often as necessary to obtain the corollary. □

Theorem 2.3 (Cauchy's inequalities). *Let $P = P(a,r)$ be a polydisc in $\mathbb{C}^n$ and let $f \in \mathcal{O}(P(a,r))$ be a holomorphic function on P. Then for every $\alpha \in \mathbb{N}^n$,*

$$|D^\alpha f(a)| \leqslant \frac{\alpha!}{r^\alpha} \sup_{z \in P(a,r)} |f(z)| \tag{2.4}$$

$$|D^\alpha f(a)| \leqslant \frac{\alpha!(\alpha_1 + 2)\cdots(\alpha_n + 2)}{(2\pi)^n r^{\alpha+\mathbf{2}}} \|f\|_{L^1(P(a,r))}, \tag{2.5}$$

where $\mathbf{2}$ *denotes the multi-index* $(2,2,\ldots,2) \in \mathbb{N}^n$.

Proof. Fix ρ such that $0 < \rho < r$ and apply Cauchy's formula (2.1) to f and the polydisc $P(a,\rho)$. Differentiating under the integral sign we get

$$D^\alpha f(a) = \frac{\alpha!}{(2i\pi)^n} \int_{\partial_0 P(a,\rho)} \frac{f(\zeta)d\zeta_1 \cdots d\zeta_n}{(\zeta - a)^{\alpha+\mathbf{1}}}, \tag{2.6}$$

where $\mathbf{1}$ denotes the multi-index $(1,1,\ldots,1) \in \mathbb{N}^n$. We then have the upper bound

$$|D^\alpha f(a)| \leqslant \frac{\alpha!}{\rho^\alpha}\Big(\sup_{\zeta \in P(a,r)} |f(\zeta)|\Big)\Big(\frac{1}{(2\pi)^n}\int_{[0,2\pi]^n} d\theta_1 \cdots d\theta_n\Big)$$

and we deduce formula (2.4) by letting ρ tend to r.

Multiplying the two sides of (2.6) by $\rho^{\alpha+1}$ and integrating with respect to ρ, we get

$$|D^\alpha f(a)| \int_{[0,r_1]\times\cdots\times[0,r_n]} \rho^{\alpha+1} d\rho_1 \cdots d\rho_n$$
$$\leqslant \frac{\alpha!}{(2\pi)^n} \int_{[0,r_1]\times\cdots\times[0,r_n]} d\rho_1 \cdots d\rho_n \int_{[0,2\pi]^n} |f(\zeta(\theta))| \rho_1 \cdots \rho_n d\theta_1 \cdots d\theta_n$$

so that

$$|D^\alpha f(a)| \frac{r^{\alpha+2}}{(\alpha_1+2)\cdots(\alpha_n+2)} \leqslant \frac{\alpha!}{(2\pi)^n} \int_{P(a,r)} |f(\zeta)| dV$$

using Fubini's theorem and passing to polar coordinates. □

Corollary 2.4. *For any $\alpha \in \mathbb{N}^n$, p such that $1 \leqslant p \leqslant \infty$ and open subset $\Omega \Subset D$, there is a constant $C = C(\alpha, p, \Omega, D)$ such that*

$$\|D^\alpha f\|_\Omega \leqslant C \|f\|_{L^p(D)} \quad \textit{for all } f \in \mathcal{O}(D) \cap L^p(D).$$

Proof. Fix δ such that $0 < \delta < \operatorname{dist}(\Omega, bD)$. Set $r = \delta/\sqrt{n}$. For any $a \in \Omega$, we have $P(a, r) \Subset D$ and the inequality (2.5) holds. As we also know that

$$\|f\|_{L^1(P(a,r))} \leqslant C_p \|f\|_{L^p(P(a,r))} \leqslant C_p \|f\|_{L^p(D)}$$

the corollary follows. □

One of the main applications of Cauchy's formula (2.1) is the fact that holomorphic functions in several variables are analytic. To prove this result we need to be able to define convergence for series indexed by $\mathbb{N}^n$.

Let $(a_\alpha)_{\alpha\in\mathbb{N}^n}$ be a family of complex numbers. We will say that the series $\sum_{\alpha\in\mathbb{N}^n} a_\alpha$ is absolutely convergent if

$$\sum_{\alpha\in\mathbb{N}^n} |a_\alpha| = \sup\left\{ \textstyle\sum_{\alpha\in F} |a_\alpha| \mid F \subset \mathbb{N}^n \text{ finite}\right\} < +\infty.$$

If this is the case, then there is a unique complex number A such that

$$(\forall\, \varepsilon > 0)(\exists\, F_0 \subset \mathbb{N}^n, \text{ finite})(\forall\, F \subset \mathbb{N}^n, \text{ finite})\big(F \supset F_0 \Rightarrow \big|A - \textstyle\sum_{\alpha\in F} a_\alpha\big| < \varepsilon\big).$$

We then write $A = \sum_{\alpha\in\mathbb{N}^n} a_\alpha$ and we say that A is the sum of the series $\sum_{\alpha\in\mathbb{N}^n} a_\alpha$. Moreover, if σ is a permutation of $\mathbb{N}^n$, then the series $\sum_{\alpha\in\mathbb{N}^n} a_{\sigma(\alpha)}$ is convergent and its sum is independent of the permutation σ. We also have $\sum_{\alpha\in\mathbb{N}^n} a_\alpha = \sum_{k\in\mathbb{N}} \big(\sum_{|\alpha|=k} a_\alpha\big)$. If $(f_\alpha)_{\alpha\in\mathbb{N}^n}$ is a family of continuous functions on an open set D in $\mathbb{C}^n$ and K is a compact subset of D then we will say that the series $\sum_{\alpha\in\mathbb{N}^n} f_\alpha$ converges uniformly on K if the series $\sum_{\alpha\in\mathbb{N}^n} \sup_{z\in K} |f_\alpha(z)|$ converges.

We now consider the special case where $f_\alpha(z)$ is a monomial, i.e. $f_\alpha(z) = a_\alpha z^\alpha = a_\alpha z_1^{\alpha_1} \cdots z_n^{\alpha_n}$. As for single-variable analysis, we have Abel's Lemma:

Lemma 2.5. *Suppose there is a $\xi \in \mathbb{C}^n$ such that $\xi_j \neq 0$ for all j and the series $\sum_{\alpha \in \mathbb{N}^n} a_\alpha \xi^\alpha$ is absolutely convergent. Then the series $\sum_{\alpha \in \mathbb{N}^n} a_\alpha z^\alpha$ converges uniformly on the polydisc $\{z \in \mathbb{C}^n \mid |z_j| \leqslant |\xi_j|\}$.*

Theorem 2.6. *Let $D \subset \mathbb{C}^n$ be an open set and let $f \in \mathcal{O}(D)$ be a holomorphic function on D. The function f then has a Taylor series expansion in a neighbourhood of any point in D. In other words, for any $\xi \in D$ there is a neighbourhood V of ξ in $\mathbb{C}^n$ such that for any $z \in V$*

$$f(z) = \sum_{\alpha \in \mathbb{N}^n} \frac{D^\alpha f(\xi)}{\alpha!}(z - \xi)^\alpha.$$

Moreover, the series on the right-hand side converges uniformly to f on every closed polydisc $\overline{P} \subset D$ with centre ξ.

Proof. Consider $\xi \in D$ and let $P \Subset D$ be a polydisc with centre ξ. For any $z \in P$ and $\zeta \in \partial_0 P$

$$\frac{1}{(\zeta_1 - z_1)\cdots(\zeta_n - z_n)} = \sum_{\alpha \in \mathbb{N}^n} \frac{(z-\xi)^\alpha}{(\zeta - \xi)^{\alpha + \mathbf{1}}}, \quad \text{where } \mathbf{1} = (1, \dots, 1).$$

The right-hand side converges uniformly with respect to ζ on $\partial_0 P$. We can therefore integrate termwise in Cauchy's formula (2.1) applied to f and the polydisc P. If $z \in P$ this gives us that

$$f(z) = \sum_{\alpha \in \mathbb{N}^n} \Big[\frac{1}{(2i\pi)^n} \int_{\partial_0 P} \frac{f(\zeta)}{(\zeta - \xi)^{\alpha+\mathbf{1}}} d\zeta_1 \cdots d\zeta_n\Big](z - \xi)^\alpha.$$

Moreover, on differentiating under the integral sign in formula (2.1) we get

$$D^\alpha f(\xi) = \alpha! \Big[\frac{1}{(2i\pi)^n} \int_{\partial_0 P} \frac{f(\zeta)}{(\zeta - \xi)^{\alpha+\mathbf{1}}} d\zeta_1 \cdots d\zeta_n\Big].$$

The second statement of the theorem follows from Lemma 2.5. □

Corollary 2.7 (The analytic continuation principle). *Let $D \subset \mathbb{C}^n$ be a connected open set. If $f \in \mathcal{O}(D)$ and there is an $a \in D$ such that $D^\alpha f(a) = 0$ for all $\alpha \in \mathbb{N}^n$, then $f(z) = 0$ for all $z \in D$. In particular, if there is a non-empty open set $U \subset D$ such that $f(z) = 0$ for all $z \in U$ then $f \equiv 0$ on D.*

Proof. By Theorem 2.6, the set

$$\Omega = \{z \in D \mid D^\alpha f(z) = 0 \quad \text{for all } \alpha \in \mathbb{N}^n\}$$

is open. Ω is also closed, by the continuity of the functions $D^\alpha f$. Since the open set D is connected and $\Omega \neq \varnothing$, we see that $\Omega = D$. □

It is easy to deduce the following characterisation of holomorphic functions from Definition 1.1 and Theorems 2.1 and 2.6.

Corollary 2.8. *Let D be an open set in $\mathbb{C}^n$ and let f be a complex-valued function on D. The following are equivalent.*

1) $f \in \mathcal{O}(D)$,
2) $f \in \mathcal{C}(D)$ *and f is holomorphic with respect to each of the variables* $z_1, \dots, z_n$ *separately.*
3) *f satisfies the Cauchy formula (2.1) for every polydisc $P \Subset D$.*
4) *f has a Taylor series expansion in a neighbourhood of z for every $z \in D$.*

The hypothesis $f \in \mathcal{C}(D)$ is not necessary in 2). In fact, Hartogs' theorem tells us that the equivalence of 1) and 2) does not need any regularity hypothesis with respect to the whole variable set [Har].

Theorem 2.9. *Let D be an open set in $\mathbb{C}^n$. Then $f \in \mathcal{O}(D)$ if and only if f is holomorphic with respect to each variable separately.*

The following counter-example shows that no analogous result holds for functions of several real variables.

Let $f : \mathbb{R}^2 \to \mathbb{R}$ be defined by

$$\begin{cases} f(0) = 0 \\ f(x,y) = \dfrac{xy}{x^4 + y^4} & \text{if } (x,y) \neq (0,0). \end{cases}$$

The function f is real analytic with respect to each variable separately, but it is not bounded in a neighbourhood of 0.

3 The open mapping theorem

The open mapping theorem and the maximum principle that follows from it can be easily extended to holomorphic functions of several variables.

Theorem 3.1. *Let D be a connected open set in $\mathbb{C}^n$ and let f be a holomorphic function on D. If f is not a constant function then the map $f : D \to \mathbb{C}$ is an open map (i.e. the image of an open set in D is an open set in $\mathbb{C}$).*

Proof. Consider $a \in D$ and let U be a convex neighbourhood of a contained in D. By the analytic continuation principle $f|_U \not\equiv f(a)$ since f is not constant and D is connected. Consider $b \in U$ such that $f(b) \neq f(a)$, consider the set $\Omega = \{z \in \mathbb{C} \mid a + z(b-a) \in U\}$ and set $g(z) = f(a + z(b-a))$ for all $z \in \Omega$. The open set Ω is convex and contains 0 and 1; moreover, $g(0) = f(a) \neq f(b) = g(1)$. By the classical one-dimensional open mapping theorem, $g(\Omega)$ is a neighbourhood of $f(a)$. Since $f(U) \supset g(\Omega)$ the theorem follows. □

Corollary 3.2 (Maximum principle). *Let D be a bounded open set in $\mathbb{C}^n$ and let f be a function which is holomorphic on D and continuous on $\overline{D}$. If f is non-constant then for every $z \in D$*

$$|f(z)| < \sup_{\zeta \in \partial D} |f(\zeta)|.$$

Proof. The continuity of f on $\overline{D}$ and the open mapping theorem imply that $f(D) = U$ is a bounded open set. Consider $w \in \partial U$: we then have $w = \lim_{\nu \to \infty} f(z_\nu)$, where $(z_\nu)_{\nu \in \mathbb{N}}$ is a sequence of points in D from which we can choose a sub-sequence which converges to a point in ∂D, since f is an open mapping. It follows that $\partial U \subset \{w \in \mathbb{C} \mid |w| \leqslant \sup_{\zeta \in \partial D} |f(\zeta)|\}$ and since U is a bounded open set $U \subset \{w \in \mathbb{C} \mid |w| < \sup_{\zeta \in \partial D} |f(\zeta)|\}$. □

Corollary 3.3. *Let f be a holomorphic function on a connected open set D in $\mathbb{C}^n$. If $|f|$ has a local maximum at $z_0 \in D$ then f is constant on D.*

Proof. Let V be a connected open neighbourhood of z_0 in D such that $|f(z)| \leqslant |f(z_0)|$ for any $z \in V$. If f is non-constant on V then by Theorem 3.1 $f(V)$ is an open set contained in the closed disc of centre 0 and radius $|f(z_0)|$ and is therefore contained in the open disc, which is impossible. It follows that f is constant on V and as D is connected f is constant on D by analytic continuation. □

Let us finish this section with a version of Schwarz's Lemma for holomorphic functions in several variables.

Theorem 3.4. *Let f be a holomorphic function on the open ball $B(0, R) = \{z \in \mathbb{C}^n \mid |z| < R\}$. Assume that f has a zero of order k at the origin, i.e. the Taylor expansion of f at 0 has no terms of order strictly less than k, and $|f|$ is bounded by a constant M on $B(0, R)$. Then*

$$|f(z)| \leqslant M \left|\frac{z}{R}\right|^k \quad \text{if } z \in B(0, R).$$

Proof. For any $z \neq 0$ in $B(0, R)$ and $u \in \mathbb{C}$ such that $|u| < R$ we set $\varphi(u) = f(uz/|z|)$. The function φ thus defined is holomorphic on the disc $D(0, R)$ in $\mathbb{C}$ and $|\varphi|$ is bounded by M on this disc. Moreover, φ vanishes to order k at 0. We can therefore consider the function $\Psi(u) = \varphi(u)/u^k$, which is holomorphic on $D(0, R)$ and if $0 < r < R$ then $|\Psi(u)| \leqslant M/r^k$ if $|u| = r$. By the maximum principle applied to Ψ we have $|\Psi(u)| \leqslant M/r^k$ if $|u| \leqslant r$. In particular, $|\Psi(|z|)| \leqslant M/r^k$ or in other words $|f(z)| \leqslant M|z/r|^k$ for any $r \geqslant |z|$ and $u = |z|$. This inequality holds for any r such that $|z| \leqslant r < R$, and passing to the limit we see that $|f(z)| \leqslant M|z/R|^k$. □

4 Series of holomorphic functions

Let D be an open set in $\mathbb{C}^n$. We say that a series $(f_j)_{j\in\mathbb{N}} \subset \mathcal{C}(D)$ converges uniformly on the compact subsets of D if there is a function $f \in \mathcal{C}(D)$ such that the series $(f_j)_{j\in\mathbb{N}}$ converges uniformly to f on any compact set K in D.

We will now define a metrisable topological vector space structure on $\mathcal{C}(D)$ such that convergence with respect to this topology is simply uniform convergence on all compact sets. A fundamental system of neighbourhoods of 0 for this topology is given by the sets $V_{K,\varepsilon} = \{f \in \mathcal{C}(D) \mid \sup_{z\in K} |f(z)| < \varepsilon\}$, where K runs over the set of compact subsets of D and ε runs over $\mathbb{R}_+^*$. As D is a countable union of compact sets, the topology thus defined is metrisable. More precisely, let $(K_j)_{j\in\mathbb{N}}$ be a exhaustion of D by compact sets, i.e. $D = \bigcup_{j=1}^{\infty} K_j$ and $K_j \subset \overset{\circ}{K}_{j+1}$ (where $K_j = \{z \in D \mid d(z, \partial D) \geqslant 1/j,\ |z| \leqslant j\}$, for example). For any $f, g \in \mathcal{C}(D)$, we set

$$\delta(f,g) = \sum_{j=1}^{\infty} 2^{-j} \frac{\|f-g\|_{K_j}}{1+\|f-g\|_{K_j}}.$$

Then δ is a distance defining the above topology.

With this topology the space $\mathcal{C}(D)$ is complete and is therefore a Fréchet space. We equip the subspace $\mathcal{O}(D)$ in $\mathcal{C}(D)$ with the topology induced by the topology on $\mathcal{C}(D)$.

Theorem 4.1. *$\mathcal{O}(D)$ is a closed sub-space of $\mathcal{C}(D)$ and for every $\alpha \in \mathbb{N}^n$ the operators D^α are continuous maps from $\mathcal{O}(D)$ to itself. More precisely, if $(f_j)_{j\in\mathbb{N}} \subset \mathcal{O}(D)$ converges uniformly on all compact sets in D to a function $f \in \mathcal{C}(D)$, then $f \in \mathcal{O}(D)$ and for every $\alpha \in \mathbb{N}^n$, the sequence $(D^\alpha f_j)_{j\in\mathbb{N}}$ converges uniformly on every compact set in D to $D^\alpha f$.*

Proof. The proof of this result simply repeats the one-dimensional proof. The only change we have to make is replacing the classical Cauchy formula by the polydisc Cauchy formula. □

Definition 4.2. A subset S in $\mathcal{O}(D)$ is *bounded* if and only if, for any compact set $K \subset D$,

$$\sup_{f\in S} \|f\|_K < +\infty.$$

We now give a characterisation of compact subsets of $\mathcal{O}(D)$ similar to the characterisation of compact subsets of finite-dimensional topological vector spaces.

Theorem 4.3 (Montel's theorem). *A subset S in $\mathcal{O}(D)$ is compact if and only if it is both closed and bounded.*

Proof.

1) *Sufficiency.* Assume that S is closed and bounded. Since $\mathcal{O}(D)$ is metrisable it will be enough to show that we can extract a convergent subsequence from any sequence in S. Let $(f_\nu)_{\nu\in\mathbb{N}}$ be a sequence of elements in S. Since $f_\nu \in \mathcal{O}(D)$ for any $\nu \in \mathbb{N}$, it follows from the Cauchy inequalities that the first derivatives of the elements f_ν are uniformly bounded on any compact subset of D. Ascoli's theorem then implies that the condition given is sufficient.
2) *Necessity.* Let K be a compact set in D. The map $\|\cdot\|_K$ from $\mathcal{C}(D)$ to $\mathbb{R}$ which sends f to $\sup_{z\in K}|f(z)|$ is then continuous. Since S is compact, the set $\{\|f\|_K \mid f \in S\}$ is a compact set in $\mathbb{R}$ and hence $\sup\{\|f\|_K \mid f \in S\} < +\infty$, which proves that S is bounded. Moreover, since S is compact it is automatically closed. □

5 Holomorphic maps

The study of holomorphic maps will lead us to define complex analytic submanifolds of $\mathbb{C}^n$. We will see in Chapter II that a complex analytic submanifold of $\mathbb{C}^n$ is also an abstract complex analytic manifold.

Definition 5.1. Let D be an open set in $\mathbb{C}^n$. A map $f = (f_1, \ldots, f_m) : D \to \mathbb{C}^m$ is *holomorphic* in D if $f_1, \ldots, f_m \in \mathcal{O}(D)$. We denote by $\mathcal{O}(D, \mathbb{C}^m)$ the space of holomorphic maps from D to $\mathbb{C}^m$. If $f \in \mathcal{O}(D, \mathbb{C}^m)$ and $a \in D$ then the matrix

$$J_f(a) = \begin{pmatrix} \dfrac{\partial f_1}{\partial z_1}(a) & \cdots & \dfrac{\partial f_1}{\partial z_n}(a) \\ \vdots & & \vdots \\ \dfrac{\partial f_m}{\partial z_1}(a) & \cdots & \dfrac{\partial f_m}{\partial z_n}(a) \end{pmatrix}$$

is called the *Jacobian matrix* of f at the point a.

Proposition 5.2. *Let $D \subset \mathbb{C}^n$ be an open set and $f = (f_1, \ldots, f_m) \in \mathcal{O}(D, \mathbb{C}^m)$. Then, for any $a \in D$,*

$$f(a+z) = f(a) + J_f(a)z + o(|z|) \tag{5.1}$$

when z tends to 0 in $\mathbb{C}^n$.

Proof. By Theorem 1.3, if $f \in \mathcal{O}(D, \mathbb{C}^m)$ then the differential of f at a, $df(a)$, is a $\mathbb{C}$-linear map from $\mathbb{C}^n$ to $\mathbb{C}^m$ whose matrix in the canonical bases of $\mathbb{C}^n$ and $\mathbb{C}^m$ is simply $J_f(a)$. Formula (5.1) then follows from the definition of $df(a)$. □

Theorem 5.3. *Let $f : D \subset \mathbb{C}^n \to \Omega \subset \mathbb{C}^m$ and $g : \Omega \subset \mathbb{C}^m \to \mathbb{C}^\ell$ be two holomorphic maps. The map $g \circ f : D \to \mathbb{C}^\ell$ is then a holomorphic map and, for all $a \in D$,*

$$J_{g\circ f}(a) = J_g(f(a))J_f(a). \tag{5.2}$$

Proof. The differential of $g \circ f$ at the point a is given by

$$d(g \circ f)(a) = dg(f(a)) \circ df(a) \tag{5.3}$$

by classical results of differential calculus But now, since f and g are holomorphic maps their differentials $df(a)$ and $dg(f(a))$ are $\mathbb{C}$-linear maps whose matrices in the canonical bases are $J_f(a)$ and $J_g(f(a))$ respectively. It follows that $d(g \circ f)(a)$ is $\mathbb{C}$-linear for any $a \in D$, which implies that $g \circ f$ is a holomorphic map, and formula (5.2) then follows from (5.3). □

Definition 5.4. Let D be an open set in $\mathbb{C}^n$, let f be a holomorphic map from D to $\mathbb{C}^n$ and let a be a point of D. We say that f is *biholomorphic in a neighbourhood of* a if there is a neighbourhood U of a such that $f|_U$ is a bijection from U to $f(U)$ and $(f|_U)^{-1}$ is a holomorphic map from $f(U)$ to U.

Theorem 5.5 (Local inverse theorem). *Let U be a neighbourhood of $a \in \mathbb{C}^n$ and consider $f \in \mathcal{O}(U, \mathbb{C}^n)$. Then f is biholomorphic in a neighbourhood of a if and only if* $\det(J_f(a)) \neq 0$.

Proof. The hypothesis that $\det(J_f(a)) \neq 0$ implies that $df(a)$ is invertible. By the classical local inversion theorem f is a local $\mathcal{C}^1$-diffeomorphism in a neighbourhood of a and $df^{-1}(f(y)) = (df(y))^{-1}$ for any y sufficiently close to a. As $df(y)$ is $\mathbb{C}$-linear since f is holomorphic its inverse $(df(y))^{-1}$ is also $\mathbb{C}$-linear which implies that f^{-1} is holomorphic in a neighbourhood of $f(a)$ since $df^{-1}(f(y)) = (df(y))^{-1}$. □

Corollary 5.6. *If X is a subset of $\mathbb{C}^n$ then for any $k \in \{1, 2, \dots, n-1\}$ the following conditions are equivalent:*

i) *For any point $\xi \in X$ there is a biholomorphic map $f = (f_1, \dots, f_n)$ defined on a neighbourhood U of ξ such that*

$$X \cap U = \{z \in U \mid f_{k+1}(z) = \cdots = f_n(z) = 0\}.$$

ii) *For any point $\xi \in X$ there is a neighbourhood V of ξ in $\mathbb{C}^n$ and a holomorphic map $g : V \to \mathbb{C}^{n-k}$ such that*

$$\operatorname{rk} J_g(\xi) = n - k \quad \textit{and} \quad X \cap V = \{z \in V \mid g(z) = 0\}.$$

iii) *For any point $\xi \in X$ there is a neighbourhood W of ξ in $\mathbb{C}^n$, an open set Ω in $\mathbb{C}^k$ and a holomorphic map $h : \Omega \to \mathbb{C}^n$ such that*

$$\operatorname{rk} J_h(\xi) = k \quad \textit{and} \quad X \cap W = h(\Omega).$$

Proof. It is obvious that i) $\Rightarrow$ ii) and iii).

We now prove that ii) $\Rightarrow$ i). Consider V and g as in ii), and denote by G the linear map $\mathbb{C}^n \to \mathbb{C}^{n-k}$ defined by $J_g(\xi)$. Since $\operatorname{rk} J_g(\xi) = n - k$, the map G is surjective and there is a map $A : \mathbb{C}^n \to \mathbb{C}^k$ such that the map $A \oplus G : \mathbb{C}^n \to \mathbb{C}^n$ defined by $A \oplus G(z) = (A(z), G(z))$ is bijective (we can define A to be the projection from $\mathbb{C}^n$ to $\ker G$ after identifying $\ker G$ and $\mathbb{C}^k$ by a choice of basis). Set $f(z) = (A(z), g(z))$ for any $z \in V$. Then $\det J_f(\xi) = \det(A \oplus G) \neq 0$, so by Theorem 5.5 f is biholomorphic in an neighbourhood U of ξ contained in V. Moreover, by definition of f we know that

$$X \cap U = \{z \in U \mid f_{k+1}(z) = \cdots = f_n(z) = 0\}.$$

To complete the proof of the corollary, we show that iii) $\Rightarrow$ ii). Consider W and h as in iii), and denote by H the linear map $\mathbb{C}^k \to \mathbb{C}^n$ defined by $J_h(\xi)$. Since $\operatorname{rk} J_h(\xi) = k$, the map H is injective. If $z \in \mathbb{C}^n = \mathbb{C}^k \times \mathbb{C}^{n-k}$ we set $z = (z', z'')$, and the map Φ defined by $\Phi(z) = H(z') + (0, z'')$ is then a bijection from $\mathbb{C}^n$ to $\mathbb{C}^n$. Set $\varphi(z) = h(z') + (0, z'')$ for $z \in \Omega \times \mathbb{C}^{n-k}$. The map $J_\varphi(h^{-1}(\xi), 0) = \Phi$ is then invertible, so by Theorem 5.5 φ is biholomorphic in a neighbourhood of $(h^{-1}(\xi), 0)$. Set $f = \varphi^{-1}$: the function f is defined and biholomorphic in a neighbourhood V of ξ contained in W and satisfies $f(h(z')) = (z', 0)$. We then simply set $g = p \circ f$, where p is the projection from $\mathbb{C}^n$ to $\mathbb{C}^{n-k}$ sending z to z''. □

Definition 5.7. Let D be an open set in $\mathbb{C}^n$. A subset X of D is a *complex analytic submanifold* of $\mathbb{C}^n$ if the equivalent conditions i), ii) and iii) of Corollary 5.6 are satisfied for X. If moreover X is closed in D, then X is a closed submanifold of D.

6 Some holomorphic extension theorems

A. Riemann's extension theorem

We want to generalise the result on the holomorphic extension of bounded holomorphic functions of a complex variable over a punctured disc to the several variable case. A point of $\mathbb{C}$ can be considered as the set of zeros of a holomorphic function. We consider the extension of holomorphic bounded functions defined on an open set in $\mathbb{C}^n$ minus the set of zeros of a finite number of holomorphic functions.

Definition 6.1. Let D be an open subset in $\mathbb{C}^n$ and consider $A \subset D$. We say that A is an *analytic set* if for every $a \in D$, there is a neighbourhood U of a and a finite number of holomorphic functions $f_1, \ldots, f_p$ on U such that $A \cap U = \{z \in U \mid f_1(z) = \cdots = f_p(z) = 0\}$.

Proposition 6.2. *Let D be a connected open set in $\mathbb{C}^n$ and let $A \subset D$ be an analytic set. The following then hold.*

i) *A is closed in D,*
ii) *if $A \neq D$ then $D \setminus A$ is dense in D,*
iii) *$D \setminus A$ is connected.*

Proof.

i) By definition, any point $a \in D$ has a neighbourhood U such that $A \cap U$ is closed since the functions f_j, $j = 1, \ldots, p$, are continuous. It follows that A is closed in D.

ii) We argue by contradiction. Suppose that $B = \overset{\circ}{A}$ is not empty. We will prove that B is closed in D and since B is open and non-empty and D is connected this will imply $A = D$.
Consider a point $a \in \overline{B}$, an open connected neighbourhood U of a in D and elements $f_1, \ldots, f_p$ of $\mathcal{O}(U)$ such that

$$A \cap U = \{z \in U \mid f_1(z) = \cdots = f_p(z) = 0\}.$$

Each function f_j, $j = 1, \ldots, p$, vanishes on the open set $B \cap U$; and since U is connected, analytic continuation says that, for any $j = 1, \ldots, p$, f_j vanishes everywhere on U and hence $U \subset A$. Since U is open, $U \subset \overset{\circ}{A} = B$, and since $a \in U$, we have $a \in B$ and hence $B = \overline{B}$.

iii) Assume that $D \setminus A \neq \varnothing$ and let us prove first that:
$(*)$ every point $a \in D$ has a connected neighbourhood U such that $U \setminus A$ is connected.
Fix $a \in D$. Let U be a convex neighbourhood of a on which there are holomorphic functions $f_1, \ldots, f_p$ such that

$$U \cap A = \{z \in U \mid f_1(z) = \cdots = f_p(z) = 0\}.$$

Consider $x_0, x_1 \in U \setminus A$ and set $V = \{\lambda \in \mathbb{C} \mid \lambda x_0 + (1-\lambda)x_1 \in U\}$; V is a convex subset of $\mathbb{C}$ because U is convex and there is a $j \in \{1, \ldots, p\}$ such that the function $g_j(\lambda) = f_j(\lambda x_0 + (1-\lambda)x_1)$ does not vanish identically on V and in particular $g_j(0) \neq 0$. Since the function g_j is holomorphic, the set of its zeros A' is discrete in V. It follows that $V \setminus A'$ is connected. But A' contains the set $\{\lambda \in \mathbb{C} \mid (\lambda x_0 + (1-\lambda)x_1) \in A \cap U\}$. Suppose that $f_j(x_1) \neq 0$: 0 and 1 are then contained in $V \setminus A'$, which is connected, so there is a path $\gamma_\lambda : [0,1] \mapsto \mathbb{C}$ linking 0 and 1 in $V \setminus A'$. If $f_j(x_1) = 0$ then $1 \in A'$, but since A' is discrete there is a $\varepsilon > 0$ such that the segment $[1, 1+\varepsilon] \cap A' = \{1\}$. We then construct a path γ linking 0 to 1 which is contained in $(V \setminus A') \cup \{1\}$ by linking 0 to $1+\varepsilon$ via a path $\gamma_1 : [0, \frac{1}{2}] \to V \setminus A'$ and then linking $1+\varepsilon$ to 1 by $\gamma_2 : [\frac{1}{2}, 1] \to (V \setminus A') \cup \{1\}$, where $\gamma_2(t) = -2\varepsilon t + 1 + 2\varepsilon$. The map $\widetilde{\gamma} : t \mapsto \gamma(t)x_0 + (1-\gamma(t))x_1$ is then a path joining x_0 to x_1 in $U \setminus A$ and hence $U \setminus A$ is connected.
Let us now show that $(*)$ implies iii). We argue by contradiction. Assume that $D \setminus A = U_1 \cup U_2$, where the U_j, $j = 1, 2$, are non-empty disjoint open sets. By ii) $D = \overline{D \setminus A} = \overline{U}_1 \cup \overline{U}_2$. Since D is connected $\overline{U}_1 \cap \overline{U}_2 \neq \varnothing$.

Consider $a \in \overline{U}_1 \cap \overline{U}_2$ and let U be a neighbourhood of a such that $U \setminus A$ is connected (such a neighbourhood exists by $(*)$). Then $U \setminus A = (U \cap U_1) \cup ((U \cap U_2)$. Since the set $U \setminus A$ is connected and A is closed, one of the sets $(U \cap U_j)$, $j = 1, 2$ is empty, so may assume that $(U \cap U_1) = \varnothing$, which is not possible since U is a neighbourhood of $a \in \overline{U}_1$. □

Corollary 6.3. *Let D be an open connected set in $\mathbb{C}^n$ and let f and g be holomorphic functions on D. If f and g are equal on a subset S of D and there is a connected open set V in D such that $V \setminus S$ is not connected then f is equal to g on D.*

Proof. Let A be the set of zeros of $f - g$ in D. Assume that A is not D. By Proposition 6.2, $V \setminus A$ is connected and $V \setminus A$ is dense in V and hence in $V \setminus S$. It follows that $V \setminus S$ must be connected, which is a contradiction. Hence $A = D$. □

Example. If S is a real hypersurface in D, i.e. $S = f^{-1}(0)$ for some $f \in \mathcal{C}^1(D, \mathbb{R})$ such that $\nabla f \neq 0$ on S, then S satisfies the hypotheses of Corollary 6.3. Indeed, $D \setminus S = \{x \in D \mid f(x) > 0\} \cup \{x \in D \mid f(x) < 0\}$.

Theorem 6.4 (Riemann's theorem). *Let D be an open set in $\mathbb{C}^n$ and let A be an analytic set which is not equal to D. Let f be a holomorphic function on $D \setminus A$. Assume that every point $a \in A$ has a neighbourhood U in D such that $f|_{U \setminus A}$ is bounded. There is then a unique function F on D such that $F|_{D \setminus A} = f$.*

Proof. Uniqueness follows immediately from the analytic continuation principle.

We consider first the case where $n = 1$. It is enough to prove that if f is bounded and holomorphic on the punctured disc $\{z \in \mathbb{C} \mid 0 < |z| < R\}$ then f has a holomorphic extension to the disc $\{z \in \mathbb{C} \mid |z| < R\}$, since analytic sets in dimension 1 are discrete. Consider the Laurent series for f on the punctured disc $\{z \in \mathbb{C} \mid 0 < |z| < R\}$

$$f(z) = \sum_{\nu \in \mathbb{Z}} a_\nu z^\nu, \quad a_\nu = \frac{1}{2i\pi} \int_{|z|=r} f(z) z^{-\nu-1} dz, \quad \text{for } 0 < r < R$$

If $\nu < 0$ then $a_\nu \to 0$ as $r \to 0$ since f is bounded, and since a_ν is independent of r we have $a_\nu = 0$. The result follows on setting $F(z) = \sum_{\nu \geqslant 0} a_\nu z^\nu$.

We now consider the general case. For any $a \in A$, let V be a connected neighbourhood of a such that $f|_{V \setminus A}$ is bounded and such that there are holomorphic functions $f_1, \dots, f_p$ on V such that $A \cap V = \{z \in V \mid f_1(z) = \cdots = f_p(z) = 0\}$. We can assume that $h = f_1 \not\equiv 0$ and (making an affine change of coordinates if necessary) we can assume that $a = 0$ and $h(0, \dots, 0, z_n) \not\equiv 0$ in a neighbourhood of $z_n = 0$. There is then a $\delta > 0$ such that $h(0, \dots, 0, z_n) \neq 0$

whenever $0 < |z_n| \leqslant \delta$. Set $z' = (z_1, \dots, z_{n-1})$. Let $\varepsilon > 0$ be such that $h(z', z_n) \neq 0$ if $|z'| \leqslant \varepsilon$ and $|z_n| = \delta$. We consider the function

$$g(z) = \frac{1}{2i\pi} \int_{|t|=\delta} \frac{f(z', t)}{t - z_n} dt.$$

We note that $(z', t) \in V \setminus A$ for $|z'| \leqslant \varepsilon$ and $|t| = \delta$, and that g is holomorphic for $|z'| < \varepsilon$ and $|z_n| < \delta$. Moreover, for any z' such that $|z'| < \varepsilon$, the function $t \mapsto f(z', t)$ has a holomorphic extension to the disc $\{t \in \mathbb{C} \mid |t| < \delta\}$. Indeed, since $h(z', z_n) \neq 0$ if $|z_n| = \delta$, the function $z_n \mapsto h(z', z_n)$ has only a finite number of zeros in $\{z_n \in \mathbb{C} \mid |z_n| < \delta\}$ and since f is bounded in a neighbourhood of these zeros we can simply apply the result for $n = 1$. By the Cauchy formula in one variable, $g(z) = f(z)$ if $z \in V \setminus A$, $|z'| < \varepsilon$, $|z_n| < \delta$. The function f therefore has a holomorphic extension in a neighbourhood of each point of A and the theorem follows by uniqueness. □

B. Rado's theorem

Theorem 6.5 (Rado's theorem). *Let f be a continuous function on an open set D in $\mathbb{C}^n$. If f is holomorphic on $\{z \in D \mid f(z) \neq 0\}$ then f is holomorphic on D.*

Proof. Since the function f is continuous it will be enough, by Corollary 2.8, to show that f is holomorphic in each of the variables separately. Moreover, as holomorphy is a local property it will be enough to prove the theorem for a function of one variable on the unit disc Δ in $\mathbb{C}$.

Suppose that f is a continuous function on $\overline{\Delta}$ and f is holomorphic on $U = \{z \in \Delta \mid f(z) \neq 0\}$. We shall prove that U is dense in Δ and that f is $\mathcal{C}^\infty$ on Δ: the result follows easily, since $\partial f / \partial \overline{z}$ is then a continuous function on Δ which vanishes on a dense subset of Δ (namely U) and hence vanishes on Δ, so f is then holomorphic on the whole of Δ.

Lemma 6.6. *If g is a continuous function on $\overline{U}$ which is holomorphic on U then for all $z \in U$, then $|g(z)| \leqslant \sup_{\zeta \in \partial \Delta \cap \partial U} |g(\zeta)|$.*

Proof. For any given $n \in \mathbb{N}^*$ consider the function $g^n(z) f(z)$. This is a holomorphic function on U, so by the maximum principle,

$$|g^n(z) f(z)| \leqslant \sup_{\zeta \in \partial U} |g^n(\zeta) f(\zeta)| \leqslant \sup_{\zeta \in \partial U \cap \partial \Delta} |g^n(\zeta) f(\zeta)|$$

for all $z \in U$, since f vanishes on $\Delta \cap \partial U$. The result follows on taking the nth root of both sides of the last inequality and letting n tend to infinity. □

Lemma 6.7. *The open set U is dense in Δ.*

Proof. We argue by contradiction. Suppose that U is not dense in Δ: there is then an $a \in \Delta \cap \partial U$ and a series of points $(a_\nu)_{\nu\in\mathbb{N}}$ in $\Delta \setminus \overline{U}$ which tend to a. Consider the sequence of functions $(1/(z-a_\nu))_{\nu\in\mathbb{N}}$. These functions are holomorphic on U and continuous on $\overline{U}$. Moreover, by Lemma 6.6, if $z \in U$ and ν is large enough then

$$\left|\frac{1}{z-a_\nu}\right| \leqslant \sup_{\zeta\in\partial\Delta\cap\partial U} \frac{1}{|\zeta - a_\nu|} \leqslant \frac{2}{d(a,\partial\Delta)}$$

But since $a \in \overline{U}$, the series $(1/(z-a_\nu))_{\nu\in\mathbb{N}}$ cannot be uniformly bounded on U, which gives us a contradiction. □

Lemma 6.8. *The functions* $\operatorname{Re} f$ *and* $\operatorname{Im} f$ *are harmonic on* Δ *and hence* f *is* $\mathcal{C}^\infty$ *on* Δ.

Proof. By the Stone–Weierstraß theorem any continuous function f on $\partial\Delta$ is a uniform limit on $\partial\Delta$ of trigonometric polynomials:

$$f = \lim_{n\to\infty} Q_n(\theta) \quad \text{where } Q_n = \sum_{k=-p_n}^{p_n} c_{k,n} e^{ik\theta}.$$

Then $\operatorname{Re} Q_n(\theta) = \sum_{k=0}^{p_n} a_{k,n} \cos k\theta + b_{k,n} \sin k\theta$, where $a_{k,n}, b_{k,n} \in \mathbb{R}$, and setting $P_n(z) = \sum_{k=0}^{p_n} (a_{k,n} - ib_{k,n}) z^k$, we get $\operatorname{Re}(P_n(z)) = \operatorname{Re} Q_n(\theta)$ for $z = e^{i\theta}$ and $\operatorname{Re}(f - P_n)$ converges uniformly to 0 on $\partial\Delta$.

For any given $\varepsilon > 0$ there is a k_0 such that if $k \geqslant k_0$

$$|e^{P_k - f}| \leqslant e^\varepsilon \quad \text{and} \quad |e^{f-P_k}| \leqslant e^\varepsilon$$

on $\partial\Delta$. The functions e^{f-P_k} and $e^{P_k - f}$ are continuous on $\overline{U}$ and holomorphic on U, so we can apply Lemma 6.6 and we get, for $k \geqslant k_0$,

$$|e^{P_k - f}| \leqslant e^\varepsilon \quad \text{and} \quad |e^{f-P_k}| \leqslant e^\varepsilon$$

on U. It follows that $|\operatorname{Re}(P_k - f)| \leqslant \varepsilon$ on U and since U is dense in Δ, $|\operatorname{Re}(P_k - f)|$ is bounded above by ε on Δ because f and the polynomials P_k are continuous on Δ.

We have therefore proved that $\operatorname{Re} f$ is a uniform limit on Δ of harmonic functions $\operatorname{Re} P_k$. Harnack's theorem then implies that $\operatorname{Re} f$ is harmonic on Δ. Replacing f by if in the above argument it follows that $\operatorname{Im} f$ is harmonic on Δ. The function f is therefore harmonic on Δ so in particular it is $\mathcal{C}^\infty$. □

C. Hartogs' phenomenon

In this section we will study a special case of Hartogs' phenomenon in $\mathbb{C}^n$, $n \geqslant 2$. Hartogs' phenomenon in a general open set in $\mathbb{C}^n$, $n \geqslant 2$, will be studied in detail in Chapter IV.

Theorem 6.9. *Let D be a connected open set in $\mathbb{C}^n$, and set $Q = \{w \in \mathbb{C} \mid r < |w| < R\}$, $0 \leqslant r < R$. Let f be a holomorphic function on $D \times Q$. We assume there is a point $a \in D$ such that f has a holomorphic extension to a neighbourhood of $\{a\} \times \Delta$, $\Delta = \{w \in \mathbb{C} \mid |w| < R\}$. Then f can be extended holomorphically to $D \times \Delta$.*

Proof. We denote the points of $D \times Q$ by (z, w). Note that if the extension F of f to $D \times \Delta$ exists then for any $z \in D$ and $\rho \in]r, R[$

$$F(z, w) = \frac{1}{2i\pi} \int_{|\zeta|=\rho} \frac{f(z, \zeta)}{\zeta - w} d\zeta \quad \text{if } |w| < \rho$$

by the one-variable Cauchy formula.

For any ρ such that $r < \rho < R$, we consider the function

$$\widehat{f}_\rho(z, w) = \frac{1}{2i\pi} \int_{|\zeta|=\rho} \frac{f(z, \zeta)}{\zeta - w} d\zeta.$$

This function is continuous on $D \times \{w \in \mathbb{C} \mid |w| < \rho\}$ and holomorphic in each variable, and is hence holomorphic on the whole domain by Corollary 2.8. As f has a holomorphic extension to a neighbourhood of $\{a\} \times \Delta$, if $r<r_0<R$ then there is an $\varepsilon > 0$ such that $f(z, w)$ can be extended to a holomorphic function, again denoted $f(z, w)$, on the open set $\{|z - a| < \varepsilon, |w| < r_0\}$. Choose ρ such that $r < \rho < r_0$: then $f(z, w) = \widehat{f}_\rho(z, w)$ if $|z - a| < \varepsilon$ and $|w| < \rho$ by the one-variable Cauchy formula. The open set $D \times \{w \in \mathbb{C} \mid r < |w| < \rho\}$ is connected so we can apply the analytic continuation principle to f and $\widehat{f}_\rho$ so f and $\widehat{f}_\rho$ are equal on $D \times \{w \in \mathbb{C} \mid r < |w| < \rho\}$. It follows that the function F defined by $F(z, w) = f(z, w)$ on $D \times Q$ and $F(z, w) = \widehat{f}_\rho(z, w)$ on $D \times \{w \in \mathbb{C} \mid |w| < \rho\}$ is the extension we seek. □

Examples.

1) Let D be an open set in $\mathbb{C}^n$, $n \geqslant 2$, and let a be a point of D. Any holomorphic function on $D \setminus \{a\}$ can then be extended to a holomorphic function on D.
2) Let K be a compact set in $\mathbb{C}^n$, $n \geqslant 2$, and let f be a holomorphic function on $\mathbb{C}^n \setminus K$. Then f can be extended to a holomorphic function on the whole of $\mathbb{C}^n$.
3) Let P be the polydisc $\{(z_1, \dots, z_n) \in \mathbb{C}^n \mid |z_i| < r, 1 \leqslant i \leqslant n\}$ and set $Q = \{(z_1, \dots, z_n) \in P \mid |z_i| < r/2, 1 \leqslant i \leqslant n-1 \text{ or } |z_n| \geqslant r/2\}$, $n \geqslant 2$. Then any holomorphic function on Q can be extended to a holomorphic function on P.

These results are false if $n = 1$: consider the function $f(z) = 1/z$, for example.

Corollary 6.10.

a) *Let f be a holomorphic function on an open set D in $\mathbb{C}^n$, $n \geqslant 2$. The zero set of f has no isolated points.*
b) *If f is a holomorphic function on $\mathbb{C}^n$, $n \geqslant 2$, then the zero set of f is not bounded.*

Proof. Argue by contradiction, applying 1) and 2) above to the function $1/f$. □

Remarks

The local theory of holomorphic functions in several variables originates in the work of K. Weierstraß, H. Poincaré and P. Cousin, who studied functions in product domains of $\mathbb{C}^n$ at the end of the last century. In 1906, Hartogs [Har] discovered domains in $\mathbb{C}^2$ for which every holomorphic function could be extended holomorphically to a larger domain, which never happens in single variable analysis.

The results presented in this chapter are very classical. They can be found in most books on the theory of holomorphic functions in several variables. The interested reader may wish to consult the following books: [Ho2], [Kr], [Na1], [Ra]. Hartogs' separate holomorphy theorem is proved in [Ho2] (Th. 2.2.8)

II

Currents and complex structures

In this chapter we introduce two new ideas, one coming from differential geometry (currents) and the other coming from analysis (complex analytic manifolds and their associated complex structures), which we will use frequently throughout the rest of this book. We start by defining currents, which for differential forms play the role that distributions play for functions, and then consider the regularisation problem for currents defined on a $\mathcal{C}^\infty$ differential manifold. Solving this problem, which is easy in $\mathbb{R}^n$ by means of convolution, obliges us to introduce kernels with similar properties to the convolution kernel. We will also study the Kronecker index of two currents, which generalises the pairing of a current and a differential form. This index enables us to prove a fairly general Stokes' formula which will be used in Chapters III and IV. We then introduce the notion of a complex analytic manifold and describe the natural complex structures which appear on the tangent space of such manifolds, which leads us to define (p,q) differential forms, the $\overline{\partial}$ operator, the Dolbeault complex and the associated cohomology groups. The holomorphic extension phenomena studied in Chapter V are linked to the vanishing of certain of these cohomology groups, and some vanishing theorems will be proved in Chapter VII. We end this chapter with the definition of the complex tangent space to the boundary of a domain in a complex analytic manifold which appears later in the definitions of CR functions (Chapter IV) and pseudoconvexity (Chapter VII).

1 Currents

By X we will always denote a $\mathcal{C}^\infty$ n-dimensional oriented differentiable manifold. For any p such that $0 \leqslant p \leqslant n$ we denote by $\mathcal{D}^p(X)$ the vector space of $\mathcal{C}^\infty$ degree p compactly supported differentiable forms on X. We will define a locally convex topology on $\mathcal{D}^p(X)$ and study its dual.

A. The topology on $\mathcal{D}^p(X)$

Assume first that X is an open set in $\mathbb{R}^n$ and denote by $\mathcal{E}(X)$ the vector space of $\mathcal{C}^\infty$ functions on X.

C. Laurent-Thiébaut, *Holomorphic Function Theory in Several Variables: An Introduction*, Universitext, DOI 10.1007/978-0-85729-030-4_2, © Springer-Verlag London Limited 2011

If K is a compact set in X and α is an element of $\mathbb{N}^n$ then for any $f \in \mathcal{E}(X)$ we set

$$p_{K,\alpha}(f) = \sup_{x \in K} |D^\alpha f(x)|.$$

The functions $p_{K,\alpha}$ are semi-norms on $\mathcal{E}(X)$. Consider the topology on $\mathcal{E}(X)$ defined by these semi-norms. The sets

$$V_{K,m,\varepsilon} = \{f \in \mathcal{E}(X) \mid \forall \alpha, |\alpha| \leqslant m, p_{K,\alpha}(f) < \varepsilon\}$$

form a fundamental system of neighbourhoods of zero for this topology. As X is an open set in $\mathbb{R}^n$ there is an exhaustion of X by compact sets $(K_p)_{p \in \mathbb{N}}$ and the family $(V_{K_{p,m,1/n}})_{p,m \in \mathbb{N}, n \in \mathbb{N}^*}$ is a fundamental basis of the topology in a neighbourhood of 0. This topology is therefore metrisable. It is easy to check that a sequence of elements in $\mathcal{E}(X)$ converges to 0 in the above topology if and only if both the sequence and all its derivatives converge uniformly to 0 on any compact set in X. The vector space $\mathcal{E}(X)$ equipped with this topology is a Fréchet space (i.e. a locally convex, complete, metrisable topological vector space).

If φ is a $\mathcal{C}^\infty$ differential form of degree p on X then φ can be written in the following form

$$\varphi = \sum_{\substack{|I|=p \\ i_1 < i_2 < \cdots < i_p}} \varphi_I dx_I,$$

where $\varphi_I \in \mathcal{E}(X)$ and for any $I = (i_1, \ldots, i_p) \in \{1, \ldots, n\}^p$, we set $dx_I = dx_{i_1} \wedge \cdots \wedge dx_{i_p}$. For any compact set K in X and $\alpha \in \mathbb{N}^n$ we set

$$\widetilde{p}_{K,\alpha}(\varphi) = \sup\{p_{K,\alpha}(\varphi_I) \mid I = (i_1, \ldots, i_p) \in \{1, \ldots, n\}^p, i_1 < i_2 < \cdots < i_p\}.$$

The set of semi-norms $\widetilde{p}_{K,\alpha}$, where K is a compact set in X and α is an element of $\mathbb{N}^n$ defines a Fréchet space topology on the vector space $\mathcal{E}^p(X)$ of $\mathcal{C}^\infty$ degree p differential forms on X.

If Y is another open set in $\mathbb{R}^n$ and f is a $\mathcal{C}^\infty$ diffeomorphism from X to Y then the map

$$f^* : \mathcal{E}^p(Y) \longrightarrow \mathcal{E}^p(X)$$
$$\varphi \longmapsto f^*\varphi$$

is a linear homeomorphism.

We now consider the case where X is a manifold. Let $\mathcal{A}$ be an atlas on X. We define a topology on $\mathcal{E}^p(X)$ using the semi-norms $\widetilde{p}_{U,K,\alpha}$ which are defined by

$$\widetilde{p}_{U,K,\alpha}(\varphi) = \widetilde{p}_{K,\alpha}\big((h^{-1})^* \varphi|_U\big) \quad \text{for any } \varphi \in \mathcal{E}(X)$$

for any set U which is the domain of a chart $(U, h) \in \mathcal{A}$, any compact set K in U and any $\alpha \in \mathbb{N}^n$. This topology is independent of the choice of $\mathcal{A}$: it is the coarsest topology such that the maps $(k^{-1})^* : \mathcal{E}(X) \to \mathcal{E}(k(V))$ are continuous for any chart (V, k) on X. As X is countable at infinity we can assume that the atlas $\mathcal{A}$ contains a countable number of charts and the topology defined above is therefore metrisable. It is easy to check that $\mathcal{E}^p(X)$ is a Fréchet space with this topology.

If K is a compact set in X then we denote by $\mathcal{D}_K^p(X)$ the subspace of $\mathcal{E}^p(X)$ consisting of $\mathcal{C}^\infty$ degree p differential forms supported on K. This is a closed subset in $\mathcal{E}^p(X)$ and it follows that if we equip it with the restriction of the above topology on $\mathcal{E}^p(X)$ we get a Fréchet space. We then equip $\mathcal{D}^p(X) = \bigcup_K \mathcal{D}_K^p(X)$ with the finest locally convex vector space topology for which all the inclusions

$$\mathcal{D}_K^p(X) \hookrightarrow \mathcal{D}^p(X), \quad K \text{ a compact set in } X$$

are continuous.

Remarks. A sequence of differential forms $(\varphi_j)_{j\in\mathbb{N}} \subset \mathcal{D}^p(X)$ converges to $\varphi \in \mathcal{D}^p(X)$ in the above topology if and only if

1) the forms φ_j are all supported on some fixed compact set K in X.
2) $(\varphi_j)_{j\in\mathbb{N}}$ converges to φ in $\mathcal{D}_K^p(X)$.

B. Currents

Definition 1.1. A *p-dimensional current on X* is a continuous linear form on $\mathcal{D}^p(X)$. We denote by $\mathcal{D}'_p(X)$ the set of p-dimensional currents on X. It is a $\mathbb{C}$-vector space, the topological dual of $\mathcal{D}^p(X)$.

Consider $T \in \mathcal{D}'_p(X)$: this is a linear form on $\mathcal{D}^p(X)$ and hence, for any pair of forms $\varphi_1, \varphi_2 \in \mathcal{D}^p(X)$,

$$T(\varphi_1 + \varphi_2) = T(\varphi_1) + T(\varphi_2)$$

and for any $\lambda \in \mathbb{C}$ and $\varphi \in \mathcal{D}^p(X)$

$$T(\lambda\varphi) = \lambda T(\varphi).$$

The current T is also continuous on $\mathcal{D}^p(X)$. In other words, $T\big|_{(\mathcal{D}_K^p)(X)}$ is continuous for any compact set K in X. This is equivalent to the following statement: For any sequence $(\varphi_j)_{j\in\mathbb{N}}$ of elements in $\mathcal{D}^p(X)$ which tends to 0, the sequence $T(\varphi_j)$ also tends to 0 in $\mathbb{C}$.

Throughout the following we will write $\langle T, \varphi\rangle$ for $T(\varphi)$.

Examples of currents.

i) The Dirac delta function δ_x, $x \in X$, defined by $\delta_x(\varphi) = \varphi(x)$ for any $\varphi \in \mathcal{D}^0(X)$ is a 0-dimensional current.

ii) If ω is a locally integrable differential form of degree q on X then we define an $(n-q)$-dimensional current T_ω by

$$\langle T_\omega, \varphi\rangle = \int_X \omega \wedge \varphi \quad \text{for any } \varphi \in \mathcal{D}^{n-q}(X).$$

This definition is only possible on an oriented manifold X.

iii) If Y is a $\mathcal{C}^\infty$ closed and oriented p-dimensional submanifold of X then we define a p-dimensional current $[Y]$ on X by

$$\langle [Y], \varphi\rangle = \int_Y \varphi = \int_Y i^*\varphi \quad \text{for any } \varphi \in \mathcal{D}^p(X),$$

where i is the inclusion $Y \hookrightarrow X$. The current $[Y]$ is called the *integration current on* Y.

If K is a compact set in X and $k \in \mathbb{N}$ is an integer then we denote by $(\mathcal{C}^k_K)^p(X)$ the space of $\mathcal{C}^k$ degree p differential forms supported on K. We equip this space with the topology defined by the semi-norms $\widetilde{p}_{K,\alpha}$, $|\alpha| \leqslant k$. We define $(\mathcal{C}^k_c)^p(X)$ to be the union of the spaces $(\mathcal{C}^k_K)^p(X)$ for all compact sets K in X: it is the space of $\mathcal{C}^k$ degree p compactly supported differential forms on X. We equip this space with the finest locally convex vector space topology for which all the inclusions

$$(\mathcal{C}^k_K)^p(X) \hookrightarrow (\mathcal{C}^k_c)^p(X)$$

are continuous. Consider the inclusion $\mathcal{D}^p(X) \hookrightarrow (\mathcal{C}^k_c)^p(X)$: it is a continuous map with dense image. We denote the topological dual of $(\mathcal{C}^k_c)^p(X)$ by $(\mathcal{C}^k_c)'_p(X)$: it is a subspace of $\mathcal{D}'_p(X)$ and its elements are called *currents of order* k *and dimension* p *on* X.

Example. If Y is a $\mathcal{C}^1$ closed oriented submanifold of X then the integration current on Y is a current of order 0.

C. Support of a current

In this section we will see that it is possible to obtain global information on a current by gluing local information.

If Ω is an open set in X and $T \in \mathcal{D}'_p(X)$ then we can define $T|_\Omega$ (or, more correctly, $T|_{\mathcal{D}^p(\Omega)}$) by $\langle T|_\Omega, \varphi\rangle = \langle T, \widetilde{\varphi}\rangle$ for any $\varphi \in \mathcal{D}^p(\Omega)$, where $\widetilde{\varphi} \in \mathcal{D}^p(X)$ is defined by $\widetilde{\varphi} = \varphi$ on Ω and $\widetilde{\varphi} = 0$ on $X \setminus \Omega$.

Proposition 1.2. *Let* $(\Omega_i)_{i\in I}$ *be a open cover of* X, *and for every* $i \in I$ *let* T_i *be an element of* $\mathcal{D}'_p(\Omega_i)$. *Assume that* $T_i|_{\Omega_i\cap\Omega_j} = T_j|_{\Omega_i\cap\Omega_j}$ *for any pair* (i, j). *There is then a unique current* $T \in \mathcal{D}'_p(X)$ *such that* $T|_{\Omega_i} = T_i$ *for all* $i \in I$.

Proof. Consider a locally finite partition of unity $(\alpha_i)_{i\in I}$ subordinate to the open cover $(\Omega_i)_{i\in I}$. If $\varphi \in \mathcal{D}^p(X)$ then $\varphi = \sum_{i\in I} \alpha_i\varphi$ and the right-hand sum has only a finite number of non-zero terms. If T exists then it must have the property that, for any $\varphi \in \mathcal{D}^p(X)$,

$$\langle T, \varphi\rangle = \sum_{i\in I}\langle T, \alpha_i\varphi\rangle = \sum_{i\in I}\langle T_i, \alpha_i\varphi\rangle \tag{1.1}$$

since α_i is supported on Ω_i.

Conversely, this formula defines a continuous linear form on $\mathcal{D}^p(X)$. Indeed, let $(\varphi_j)_{j\in\mathbb{N}}$ be a sequence of elements in $\mathcal{D}^p(X)$ converging to 0. There is then a compact set K such that $\operatorname{supp}\varphi_j \subset K$ for all $j \in \mathbb{N}$ and for any $i \in I$ the sequence $(\alpha_i\varphi_j)_{j\in\mathbb{N}}$ converges to 0 in $(\mathcal{C}^\infty_{K\cap\operatorname{supp}\alpha_i})^p(X)$. It follows that $T_i(\alpha_i\varphi_j)$ converges to 0 and as only a finite number of the forms $\alpha_i\varphi_j$ are non-zero on K for any j, $T(\varphi_j) = \sum T_i(\alpha_i\varphi_j)$ tends to zero as j tends to infinity. Formula (1.1) therefore defines a p-dimensional current T on X and we now check that $T|_{\Omega_i} = T_i$. Consider $\varphi \in \mathcal{D}(\Omega_i)$. Then:

$$\langle T_i, \varphi\rangle = \sum_{k\in I}\langle T_i, \alpha_k\varphi\rangle$$

but $\operatorname{supp}\alpha_k\varphi \subset \Omega_k \cap \Omega_i$ and it follows that $\langle T_i, \alpha_k\varphi\rangle = \langle T_k, \alpha_k\varphi\rangle$ which implies that $\langle T_i, \varphi\rangle = \sum_{k\in I}\langle T_k, \alpha_k\varphi\rangle = \langle T, \varphi\rangle$. □

Corollary 1.3. *If T is a p-dimensional current on X then there is a largest possible open set Ω in X such that $T|_\Omega = 0$.*

Definition 1.4. If $T \in \mathcal{D}'_p(X)$ then the *support* of T is the complement of the largest open set on which T is identically zero.

Example. If Y is a $\mathcal{C}^\infty$ closed oriented submanifold of X then the support of the current $[Y]$ is Y.

Remark. Note that if $T \in \mathcal{D}'_p(X)$ is a current with compact support then the expression $\langle T, \psi\rangle$ is meaningful for any $\mathcal{C}^\infty$ differential form ψ on X of degree p. Indeed, let χ be a compactly supported $\mathcal{C}^\infty$ function on X such that χ is identically 1 on a neighbourhood of the support of T. We then set $\langle T, \psi\rangle = \langle T, \chi\Psi\rangle$.

Local expressions of currents. Let (U, h) be a chart of X and let $(x_1, \ldots, x_n)$ be the associated local coordinates. Consider the expression

$$T = \sum_{|I|=p} T_I dx_I, \tag{1.2}$$

where T_I is an n-dimensional current on U and $dx_I = dx_{i_1} \wedge \cdots \wedge dx_{i_p}$ for any $I = (i_1, \ldots, i_p) \in \{1, \ldots, n\}^p$. This defines an $(n-p)$-dimensional current on U in the following way: if φ in $\mathcal{D}^{n-p}(U)$ can be written in the form $\varphi = \sum_{|J|=n-p} \varphi_J dx_J$ then we set

$$\langle T, \varphi \rangle = \sum_{|I|=p} \varepsilon(I, \complement I) \langle T_I, \varphi_{\complement I} dx_1 \wedge \cdots \wedge dx_n \rangle,$$

where $I = (i_1, \ldots, i_p)$ and $\complement I = (j_1, \ldots, j_{n-p})$ have the property that $\{i_1, \ldots, i_p, j_1, \ldots, j_{n-p}\} = \{1, \ldots, n\}$ and $\varepsilon(I, \complement I)$ is the sign of the permutation sending $(1, \ldots, n)$ to $(i_1, \ldots, i_p, j_1, \ldots, j_{n-p})$.

Conversely, if T is any current on X then the current $T\big|_U$ can be written in the form (1.2). Indeed, if we set

$$\langle T_I, \varphi dx_1 \wedge \cdots \wedge dx_n \rangle = \varepsilon(I, \complement I) \langle T, \varphi dx_{\complement I} \rangle$$

for any $I = (i_1, \ldots, i_p)$ and any $\mathcal{C}^\infty$ function with compact support on U φ, then this formula defines a set of n-dimensional currents T_I and the current T can be written in the form (1.2) using these currents T_I.

Definition 1.5. If T is a p-dimensional current on a differentiable manifold of dimension n then the number $(n-p)$ is called the *degree* of the current T. We denote the set of degree q currents on X by $\mathcal{D}'^q(X)$.

We have just proved that a degree q current on X can be locally written as a degree q differential form whose coefficients are degree 0 currents.

Example. Degree n currents in $\mathbb{R}^n$ are simply distributions and can be naturally identified with degree 0 currents.

D. Operations on currents

We now show how to extend the classical operations on differential forms to currents and define some new operations.

Wedge product with a $\mathcal{C}^\infty$ differential form. Consider a current $T \in \mathcal{D}'^p(X)$ and a differential form $\alpha \in \mathcal{E}^q(X)$ such that $0 \leqslant p+q \leqslant n$. We define the wedge product $T \wedge \alpha$ by

$$\langle T \wedge \alpha, \varphi \rangle = \langle T, \alpha \wedge \varphi \rangle \quad \text{for any } \varphi \in \mathcal{D}^{n-p-q}(X).$$

This is a degree $p+q$ current on X. If $T = T_\omega$ is the current defined by a $\mathcal{C}^\infty$ differential form of degree p then

$$T_\omega \wedge \alpha = T_{\omega \wedge \alpha} = (-1)^{pq} T_{\alpha \wedge \omega}.$$

For any $T \in \mathcal{D}'^p(X)$ and $\alpha \in \mathcal{D}^q(X)$ we set

$$\alpha \wedge T = (-1)^{pq} T \wedge \alpha.$$

If T is a current of order k then we can define its wedge product with a $\mathcal{C}^k$ differential form in a similar way.

Boundary and differential of a current. If $T \in \mathcal{D}'^p(X)$, then we define the *boundary* bT of the current T by

$$\langle bT, \varphi \rangle = \langle T, d\varphi \rangle, \quad \text{for any } \varphi \in \mathcal{D}^{n-p-1}(X).$$

This is a $p+1$ degree current on X. The *differential* dT of the current $T \in \mathcal{D}'^p(X)$ is then defined by the formula

$$dT = (-1)^{p-1} bT.$$

Examples.

1) Let $D \Subset X$ be a open set with $\mathcal{C}^1$ boundary which is relatively compact in X. We denote by $[D]$ the degree 0 current defined by

$$\langle [D], \varphi \rangle = \int_D \varphi \quad \text{for any } \varphi \in \mathcal{D}^n(X).$$

Stokes' theorem then says that $b[D] = [bD]$.

2) Let ω be an element of $\mathcal{E}^p(X)$ and let us calculate dT_ω. For any $\varphi \in \mathcal{D}^{n-p-1}(X)$,

$$\langle dT_\omega, \varphi \rangle = (-1)^{p-1} \langle bT_\omega, \varphi \rangle = (-1)^{p-1} \langle T_\omega, d\varphi \rangle = (-1)^{p-1} \int_X \omega \wedge d\varphi;$$

but now

$$d(\omega \wedge \varphi) = d\omega \wedge \varphi + (-1)^p \omega \wedge d\varphi,$$

so that

$$\langle dT_\omega, \varphi \rangle = \int_X d\omega \wedge \varphi - \int_X d(\omega \wedge \varphi).$$

Since $\omega \wedge \varphi$ is a compactly supported form, Stokes' formula now says that $\int_X d(\omega \wedge \varphi) = 0$ and hence $dT_\omega = T_{d\omega}$.

Remark. If $T \in \mathcal{D}'^p(X)$ then $d(dT) = 0$.

Direct image of a current under a proper map. Let X and Y be two oriented $\mathcal{C}^\infty$ differentiable manifolds and let f be a $\mathcal{C}^\infty$ map from X to Y. We say that f is proper if and only if for any compact set K in Y $f^{-1}(K)$ is a compact set in X. If $T \in \mathcal{D}'_p(X)$ then the *direct image* of T under the proper map f is the current f_*T defined by

$$\langle f_*T, \varphi \rangle = \langle T, f^*\varphi \rangle, \quad \text{for any } \varphi \in \mathcal{D}'^p(Y).$$

(This definition is meaningful because $\operatorname{supp} f^*\varphi \subset f^{-1}(\operatorname{supp} \varphi)$ is compact for any proper f). The current f_*T is a p-dimensional current on Y. It follows from the definition of the operator f_* and Proposition 5.4 ii) of Appendix A that if T is contained in $\mathcal{D}'_p(X)$ then

$$f_* dT = d f_* T.$$

Inverse image of a current under projection. Let Y and Z be two oriented $\mathcal{C}^\infty$ differentiable manifolds and let f be the projection from $X = Y \times Z$ to Z. If $\varphi \in \mathcal{D}^\bullet(Y \times Z)$ then we define the integral $\int_Y \varphi$ to be the unique differential form on Z such that, for any $\psi \in \mathcal{D}^\bullet(Z)$,

$$\left\langle \int_Y \varphi, \psi \right\rangle_Z = \langle \varphi, f^*\psi \rangle_{Y \times Z} = \int_{Y \times Z} \varphi \wedge f^*\psi.$$

If $T = T_\varphi$ is the current defined by a compactly supported $\mathcal{C}^\infty$ differential form φ on X then the current $f_* T_\varphi$ is the current defined by the $\mathcal{C}^\infty$ compactly supported differential form $\psi(z) = \int_Y \varphi$ on Z. (This follows from the definition of f_* and Fubini's theorem). Consider $T \in \mathcal{D}'^p(Z)$: we define the *inverse image* of the current T under the projection f by

$$\langle f^*T, \varphi \rangle = \langle T, f_* T_\varphi \rangle, \quad \text{for any } \varphi \in \mathcal{D}^{\dim X - p}(X).$$

This is a degree p current on X. It is clear that if $T = T_\omega$ is the current defined by a $\mathcal{C}^\infty$ differential form on Y then $f^*T_\omega = T_{f^*\omega}$, where $f^*\omega$ is the inverse image of the differential form ω.

2 Regularisation

Let X be an n-dimensional oriented $\mathcal{C}^\infty$ differentiable manifold. We denote by $\mathcal{D}'^\bullet(X) = \bigoplus_{p=0}^n \mathcal{D}'^p(X)$ the vector space of currents on X – this is the topological dual of the vector space $\mathcal{D}^\bullet(X)$ of compactly supported $\mathcal{C}^\infty$ differential forms on X. Traditionally, we consider two topologies on $\mathcal{D}'^\bullet(X)$:

1) The weak topology, or the topology of simple convergence on $\mathcal{D}^\bullet(X)$. More precisely, a family $(T_\varepsilon)_{\varepsilon \in \mathbb{R}^+} \subset \mathcal{D}'^\bullet(X)$ converges weakly to $T \in \mathcal{D}'^\bullet(X)$ as ε tends to 0 if for every $\varphi \in \mathcal{D}^\bullet(X)$

$$\lim_{\varepsilon \to 0} \langle T_\varepsilon, \varphi \rangle = \langle T, \varphi \rangle.$$

2) The strong topology, or the topology of uniform convergence on bounded sets in $\mathcal{D}'(X)$. We recall that a subset B in $\mathcal{D}^\bullet(X)$ is bounded if the elements φ in B are all supported in some given compact set K and if for any $\alpha \in \mathbb{N}^n$ and any chart domain U of an atlas $\mathcal{A}$, $\sup_{\varphi \in B}\{\widetilde{p}_{U,K,\alpha}(\varphi)\} < +\infty$, where the $\widetilde{p}_{U,K,\alpha}$ are the semi-norms defined in § 1.A.

It is easy to see that the strong topology is finer than the weak topology.

The aim of this section is to prove that $\mathcal{E}^\bullet(X) = \bigoplus_{p=0}^n \mathcal{E}^p(X)$ is dense in $\mathcal{D}'^\bullet(X)$ with respect to either the weak or the strong topology and give a method for constructing families of $\mathcal{C}^\infty$ differential forms converging to a given current in either topology.

A current of degree 0 on X is called a *distribution* on X.

A. Regularising distributions on $\mathbb{R}^n$

We consider a positive $\mathcal{C}^\infty$ function θ which is compactly supported in a neighbourhood of 0 in $\mathbb{R}^n$ and which has the property that $\int_{\mathbb{R}^n} \theta(x)dx = 1$. For example, we can take the function defined by

$$\theta(x) = \begin{cases} c\, e^{-1/(1-\|x\|^2)} & \text{for } \|x\| \leqslant 1 \\ 0 & \text{for } \|x\| \geqslant 1, \end{cases}$$

where the constant c is chosen in such a way that $\int_{\mathbb{R}^n} \theta(x)dx = 1$. We set $\theta_\varepsilon(x) = \frac{1}{\varepsilon^n}\theta(x/\varepsilon)$ for any $\varepsilon > 0$ and we set $K_\varepsilon(x, y) = \theta_\varepsilon(x - y)$ for any $x, y \in \mathbb{R}^n$. If u is a continuous function on $\mathbb{R}^n$ then we define the regularisations u_ε of u by

$$u_\varepsilon(x) = \int_{\mathbb{R}^n} K_\varepsilon(x, y)u(y)dy.$$

The following classical proposition and corollary will be proved in a more general form in Section B.

Proposition 2.1. *If $u \in \mathcal{D}^0(\mathbb{R}^n)$ is a compactly supported $\mathcal{C}^\infty$ function on $\mathbb{R}^n$, then the family $(u_\varepsilon)_{\varepsilon\in\mathbb{R}^+}$ of regularisations of u converges to u in $\mathcal{D}^0(\mathbb{R}^n)$ as ε tends to 0. Moreover, the convergences of these series to u is uniform with respect to u over any bounded sets in $\mathcal{D}^0(\mathbb{R}^n)$.*

Definition 2.2. Let $T \in \mathcal{D}'^0(\mathbb{R}^n)$ be a distribution on $\mathbb{R}^n$. We then define *the family $(T_\varepsilon)_{\varepsilon\in\mathbb{R}^+}$ of regularisations* of T in the following way: for any $\varphi \in \mathcal{D}^0(\mathbb{R}^n)$ we set

$$\langle T_\varepsilon, \varphi dx_1 \wedge \cdots \wedge dx_n\rangle = \langle T, \varphi_\varepsilon dx_1 \wedge \cdots \wedge dx_n\rangle,$$

where $\varphi_\varepsilon(x) = \int_{\mathbb{R}^n} K_\varepsilon(x, y)\varphi(y)dy$.

Corollary 2.3. *The family $(T_\varepsilon)_{\varepsilon\in\mathbb{R}^+}$ of regularisations of the distribution T is a family of $\mathcal{C}^\infty$ functions on $\mathbb{R}^n$ which converges both weakly and strongly to T when ε tends to 0.*

The interested reader may consult [Sc, Chap. 6] for more information on regularisation in $\mathbb{R}^n$.

Let us consider the main properties of the function K_ε defined on $\mathbb{R}^n \times \mathbb{R}^n$:

1) K_ε is a $\mathcal{C}^\infty$ function,
2) K_ε is supported in a strip containing the diagonal in $\mathbb{R}^n \times \mathbb{R}^n$ whose width is of order ε,
3) $\displaystyle\int_{\mathbb{R}^n} K_\varepsilon(x, y)dy = \int_{\mathbb{R}^n} K_\varepsilon(x, y)dx = 1,$
4) $\displaystyle\int_{\mathbb{R}^n} \Big(\frac{\partial^\alpha}{\partial x_\alpha} + (-1)^{|\alpha|+1}\frac{\partial^\alpha}{\partial y_\alpha}\Big)K_\varepsilon(x, y)dy = 0$ for any $\alpha \in \mathbb{N}^n$.

These properties are central to the proofs of Proposition 2.1 and Corollary 2.3.

B. Regularising distributions on manifolds

Let X be an n-dimensional oriented $\mathcal{C}^\infty$ differentiable manifold.

As a manifold does not have a group law in general, we can no longer use convolution to regularise distributions as in Section A. The idea is to use kernels $(K_\varepsilon)_{\varepsilon\in\mathbb{R}^+}$ which are functions defined on $X\times X$ with properties similar to the four properties mentioned in Section A.

Definition 2.4. Let π_1 and π_2 be the two projections from $X \times X$ to X. We say that a subset A in $X \times X$ is *proper* if for any compact set K in X the sets $\pi_1(\pi_2^{-1}(K) \cap A)$ and $\pi_2(\pi_1^{-1}(K) \cap A)$ are relatively compact in X.

We consider a family of nested neighbourhoods of the diagonal $\Delta \subset X\times X$ which we denote by $(U_\varepsilon)_{\varepsilon\in\mathbb{R}^+}$ and which we construct in the following way. Consider a locally finite cover $\mathcal{U}$ of Δ by open sets $\widetilde{U} = (U \times U)$ such that U is the domain of a chart (U,h) on X. We then define U_ε by

$$U_\varepsilon = \bigcup_{\widetilde{U}\in\mathcal{U}} \{(x,y) \in \widetilde{U} \mid \|h(x) - h(y)\| < \varepsilon\}.$$

Let ω be a $\mathcal{C}^\infty$ degree n nowhere vanishing differential form on X defining the orientation of X.

Definition 2.5. A *family of regularising kernels* on $X \times X$ is a family $(K_\varepsilon(x,y))_{\varepsilon\in\mathbb{R}^+}$ of positive $\mathcal{C}^\infty$ functions on $X\times X$ which has the following two properties. Firstly, for any $\varepsilon > 0$ the support of K_ε must be proper, contained in U_ε and contain the diagonal $\Delta \subset X \times X$. Secondly, as ε tends to 0 in $\mathbb{R}^+$ the family of functions $(x \mapsto \int_X K_\varepsilon(x,y)\omega(y))_{\varepsilon\in\mathbb{R}^+}$ must converge uniformly on any compact set in X to the constant function 1.

Definition 2.6. Let f be a continuous function on X. The *family of regularisations of* f is the family of functions $(f_\varepsilon)_{\varepsilon\in\mathbb{R}^+}$ defined by

$$f_\varepsilon(x) = \int_X K_\varepsilon(x,y)f(y)\omega(y) \quad \text{for any } x \in X.$$

Definition 2.7. Let $T \in \mathcal{D}'^0(X)$ be a distribution on the manifold X. We define the *family* $(T_\varepsilon)_{\varepsilon\in\mathbb{R}^+}$ *of regularisations of* T in the following way: for any φ in $\mathcal{D}^0(V)$ we set

$$\langle T_\varepsilon, \varphi\omega\rangle = \langle T, \varphi_\varepsilon\omega\rangle, \quad \text{where } \varphi_\varepsilon(x) = \int_X K_\varepsilon(x,y)\varphi(y)\omega(y).$$

Definition 2.7 is meaningful because K_ε is $\mathcal{C}^\infty$ on $X\times X$ so φ_ε is $\mathcal{C}^\infty$ on X and the support of φ_ε is contained in the set $\pi_1(\pi_2^{-1}(\operatorname{supp}\varphi)\cap\operatorname{supp}K_\varepsilon)$ which is compact because K_ε is assumed to have proper support.

Proposition 2.8. *Let $T \in \mathcal{D}'^0(X)$ be a distribution on X. The regularisations T_ε of T are then $\mathcal{C}^\infty$ functions on X and for every $y \in X$*

$$T_\varepsilon(y) = \langle T, K_\varepsilon(x,y)\omega(x)\rangle.$$

Proof. Consider $\varphi \in \mathcal{D}^0(X)$. By definition of regularisations,

$$\langle T_\varepsilon, \varphi\omega\rangle = \langle T, \varphi_\varepsilon\omega\rangle = \Big\langle T, \Big(\int_X K_\varepsilon(x,y)\varphi(y)\omega(y)\Big)\omega(x)\Big\rangle.$$

The function $x \mapsto K_\varepsilon(x,y)$ is a $\mathcal{C}^\infty$ compactly supported function on X for any given y and its dependence on y is also $\mathcal{C}^\infty$. Moreover, as ω is a $\mathcal{C}^\infty$ differential form on X, $\langle T, K_\varepsilon(x,y)\omega(x)\rangle$ is a well-defined $\mathcal{C}^\infty$ function on X. Using the density of the vector space generated by functions of the form $u(x)v(y), u, v \in \mathcal{C}^\infty(X)$ in $\mathcal{D}^0(X \times X)$ we get

$$\Big\langle T, \Big(\int_X K_\varepsilon(x,y)\varphi(y)\omega(y)\Big)\omega(x)\Big\rangle = \langle\langle T, K_\varepsilon(x,y)\omega(x)\rangle, \varphi(y)\omega(y)\rangle$$

and hence it follows by definition of T_ε that $T_\varepsilon(y) = \langle T, K_\varepsilon(x,y)\omega(x)\rangle$. □

Let us now study the convergence of the family $(T_\varepsilon)_{\varepsilon\in\mathbb{R}^+}$ in the weak and strong topologies on $\mathcal{D}'(X)$. Let $\psi \in \mathcal{D}^n(X)$ be a $\mathcal{C}^\infty$ compactly supported differential form of degree n on X. Since we have assumed that ω does not vanish on X, there is a $\varphi \in \mathcal{D}^0(X)$ such that $\psi = \varphi\omega$. Then,

$$\langle T - T_\varepsilon, \psi\rangle = \langle T - T_\varepsilon, \varphi\omega\rangle = \langle T, (\varphi - \varphi_\varepsilon)\omega\rangle.$$

To prove that the family $(T_\varepsilon)_{\varepsilon\in\mathbb{R}^+}$ converges weakly to T as ε tends to 0 it will be enough to show that $(\varphi_\varepsilon)_{\varepsilon\in\mathbb{R}^+}$ tends to φ in $\mathcal{D}^0(X)$. To prove that the family $(T_\varepsilon)_{\varepsilon\in\mathbb{R}^+}$ converges strongly to T as ε tends to 0, it will be enough to prove that the convergence of the family $(\varphi_\varepsilon)_{\varepsilon\in\mathbb{R}^+}$ to φ is uniform with respect to φ over any bounded subset of $\mathcal{D}^0(X)$.

Definition 2.9. A (linear) finite order differential operator with $\mathcal{C}^\infty$ coefficients is a linear map $P : \mathcal{C}^\infty(X) \to \mathcal{C}^\infty(X)$ such that for any local chart (U,h) on X there is a differential operator $P_{(U,h)}$ with $\mathcal{C}^\infty$ coefficients on the open set $h(U)$ in $\mathbb{R}^n$ such that, for any function $f \in \mathcal{C}^\infty(X)$,

$$(Pf) \circ h^{-1} = P_{(U,h)}(f \circ h^{-1}) \quad \text{in } h(U).$$

We denote by P^* the formal adjoint of P with respect to the scalar product on $\mathcal{D}^0(X)$ defined by $(f/g) = \int_X f(y)g(y)\omega(y)$.

Throughout the following the term "differential operator" will always denote a finite order differential operator with $\mathcal{C}^\infty$ coefficients.

Remark 2.10. Let $(f_\varepsilon)_{\varepsilon\in\mathbb{R}^+}$ and f be functions in $\mathcal{D}^0(X)$ which are all supported on some fixed compact subset of X. The family $(f_\varepsilon)_{\varepsilon\in\mathbb{R}^+}$ converges to f in $\mathcal{D}^0(X)$ as ε tends to 0 if and only if for any differential operator $P = P(x, D)$,

$$\lim_{\varepsilon\to 0} \sup_{x\in X} \left|P(x, D_x)f(x) - P(x, D_x)f_\varepsilon(x)\right| = 0.$$

Proposition 2.11. *Consider $f \in \mathcal{D}^0(X)$ and set*

$$f_\varepsilon(x) = \int_X K_\varepsilon(x, y)f(y)\omega(y).$$

The following then hold.

1) *The family $(f_\varepsilon)_{\varepsilon\in\mathbb{R}^+}$ converges to f in $\mathcal{D}^0(X)$ as ε tends to 0 if and only if for any differential operator P on X*

$$(*) \quad \lim_{\varepsilon\to 0} \sup_{x\in X} \left| \int_X \big((P(x, D_x) - P^*(y, D_y))K_\varepsilon(x, y)\big) f(y)\omega(y)\right| = 0.$$

2) *The sequence $(f_\varepsilon)_{\varepsilon\in\mathbb{R}^+}$ converges to f and this convergence is uniform with respect to f on any bounded subset of $\mathcal{D}^0(X)$ if and only if the following holds:*

$$(**) \quad \begin{cases} \text{For any differential operator } P \text{ on } X \\ \sup_{x\in X} \left| \int_X \big((P(x, D_x) - P^*(y, D_y))K_\varepsilon(x, y)\big) f(y)\omega(y)\right| \to 0 \\ \text{as } \varepsilon \text{ tends to } 0. \text{ This convergence is uniform with respect to} \\ f \text{ in any bounded set in } \mathcal{D}^0(X). \end{cases}$$

Proof.

a) *Necessity.* Assume either that $(f_\varepsilon)_{\varepsilon\in\mathbb{R}^+}$ converges to f in $\mathcal{D}^0(V)$ as ε tends to 0 or that $(f_\varepsilon)_{\varepsilon\in\mathbb{R}^+}$ converges to f as ε tends to 0 and this convergence is uniform with respect to f on any bounded subset of $\mathcal{D}^0(X)$. Then,

$$\begin{aligned} &\int_X \big((P(x, D_x) - P^*(y, D_y))K_\varepsilon(x, y)\big) f(y)\omega(y) \\ &\quad = \int_X (P(x, D_x)K_\varepsilon(x, y))f(y)\omega(y) - \int_X K_\varepsilon(x, y)(P(y, D_y)f(y))\omega(y) \\ &\quad = P(x, D_x)f_\varepsilon(x) - (P(y, D_y)f(y))_\varepsilon(x) \\ &\quad \text{(differentiating the first term under the integral sign in each chart)} \\ &\quad = P(x, D_x)(f_\varepsilon(x) - f(x)) + P(x, D_x)f(x) - (P(y, D_y)f(y))_\varepsilon(x). \end{aligned}$$

Since $(f_\varepsilon)_{\varepsilon\in\mathbb{R}^+}$ converges to f in $\mathcal{D}^0(X)$ and P is continuous on $\mathcal{D}^0(X)$,

$$\lim_{\varepsilon\to 0} \sup_{x\in X} |P(x, D_x)(f_\varepsilon(x) - f(x))| = 0.$$

If the convergence of the sequence $(f_\varepsilon)_{\varepsilon\in\mathbb{R}^+}$ to f is uniform with respect to f on any bounded subset of $\mathcal{D}^0(X)$, then the above limit is also uniform with respect to f on any bounded subset of $\mathcal{D}^0(X)$.

To complete the proof it will be enough to prove the following lemma which we will then apply to the function $P(x, D_x)f$.

Lemma 2.12. *Let f be a compactly supported continuous function on X. We set $f_\varepsilon(x) = \int_X K_\varepsilon(x,y)f(y)\omega(y)$. The functions $(f_\varepsilon)_{\varepsilon\in\mathbb{R}^+}$ then converge uniformly to f on X as ε tends to 0. If moreover $\mathcal{B}$ is an equicontinuous subset of $\mathcal{C}(X)$ consisting of functions which are all supported on some fixed compact set and $\mathcal{B}$ has the property that* $\sup\{\|f\|_\infty \mid f \in \mathcal{B}\}$ *is finite then the families $(f_\varepsilon)_{\varepsilon\in\mathbb{R}^+}$ converge to f and this convergence is uniform with respect to f on $\mathcal{B}$.*

Proof. By definition of f_ε,

$$\begin{aligned} f_\varepsilon(x) - f(x) &= \int_X K_\varepsilon(x,y)f(y)\omega(y) - f(x) \\ &= \int_X K_\varepsilon(x,y)(f(y)-f(x))\omega(y) + f(x)\Big(\int_X K_\varepsilon(x,y)\omega(y) - 1\Big). \end{aligned}$$

$$\begin{aligned} \|f_\varepsilon - f\|_\infty \leqslant & \Big\| \int_X K_\varepsilon(x,y)(f(y)-f(x))\omega(y)\Big\|_\infty \\ & + \|f\|_\infty \Big\| \Big[\int_X K_\varepsilon(x,y)\omega(y) - 1\Big]\Big|_{\operatorname{supp} f}\Big\|_\infty . \end{aligned}$$

As the function f is continuous and compactly supported it is uniformly continuous and hence

$$(\forall\, \alpha > 0)(\exists\, \varepsilon_0 > 0)(\forall\, \varepsilon < \varepsilon_0)((x,y) \in U_\varepsilon \Longrightarrow |f(x) - f(y)| < \alpha).$$

Note that ε_0 is independent of f for $f \in \mathcal{B}$ since the elements of $\mathcal{B}$ are uniformly equicontinuous. It follows that if $\varepsilon < \varepsilon_0$ then

$$\Big\| \int_X K_\varepsilon(x,y)(f(x)-f(y))\omega(y)\Big\|_\infty \leqslant m_{L,\varepsilon}\alpha,$$

where $m_{L,\varepsilon} = \sup_{x\in L}\big(\int_X K_\varepsilon(x,y)\omega(y)\big)$ whenever L is a compact set containing $\operatorname{supp} f \cup \pi_1(\pi_2^{-1}(\operatorname{supp} f))$. By our assumptions on K_ε, $m_{L,\varepsilon}$ can be bounded independently of ε and if L' is a compact set in X containing the support of f then there is a $\varepsilon_0' > 0$ such that, for any $\varepsilon < \varepsilon_0'$,

$$\Big\| \int_X K_\varepsilon(x,y)\omega(y) - 1\Big\|_{\infty,L'} < \alpha.$$

We note that if $f \in \mathcal{B}$ then the compact sets L and L' can be chosen independently of f. It follows that if $\varepsilon < \min(\varepsilon_0, \varepsilon_0')$ then

$$\|f - f_\varepsilon\|_\infty \leqslant M\alpha + \|f\|_\infty\alpha.$$

This proves that $(f_\varepsilon)_{\varepsilon\in\mathbb{R}^+}$ tends to f as ε tends to 0 and this convergence is uniform with respect to f on $\mathcal{B}$ since $\sup\{\|f\|_\infty \mid f \in \mathcal{B}\}$ is finite. □

End of the proof of Proposition 2.11.

b) *Sufficiency.* To prove that the family $(f_\varepsilon)_{\varepsilon\in\mathbb{R}^+}$ tends to f in $\mathcal{D}^0(X)$ as ε tends to 0 it will be enough to prove that, for any differential operator P on X,

$$\lim_{\varepsilon\to 0}\sup_{x\in X}|P(x,D_x)f_\varepsilon(x)-P(x,D_x)f(x)|=0$$

and

$$\begin{aligned}P(x,D_x)f(x)-P(x,D_x)f_\varepsilon(x)&=P(x,D_x)f(x)-(P(y,D_y)f(y)_\varepsilon(x)\\&\quad+(P(y,D_y)f(y))_\varepsilon(x)-P(x,D_x)f_\varepsilon(x)\end{aligned}$$

but now by Lemma 2.12 $\sup_{x\in X}|P(x,D_x)f(x)-(P(y,D_y)f(y))_\varepsilon(x)|$ tends to 0 as ε tends to 0 and

$$\begin{aligned}&\sup_{x\in X}|(P(y,D_y)f(y))_\varepsilon(x)-P(x,D_x)f_\varepsilon(x)|\\&\qquad=\sup_{x\in X}\left|\int_X((P(x,D_x)-P^*(y,D_y))K_\varepsilon(x,y))f(y)\omega(y)\right|\end{aligned}$$

tends to 0 as ε tends to 0 by assumption. The theorem follows. If (**) holds then it is easy to prove that the convergence is uniform with respect to f on any closed set by repeating the proof given above mutatis mutandis. □

Consider a local chart (U,h) on X and let $\xi=(\xi_1,\dots,\xi_n)$ be the associated local coordinates. Let P be a differential operator on X. There is then an operator $P_{(U,h)}$ on $h(U)$ such that, for any $f\in\mathcal{D}^0(V)$,

$$P(f)\circ h^{-1}=P_{(U,h)}(f\circ h^{-1})=\sum_\alpha a_\alpha(\xi)D_\xi^\alpha(f\circ h^{-1}),$$

where $\alpha\in\mathbb{N}^n$, the functions a_α are $\mathcal{C}^\infty$ on $h(U)$ all except a finite number of which are zero, and $D_\xi^\alpha=\partial^{|\alpha|}/\partial\xi_1^{\alpha_1}\cdots\partial\xi_1^{\alpha_n}$.

Proposition 2.13. *Let $(K_\varepsilon(x,y))_{\varepsilon\in\mathbb{R}^+}$ be a family of regularising kernels on $X\times X$. The following are then equivalent.*

$$(*)\quad\begin{cases}\text{For any differential operator } P \text{ on } X \text{ and any function } f\in\mathcal{D}^0(X)\\ \lim\limits_{\varepsilon\to 0}\sup\limits_{x\in X}\left|\int_X((P(x,D_x)-P^*(y,D_y))K_\varepsilon(x,y))f(y)\omega(y)\right|=0.\end{cases}$$

$$(*')\quad\begin{cases}\text{For any function } f\in\mathcal{D}^0(X) \text{ which is supported in the domain}\\ \text{of some chart and any multi-index } \alpha\in\mathbb{N}^n\\ \lim\limits_{\varepsilon\to 0}\sup\limits_{x\in X}\left|\int_X((D_x^\alpha+(-1)^{|\alpha|+1}D_y^\alpha)K_\varepsilon(x,y))f(y)\omega(y)\right|=0.\end{cases}$$

Proof. It is obvious that $(*)$ implies $(*')$. Conversely, let $(U_i)_{i\in I}$ be a locally finite cover of X by chart domains and let $(\chi_i)_{i\in I}$ be a partition of unity subordinate to this cover. Consider an element $f \in \mathcal{D}^0(X)$ and let P be a differential operator on V; as the support of f meets only a finite number of the open sets U_i, $(U_{i_k})_{k=1,\dots,\ell}$,

$$\int_X ((P(x,D_x) - P^*(y,D_y))K_\varepsilon(x,y))f(y)\omega(y)$$
$$= \sum_{k=1}^{\ell} \int_X ((P(x,D_x) - P^*(y,D_y))K_\varepsilon(x,y))\chi_{i_k}(y)f(y)\omega(y).$$

Since $\chi_{i_k} f$ is supported on the domain of the chart U_{i_k} it will therefore be enough to prove $(*)$ for any function $g \in \mathcal{D}^0(X)$ which is supported in the domain of a chart of X. Moreover, by linearity, it will be enough to show that $(*)$ holds for any chart (U,h), function $g \in \mathcal{D}^0(X)$ supported on U and differential operator P on X such that $P_{(U,h)} = a(\xi)D_\xi^\alpha$. In the following calculations we identify U and the open set $h(U)$ in $\mathbb{R}^n$ in order to simplify the notation. The differential operator P can then be written in the form $P(x,D_x) = a(x)D_x^\alpha$ and

$$\int_X ((P(x,D_x) - P^*(y,D_y))K_\varepsilon(x,y))g(y)dy$$

is the sum of the three following terms

$$(\mathrm{I}) = a(x)\int_X [(D_x^\alpha + (-1)^{|\alpha|+1}D_y^\alpha)K_\varepsilon(x,y)]g(y)dy$$
$$(\mathrm{II}) = (-1)^{|\alpha|}a(x)\int_X (D_y^\alpha K_\varepsilon(x,y))g(y)dy$$
$$(\mathrm{III}) = -\int_X K_\varepsilon(x,y)(P(y,D_y)g(y))dy \quad \text{(by definition of } P^*\text{)}.$$

By $(*')$, (I) converges uniformly to 0 when ε converges to 0 since the continuous function a is bounded on the support of the integral in (I). (The support of this integral is compact because the support of K_ε is proper.)

Integrating by parts we see that (II) $= a(x)(D_y^\alpha g(y))_\varepsilon(x)$ and by Lemma 2.12 this quantity converges uniformly to $a(x)D_x^\alpha g(x)$ on X.

And finally, (III) $= -(a(y)D_y^\alpha g(y))_\varepsilon(x)$ which converges uniformly to $-a(x)D_x^\alpha g(x)$ by Lemma 9.2.9. This completes the proof of the proposition. □

Proposition 2.14. *Let $(K_\varepsilon(x,y))_{\varepsilon\in\mathbb{R}^+}$ be a family of regularising kernels on $X\times X$. The following are then equivalent.*

$$(**)\quad \begin{cases} \text{For any differential operator } P \text{ on } X \\ \sup_{x\in X}\left|\int_X \big((P(x,D_x)-P^*(y,D_y))K_\varepsilon(x,y)\big)f(y)\omega(y)\right| \longrightarrow 0 \\ \text{as } \varepsilon \text{ tends to } 0. \text{ This convergence is uniform with respect to } f \\ \text{on any bounded set in } \mathcal{D}^0(X). \end{cases}$$

$$(**')\quad \begin{cases} \text{For any chart domain and any multi-index } \alpha \\ \sup_{x\in X}\left|\int_X \big((D_x^\alpha+(-1)^{|\alpha|+1}D_y^\alpha)K_\varepsilon(x,y)\big)f(y)\omega(y)\right| \longrightarrow 0 \\ \text{as } \varepsilon \text{ tends to } 0. \text{ This convergence is uniform with respect to } f \\ \text{on any bounded set in } \mathcal{D}^0(X) \text{ whose functions are supported} \\ \text{on the domain of a local chart.} \end{cases}$$

Proof. Repeat the proof of Proposition 2.13, noting that it is still possible to use a partition of unity because the functions in a bounded set are all supported on some fixed compact set, and use the results of Lemma 2.12 on the uniformity of the convergence. □

We have therefore proved the following result.

Theorem 2.15. *Let $(K_\varepsilon(x,y))_{\varepsilon\in\mathbb{R}^+}$ be a family of regularising kernels on $X\times X$.*

- *If the equivalent conditions $(*)$ and $(*')$ hold then the family $(T_\varepsilon)_{\varepsilon\in\mathbb{R}^+}$ of regularisations of T converges weakly to T in $\mathcal{D}'^0(X)$.*
- *If the equivalent conditions $(**)$ and $(**')$ hold then the family $(T_\varepsilon)_{\varepsilon\in\mathbb{R}^+}$ of regularisations of T converges strongly to T in $\mathcal{D}'^0(X)$.*

We will finish this section by constructing regularising operators whose kernels are regularising kernels as in Definition 2.5 which satisfy conditions $(*)$ and $(**)$. This construction is due to de Rham ([Rh], § 15).

Consider a locally finite countable cover of X by chart domains $(U_i)_{i\in\mathbb{N}}$ which are homeomorphic to $\mathbb{R}^n$. Let h_i be a homeomorphism from U_i to $\mathbb{R}^n$. We can then find an open cover of X by open sets $V_i \Subset U_i$ and $\mathcal{C}^\infty$ functions f_i with compact support contained in U_i such that $f_i\equiv 1$ on $\overline{V}_i$ (cf. Appendix A, Lemmas 2.1 and 2.2). If T is a distribution on X then set $R_{i,\varepsilon}T=\overline{R}_{i,\varepsilon}f_iT+(1-f_i)T$, where $\overline{R}_{i,\varepsilon}=h_i^*r_\varepsilon h_{i_*}$ and r_ε is the convolution on $\mathbb{R}^n$ by the function θ_ε of Section 2.A. In a neighbourhood of any compact set of X the sequence of operators $R^i(\varepsilon)=R_{i,\varepsilon}\circ\cdots\circ R_{1,\varepsilon}$ is stationary. We set $R_\varepsilon=\lim_{i\to\infty}R^i(\varepsilon)$.

The regularising operators constructed by this method are called the *de Rham regularising operators*.

As an example, consider a manifold with an atlas consisting of two charts. Then,

$$R_\varepsilon T = R_{1,\varepsilon} R_{2,\varepsilon} T = \overline{R}_{1,\varepsilon} f_1 \overline{R}_{2,\varepsilon} f_2 T + \overline{R}_{1,\varepsilon} f_1 (1-f_2) T \\ + (1-f_1)\overline{R}_{2,\varepsilon} f_2 T + (1-f_1)(1-f_2)T,$$

where the last term in the sum vanishes. The kernel associated to the operator R_ε can be written as

$$K_\varepsilon(x,z) = \int_X K_{1,\varepsilon}(x,y) f_1(y) K_{2,\varepsilon}(y,z) f_2(z) dy \\ + K_{1,\varepsilon}(x,z) f_1(z)(1-f_2(z)) + (1-f_1(x)) K_{2,\varepsilon}(x,z) f_2(z),$$

where K_{i,ε_i} is the kernel associated to the image under h_i of convolution with θ_{ε_i}.

The following theorem was proved by de Rham (Theorem 12, [Rh], § 15).

Theorem 2.16. *Let $(R_\varepsilon)_{\varepsilon>0}$ be a family of de Rham regularising operators and let $T \in \mathcal{D}'^0(X)$ be a distribution on X. Then:*

1) *$R_\varepsilon T$ is a $\mathcal{C}^\infty$ function on X,*
2) *The support of $R_\varepsilon T$ is contained in any given neighbourhood of the support of T for small enough ε,*
3) *$R_\varepsilon T$ converges both weakly and strongly to T as ε tends to 0.*

We leave it to the reader to check that the kernels K_ε, $\varepsilon \in \mathbb{R}^+$, associated to the operators R_ε form a regularising family of operators satisfying conditions $(*)$ and $(**)$.

C. Regularising currents

To regularise currents on X we simply replace the kernels in Section B by $\mathcal{C}^\infty$ differential forms on $X \times X$ supported on a fundamental system of neighbourhoods $(U_\varepsilon)_{\varepsilon>0}$ of the diagonal $\Delta \subset X \times X$. Let $(\psi_\varepsilon)_{\varepsilon\in\mathbb{R}^+}$ be such a family. If $\varphi \in \mathcal{D}^p(X)$ is a compactly supported $\mathcal{C}^\infty$ differential form on X and π_1 and π_2 are the two projections $X \times X \to X$ then we set

$$\varphi_\varepsilon = (-1)^{np}(\pi_1)_*(\psi_\varepsilon \wedge \pi_2^*\varphi).$$

If, moreover, the support of φ is contained in a chart domain U and ε is chosen small enough that $\pi_2^{-1}(U) \cap U_\varepsilon$ is contained in a chart domain of $X \times X$ then

$$\varphi(x) = \sum_{|I|=p} \varphi_I(x) dx_I \quad \text{for any } x \in U$$

$$\psi_\varepsilon(x,y) = \sum_{I,J} K_{\varepsilon,I,J}(x,y) dx_I \wedge dy_J \quad \text{on } \pi_2^{-1}(U) \cap U_\varepsilon,$$

and hence

$$\varphi_\varepsilon(x) = \sum_{\substack{|I|=p, I\cap J=\varnothing \\ I\cup J=\{1,\dots,n\}}} \sigma(\tau)\Big(\int_U K_{\varepsilon,I,J}(x,y)\varphi_I(y)dy\Big)dx_I,$$

where $\sigma(\tau)$ is the signature of the permutation $(I, J) \mapsto (1, \dots, n)$ and dy is the differential form $dy_1 \wedge \cdots \wedge dy_n$.

Reasoning as in Section B we get the following theorem.

Theorem 2.17. *Let $(\psi_\varepsilon)_{\varepsilon>0}$ be a family of $\mathcal{C}^\infty$ differential forms on $X \times X$ with proper support in a fundamental system of neighbourhoods of the diagonal $\Delta \subset X \times X$ such that, for every chart on $X \times X$,*

$$\psi_\varepsilon(x,y) = \sum_{\substack{I\cup J=\{1,\dots,n\} \\ I\cap J=\varnothing}} K_{\varepsilon,I,J}(x,y)dx_I \wedge dy_J,$$

where the functions $K_{\varepsilon,I,J}$ have the two following properties: firstly, condition $(**')$ of Proposition 2.13 holds and secondly, the functions $(x \mapsto \int_X K_{\varepsilon,I,J}(x,y)dy)$ converge uniformly on any compact subset of the chart of definition to the constant function 1 as ε tends to 0 in $\mathbb{R}^+$. If $T \in \mathcal{D}'^p(X)$ is a degree p current on the manifold X then consider the family $(T_\varepsilon)_{\varepsilon\in\mathbb{R}^+}$ of regularisations of T defined by $\langle T_\varepsilon, \varphi\rangle = \langle T, \varphi_\varepsilon\rangle$ for any $\varphi \in \mathcal{D}^{n-p}(X)$, where φ_ε is the regularisation of φ by ψ_ε. This family then converges to T in both the weak and strong topologies on $\mathcal{D}'^p(X)$ as ε tends to 0 in $\mathbb{R}^+$.

Proof. By definition of the strong topology on $\mathcal{D}'^p(X)$ it will be enough to prove that the family $(\varphi_\varepsilon)_{\varepsilon\in\mathbb{R}^+}$ of regularisations of φ converges to φ and this convergence is uniform with respect to φ over any bounded set in $\mathcal{D}^{n-p}(X)$. Let $(U_i)_{i\in I}$ be a cover of X by chart domains in X and let $(\chi_i)_{i\in I}$ be a partition of unity subordinate to this cover. If $\mathcal{B}$ is a bounded set in $\mathcal{D}^{n-p}(X)$ and $\varphi \in \mathcal{B}$ then we set $\varphi_i = \chi_i\varphi$. As the functions φ are all supported on the same compact subset of X we can write $\varphi = \sum_{i\in I'} \varphi_i$, where I' is a finite subset of I which is independent of $\varphi \in \mathcal{B}$. We obtain the desired result by linearity on applying Proposition 2.11 to each of the functions φ_i. □

Definition 2.18. If $(\psi_\varepsilon)_{\varepsilon\in\mathbb{R}^+}$ is a family of double differential forms satisfying the hypotheses of Theorem 2.17 then the operators R_ε mapping $\mathcal{D}'^\bullet(X)$ to itself defined by $\langle R_\varepsilon T, \varphi\rangle = \langle T, \varphi_\varepsilon\rangle$ for any $T \in \mathcal{D}'^\bullet(X)$ and $\varphi \in \mathcal{D}^\bullet(X)$ are called *regularising operators.* Here, the functions φ_ε are the regularisations of φ obtained using the kernels ψ_ε.

3 Kronecker index of two currents

Throughout the following, X is an n-dimensional $\mathcal{C}^\infty$ differentiable manifold.

Definition 3.1. If T and S are two currents on X such that $d^\circ T + d^\circ S = n$, we will say that the *Kronecker index of* T *and* S, $\mathcal{K}(T,S)$, is defined in de Rham's sense if for any choice of families of regularising operators $(R_\varepsilon)_{\varepsilon>0}$ and $(R'_{\varepsilon'})_{\varepsilon'>0}$ commuting with the operator d the quantity $\langle R_\varepsilon T \wedge R'_{\varepsilon'} S, 1\rangle$ has a limit as ε and ε' tend to 0 which is independent of the choice of $(R_\varepsilon)_{\varepsilon>0}$ and $(R'_{\varepsilon'})_{\varepsilon'>0}$. If this is the case then we denote this limit by $\mathcal{K}(T,S)$.

If $X = \mathbb{C}^n$ then we can choose our regularising operators to be the operators associated to the convolution kernels defined in Section 2.A. If X is a manifold then the de Rham regularising operators commute with the operator d (cf. [Rh], § 15, Prop. 1).

Remarks.

1) The map $(T,S) \mapsto \mathcal{K}(T,S)$ is bilinear.
2) If one of the two currents T or S is a $\mathcal{C}^\infty$ differential form and either the support of T or the support of S is compact then the Kronecker index $\mathcal{K}(T,S)$ of T and S exists and is equal to $\langle T \wedge S, 1\rangle$. (This follows from the convergence properties of regularisations in $\mathcal{D}^\bullet(X)$ and $\mathcal{D}'^\bullet(X)$ for the strong topology.)
3) If the supports of T and S do not meet and T or S has compact support then $\mathcal{K}(T,S)$ is well defined and is equal to 0. Indeed, if ε and ε' are small enough then $R_\varepsilon T$ and $R'_{\varepsilon'} S$ will have disjoint support and it follows that $R_\varepsilon T \wedge R'_{\varepsilon'} S = 0$.

We will now give some sufficient conditions for the Kronecker index of two currents to exist.

Definition 3.2. The *singular support* of a current is the complement of the set of points which have a neighbourhood in which the current is defined by a $\mathcal{C}^\infty$ differential form. We denote by $SS(T)$ the singular support of the current T.

If $T \in \mathcal{D}'^\bullet(X)$ is a current on X and U is a neighbourhood of the singular support of T then we can write $T = T' + T''$, where T' is a current supported on U and T'' is a $\mathcal{C}^\infty$ differential form on X – simply set $T' = \rho T$ and $T'' = (1-\rho)T$, where ρ is a positive $\mathcal{C}^\infty$ function supported on U and equal to 1 in a neighbourhood of the singular support of T.

Proposition 3.3. *If T and S are two currents on X such that $d^\circ T + d^\circ S = n$, at least one of which is of compact support and whose singular supports do not meet, then the Kronecker index $\mathcal{K}(T,S)$ of T and S is well defined.*

Proof. Under the hypotheses of the proposition there are decompositions $T = T' + T''$ and $S = S' + S''$ such that T'' and S'' are $\mathcal{C}^\infty$ differential forms and the supports of T' and S' do not meet. We can then apply 1), 2) and 3) of the above remark to get $\mathcal{K}(T,S) = \mathcal{K}(T',S'') + \mathcal{K}(T'',S') + \mathcal{K}(T'',S'')$. □

Proposition 3.4. *Let T and S be two currents on X such that $d^\circ T + d^\circ S = n-1$, at least one of which has compact support. If $\mathcal{K}(bT, S)$ or $\mathcal{K}(T, dS)$ exists then then other also exists and they are equal, i.e.*

$$\mathcal{K}(bT, S) = \mathcal{K}(T, dS).$$

Proof. Since the regularising operators R_ε and $R'_{\varepsilon'}$ commute with d and therefore with b,

$$\langle R_\varepsilon bT, R'_{\varepsilon'} S\rangle = \langle bR_\varepsilon T, R'_{\varepsilon'} S\rangle = \langle R_\varepsilon T, dR'_{\varepsilon'} S\rangle = \langle R_\varepsilon T, R'_\varepsilon dS\rangle$$

and the result follows. □

Application. Proposition 3.4 enables us to extend Stokes' theorem to a domain $D \Subset X$ with $\mathcal{C}^1$ boundary and a differential form $\omega \in \mathcal{C}_{n-1}(\overline{D})$ such that $d\omega$, calculated as a current, is continuous on $\overline{D}$. Indeed, setting $T = [D]$ and $S = \omega$ we see that $\mathcal{K}(bT, S)$ exists and is equal to $\int_{bD} \omega$ and it follows that $\int_{bD} \omega = \int_D d\omega$.

Theorem 3.5. *Let T and S be two currents on X such that $d^\circ T + d^\circ S = n$, at least one of which has compact support. The Kronecker index $\mathcal{K}(T, S)$ of the currents T and S exists if*

$$SS(T) \cap SS(bS) = \varnothing \quad \textit{and} \quad SS(bT) \cap SS(S) = \varnothing.$$

Remark. We deduce from Theorem 3.5 that the Kronecker index of T and S exists whenever T is closed with compact support and S is closed.

To prove Theorem 3.5, we need a parametrix of the operator d which does not increase the singular support. The parametrix presented below is due to J.B. Poly.

Proposition 3.6. *If T is a current on X then there are operators A and R such that*

1) $T - RT = dAT + AdT$
2) *A does not increase the singular support and R is regularising.*

Proof. We start with the case $X = \mathbb{R}^n$. Denote by δ the operator on $\mathcal{D}'(\mathbb{R}^n)$ defined as follows: if the current $T \in \mathcal{D}'(\mathbb{R}^n)$, considered as a differential form with distribution coefficients can be written as $T = \sum'_I T_I dx_I$, where $\sum'_I$ denotes $\sum_{\substack{I=(i_1,\dots,i_k)\\ i_1<\dots<i_k}}$ then we set

$$\delta T = -\sum_J{}' \sum_i \frac{\partial}{\partial x_i} T_{iJ} dx_J.$$

Note that $d\delta + \delta d = -\Delta$, where Δ is the usual Laplacian $\Delta = \sum_{i=1}^{n} \partial^2/\partial x_i^2$. Let E be the elementary solution to the Laplacian Δ:

$$E = \begin{cases} \dfrac{1}{(n-2)s_n r^{n-2}} & \text{if } n \geqslant 3, \\ \frac{1}{2\pi}\log r & \text{if } n = 2, \\ r/2 & \text{if } n = 1, \end{cases}$$

where $r = (x_1^2 + \cdots + x_n^2)^{1/2}$ and s_n is the area of the unit sphere in $\mathbb{R}^n$.

For any compactly supported current S on $\mathbb{R}^n$ we set $GS = -E * S$ and $KS = \delta GS$. If S (considered as a differential form with distribution coefficients) can be written as $S = \sum_I' S_I dx_I$ then the convolution product $GS = E * S$ can be written as $GS = -\sum_I' E * S_I dx_I$ whence we get the following expression for KS

$$KS = \sum_J{}' \sum_i \frac{\partial E}{\partial x_i} * S_{iJ} dx_J.$$

The operator $K : \mathcal{E}'(\mathbb{R}^n) \to \mathcal{D}'(\mathbb{R}^n)$ thus defined has the following properties.

a) $S = dKS + KdS$ since $S = -\Delta GS = d\delta GS + \delta dGS = d\delta GS + \delta GdS$.
b) K does not increase the singular support. Indeed, Δ is elliptic and it follows that GS is $\mathcal{C}^\infty$ outside of the singular support of S.

Let us return to the case where X is a manifold. Since X is a countable union of compact sets, there is a countable locally finite cover of X by chart domains $W_i \Subset X$. By Lemmas 2.3 and 2.1 of Appendix A, we can therefore find a cover of X by open sets V_i such that $V_i \Subset W_i$ and $\mathcal{C}^\infty$ functions η_i with compact support in W_i such that $\eta_i = 1$ in a neighbourhood of V_i. For any current T on X, we set

$$A_i T = \eta_i K(\eta_i T)$$

(where by abuse of notation we consider W_i as an open set in $\mathbb{R}^n$), and

$$\begin{aligned} R_i T &= T - dA_i T - A_i dT \\ &= T - (\eta_i)^2 T + \eta_i K(d\eta_i \wedge T) - d\eta_i \wedge K(\eta_i T). \end{aligned}$$

(To be completely rigorous we should consider the chart (W_i, h_i) in X and define the operator A_i by $A_i T = \eta_i (h_i^{-1})_* (K(h_i)_*(\eta_i T))$. The reader can easily check that the identification of W_i and its image under h_i in $\mathbb{R}^n$ does not change any of the properties of A_i but greatly simplifies the notation.)

The operators A_i and R_i have the following properties

a) $T = dA_i T + A_i dT + R_i T$ by construction, from which it follows that $dR_i T = R_i dT$.
b) A_i and R_i do not increase the singular support. Moreover, $R_i T$ is $\mathcal{C}^\infty$ on V_i – indeed $R_i T$ and $K(d\eta_i \wedge T)$ are equal on V_i, and since K does not increase the singular support, $K(d\eta_i \wedge T)$ is $\mathcal{C}^\infty$ on V_i because $d\eta_i \wedge T$ vanishes on V_i.

We set $A^k = A_k R_{k-1} \cdots R_1$ and $R^k = R_k R_{k-1} \cdots R_1$. Since the cover (W_i) is locally finite, it is easy to show that

$$RT = \lim_{k \to \infty} R^k T \quad \text{and} \quad AT = \lim_{k \to \infty} A^k T$$

exist since $R^k T$ is stationary on any open set $U \Subset X$ and $A^k T$ vanishes as soon as k is large enough. The operators A and R have the desired properties:

a) We have

$$\begin{aligned} R^{k-1}T - R^k T &= (1 - R_k) R^{k-1} T \\ &= (dA_k + A_k d) R^{k-1} T \\ &= dA^k T + A^k dT \end{aligned}$$

since R_i and d commute, and summing it follows that

$$T - RT = dAT + AdT.$$

b) A does not increase the singular support since A^k does not increase the singular support. The operator R is regularising since $R^k T$ is $\mathcal{C}^\infty$ on V_i whenever $k \geqslant i$.

Proof of Theorem 3.5. By Proposition 3.6, T and S have the following decompositions:

$$\begin{aligned} T &= RT + dAT + AdT \\ S &= RS + dAS + AdS. \end{aligned}$$

We set

$$\begin{aligned} T_1 &= RT, \quad T_2 = dAT, \quad T_3 = AdT \\ S_1 &= RS, \quad S_2 = dAS, \quad S_3 = AdS. \end{aligned}$$

By linearity, $\mathcal{K}(T, S)$ exists if each of the $\mathcal{K}(T_i, S_k)$ for $i, k = 1, 2, 3$ exists. We note that T_i is a compactly supported current for $i = 1, 2, 3$. The current T_1 is a $\mathcal{C}^\infty$ form because R is regularising. It follows that $\mathcal{K}(T_1, S_k)$ is defined for $k = 1, 2, 3$. As the current S_1 is a $\mathcal{C}^\infty$ form, $\mathcal{K}(T_i, S_1)$ exists for $i = 1, 2, 3$. For $i = k = 2$, we apply Proposition 3.4 and we get $\mathcal{K}(T_2, S_2) = 0$. The cases $i = 2$ and $k = 3$, $i = 3$ and $k = 2$ and $i = k = 3$ follow from Proposition 3.3 because the operators d and A do not increase the singular support. □

Corollary 3.7 (Stokes' formula for the Kronecker index). *Let T and S be two currents on X such that $d^0 T + d^0 S = n - 1$, at least one of which has compact support. If*

$$SS(bT) \cap SS(bS) = \varnothing$$

then the Kronecker indices $\mathcal{K}(bT, S)$ and $\mathcal{K}(T, bS)$ exist and

$$\mathcal{K}(bT, S) = (-1)^{d^0 S - 1} \mathcal{K}(T, bS).$$

Proof. This follows immediately from Theorem 3.5 and Proposition 3.4. □

Example of an application: the Cauchy–Green formula in $\mathbb{C}$. Let D be a bounded open set in $\mathbb{C}$ with $\mathcal{C}^1$ boundary containing the origin. Let ψ be a $\mathcal{C}^\infty$ function on $\mathbb{C}$. Identifying $\mathbb{C}$ with $\mathbb{R}^2$ we define degree 1 differential forms dz and $d\overline{z}$ by $dz = dx + idy$ and $d\overline{z} = dx - idy$.

For any $\mathcal{C}^\infty$ function ψ on $\mathbb{C}$ we set $T = \psi[D]$ where $[D]$ is the integration current on D. Then $SS(T) = bD$.

If $S = \frac{1}{2i\pi} dz/z$ then $dS = [0]$, where $[0]$ is the integration current on the point manifold 0. Indeed, if $\varphi \in \mathcal{D}^0(\mathbb{C})$ then, by definition of d,

$$\langle dS, \varphi\rangle = \langle S, d\varphi\rangle = \frac{1}{2i\pi}\int_{\mathbb{C}} \frac{\partial\varphi}{\partial\overline{z}}(z)\frac{dz\wedge d\overline{z}}{z}$$

since S is defined by the locally integrable differential form $\frac{1}{2i\pi}dz/z$. For any $\varepsilon > 0$, we set $B_\varepsilon = \{z \in \mathbb{C} \mid |z| < \varepsilon\}$ and then

$$\langle dS, \varphi\rangle = \lim_{\varepsilon\to 0}\frac{1}{2i\pi}\int_{\mathbb{C}\setminus B_\varepsilon} \frac{\partial\varphi}{\partial\overline{z}}(z)\frac{dz\wedge d\overline{z}}{z} = -\lim_{\varepsilon\to 0}\frac{1}{2i\pi}\int_{\mathbb{C}\setminus B_\varepsilon} d\Big(\frac{\varphi(z)}{z}dz\Big).$$

Applying Stokes' theorem, we get

$$\int_{\mathbb{C}\setminus B_\varepsilon} d\Big(\frac{\varphi(z)}{z}dz\Big) = -\int_{\partial B_\varepsilon}\frac{\varphi(z)}{z}dz$$

since φ has compact support. Moreover,

$$\int_{\partial B_\varepsilon}\frac{\varphi(z)}{z}dz = \int_{\partial B_\varepsilon}\frac{\varphi(z)-\varphi(0)}{z}dz + \varphi(0)\int_{\partial B_\varepsilon}\frac{dz}{z}.$$

As the function φ is $\mathcal{C}^\infty$ and in particular $\mathcal{C}^1$ the first integral on the right-hand side tends to 0 as ε tends to 0. Moreover, we know by Cauchy's formula that $\frac{1}{2i\pi}\int_{\partial B_\varepsilon} dz/z = 1$ for any $\varepsilon > 0$. It follows that

$$\langle dS, \varphi\rangle = \varphi(0) = \langle [0], \varphi\rangle.$$

Moreover, $bT = -d\psi\wedge[D] + \psi\wedge[bD]$ and hence $SS(bT) = SS(T) = bD$: as we also have $SS(bS) = \{0\}$ it follows that $SS(bT)\cap SS(bS) = \varnothing$ since $0 \in D$.

We can therefore apply Corollary 3.7 and we get

$$\begin{aligned}
\mathcal{K}(bT, S) &= \mathcal{K}\Big(-d\psi\wedge[D] + \psi\wedge[bD], \frac{1}{2i\pi}\frac{dz}{z}\Big)\\
&= -\frac{1}{2i\pi}\int_D d\psi\wedge\frac{dz}{z} + \frac{1}{2i\pi}\int_{bD}\psi(z)\frac{dz}{z}\\
&= \frac{1}{2i\pi}\Big(\int_{bD}\psi(z)\frac{dz}{z} + \int_D\frac{\partial\psi}{\partial\overline{z}}(z)\frac{dz\wedge d\overline{z}}{z}\Big).\\
\mathcal{K}(T, dS) &= \mathcal{K}(\psi[D], [0]) = \langle\psi, [0]\rangle = \psi(0).
\end{aligned}$$

Finally, it follows that

$$\psi(0) = \frac{1}{2i\pi}\Big(\int_{bD}\frac{\psi(z)}{z}dz + \int_D\frac{\partial\psi}{\partial\overline{z}}(z)\frac{dz\wedge d\overline{z}}{z}\Big).$$

Geometric interpretation of the Kronecker index. We state the following result without proof: the interested reader will find more details and better sufficient conditions for the existence of the Kronecker index of two currents in [Rh, § 20] and [L-T1].

If Y and Z are two p- and $(n-p)$-dimensional closed oriented submanifolds of X which meet transversally such that either Y or Z is a compact submanifold of X, then the integration currents $[Y]$ and $[Z]$ on Y and Z are closed and therefore satisfy the hypotheses of Theorem 3.5. Then

$$\mathcal{K}([Y],[Z]) = \langle [Y \cap Z], 1 \rangle.$$

Here $Y \cap Z$ contains a finite number of points and $\langle [Y \cap Z], 1\rangle$ is equal to the number of points of $Y \cap Z$ where the orientations of Y and Z coincide minus the number of points where they differ. More generally, if Y and Z are two closed oriented submanifolds of X of dimensions p and q respectively which meet transversally in such a way that $Y \cap Z$ is a submanifold of X and φ is a $\mathcal{C}^\infty$ differential form of degree $p+q-n$ with compact support on X, then

$$\mathcal{K}([Z],[Y] \wedge \varphi) = \langle [Z \cap Y], \varphi \rangle.$$

4 Complex analytic manifolds

When studying holomorphic functions, it is natural to try to introduce objects which play the role in the holomorphic setting which is played by differentiable manifolds in the differential setting, that is, objects which locally inherit the analytic properties of open sets in $\mathbb{C}^n$.

Definition 4.1. Let X be a topological space. A *complex atlas* on X is a set of charts (U,φ) such that the domains U form an open cover of X and the maps φ are homeomorphisms from U to an open set in $\mathbb{C}^n$ satisfying the holomorphic compatibility condition: if $U \cap U' \neq \varnothing$ then the map

$$\varphi' \circ \varphi^{-1} : \varphi(U \cap U') \longrightarrow \varphi'(U \cap U')$$

is a biholomorphic map between two open sets in $\mathbb{C}^n$. We say that two complex atlases are compatible if their union is also a complex atlas. Compatibility is an equivalence relation.

Definition 4.2. A *complex analytic manifold* is a Hausdorff topological space which is a countable union of compact sets equipped with an equivalence class of complex atlases.

For any complex analytic manifold X, any point of $x \in X$ and any chart (U,φ) in a neighbourhood of x the map φ is a homeomorphism from U to an open set in some $\mathbb{C}^n$ and the number n is called the *complex dimension* of X at x. (Of course, exactly as for differentiable manifolds, the number n

is independent of the choice of chart (U, φ) in a neighbourhood of x.) We say that X is a *complex analytic manifold of dimension* n if for any $x \in X$ the complex dimension of X at x is n. If X and Y are two complex analytic manifolds then a map $f : X \to Y$ is said to be *holomorphic* if it is continuous and for any pair of charts (U, φ) and (V, Ψ) of X and Y such that $f(U) \subset V$, the map $\psi \circ (f|_U) \circ \varphi^{-1} : \varphi(U) \to \psi(V)$ is holomorphic. When $Y = \mathbb{C}$ such a map will be called a holomorphic function. Of course, exactly as for $\mathcal{C}^q$ maps, holomorphy can be checked on all the charts of some given atlas only. We denote the vector space of complex-valued holomorphic functions on X by $\mathcal{O}(X)$.

Definition 4.3. Let (U, φ) be a chart of a complex analytic manifold. The function φ is then a holomorphic map from U to $\mathbb{C}^n$ and we can write $\varphi(x) = (z_1(x), \dots, z_n(x))$ where the $z_j : U \to \mathbb{C}$ $(j = 1, \dots, n)$ are holomorphic functions on U. The functions $(z_1, \dots, z_n)$ are then called the *holomorphic coordinates* of X on U defined by the chart (U, φ).

Remark. It is clear that any n-dimensional complex analytic manifold X has a natural $2n$-dimensional $\mathcal{C}^\infty$ differentiable manifold structure. The tangent and cotangent spaces T_xX and T_x^*X of X at x are therefore well defined. In particular, we have a space $\mathbb{C}T_x^*X$ of complex-valued differential 1-forms at $x \in X$ which is the dual of the complexified space $\mathbb{C}T_xX$ of T_xX (cf. Appendix A, §4). Let us look at what happens in a chart (U, φ) in a neighbourhood of x whose local coordinates are $(z_1, \dots, z_n)$ where $z_j = x_j + iy_j$, $j = 1, \dots, n$. The family $\{(dx_1)_x, (dy_1)_x, \dots, (dx_n)_x(dy_n)_x\}$ is then a basis for $\mathbb{C}T_x^*X$ and the corresponding dual basis in $\mathbb{C}T_xX$ is $\{(\partial/\partial x_1)_x, (\partial/\partial y_1)_x, \dots, (\partial/\partial x_n)_x, (\partial/\partial y_n)_x\}$. It is often more convenient to consider the basis

$$\{(dz_1)_x, (d\overline{z}_1)_x, \dots, (dz_n)_x, (d\overline{z}_n)_x\}$$

in $\mathbb{C}T_x^*X$ and the associated dual basis in $\mathbb{C}T_xX$ which we denote by

$$\Big\{\Big(\frac{\partial}{\partial z_1}\Big)_x, \Big(\frac{\partial}{\partial \overline{z}_1}\Big)_x, \dots, \Big(\frac{\partial}{\partial z_n}\Big)_x, \Big(\frac{\partial}{\partial \overline{z}_n}\Big)_x\Big\}.$$

By definition,

$$(dz_j)_x\Big(\Big(\frac{\partial}{\partial z_k}\Big)_x\Big) = \delta_{jk}, \qquad (d\overline{z}_j)_x\Big(\Big(\frac{\partial}{\partial z_k}\Big)_x\Big) = 0,$$
$$(dz_j)_x\Big(\Big(\frac{\partial}{\partial \overline{z}_k}\Big)_x\Big) = 0 \quad \text{and} \quad (d\overline{z}_j)_x\Big(\Big(\frac{\partial}{\partial \overline{z}_k}\Big)_x\Big) = \delta_{jk}.$$

If f is a complex-valued $\mathcal{C}^1$ function on a neighbourhood of x in X then its differential df_x defines an element of T_x^*X which by definition of a dual basis can be written as

$$(df)_x = \sum_{j=1}^n \frac{\partial f}{\partial x_j}(x)(dx_j)_x + \frac{\partial f}{\partial y_j}(x)(dy_j)_x$$

or alternatively

$$(df)_x = \sum_{j=1}^{n} \frac{\partial f}{\partial z_j}(x)(dz_j)_x + \frac{\partial f}{\partial \overline{z}_j}(x)(d\overline{z}_j)_x$$

depending on our choice of basis. An easy calculation shows that

$$\Big(\frac{\partial}{\partial z_j}\Big)_x = \frac{1}{2}\Big(\Big(\frac{\partial}{\partial x_j}\Big)_x - i\Big(\frac{\partial}{\partial y_j}\Big)_x\Big) \quad \text{and} \quad \Big(\frac{\partial}{\partial \overline{z}_j}\Big)_x = \frac{1}{2}\Big(\Big(\frac{\partial}{\partial x_j}\Big)_x + i\Big(\frac{\partial}{\partial y_i}\Big)_x\Big)$$

which agrees with the definition given in Chapter I.

We further note that the construction of the complexifications of $T_x^* X$ and $T_x X$ – in other words the construction of complex-valued forms – does not involve the complex structure on X and can be carried out for any differentiable manifold. We have only used the complex structure on X when writing certain expressions in local coordinates.

We end this section by proving that any complex analytic manifold is orientable. Let X be a complex analytic manifold of dimension n. By Proposition 6.2 of Appendix A, X is orientable if it has a $\mathcal{C}^\infty$ atlas, $(U_i, h_i)_{i \in I}$, such that, for any $i, j \in I$,

$$d_{ij}(x) = \det \big[J(h_i \circ h_j^{-1})(h_j(x))\big] > 0 \quad \text{for any } x \in U_i \cap U_j.$$

Consider a complex atlas $(U_j, \varphi_j)_{j \in J}$ on X. If $\varphi_j = h_j + ik_j$ then the set $(U_j, (h_j, k_j))_{j \in J}$ is a $\mathcal{C}^\infty$ atlas on X such that, for any $x \in U_i \cap U_j$,

$$d_{ij}(x) = \det \begin{vmatrix} \overline{J(\varphi_i \circ \varphi_j^{-1})}(x) & 0 \\ 0 & J(\varphi_i \circ \varphi_j^{-1})(x) \end{vmatrix} = |J(\varphi_i \circ \varphi_j^{-1})|^2(x) > 0.$$

Any complex analytic manifold X is therefore orientable. Throughout the rest of this book all complex analytic manifolds will be equipped with the following orientation: given a complex atlas $(U_i, \varphi_i)_{i \in I}$ on X, we consider the holomorphic coordinates $(z_1, \dots, z_n)$ associated to the chart (U_i, φ_i) and we choose the orientation associated to the $2n$-differential form

$$d\overline{z}_1 \wedge \cdots \wedge d\overline{z}_n \wedge dz_1 \wedge \cdots \wedge dz_n$$

which is simply the orientation defined by

$$dx_1 \wedge \cdots \wedge dx_n \wedge dy_1 \wedge \cdots \wedge dy_n,$$

where $z_j = x_j + iy_j$.

5 Complex structures

Let X be a complex analytic manifold of dimension n. We will show that at any point $x \in X$ the real vector space $T_x^* X$ has a natural $\mathbb{C}$-vector space structure.

We start by considering the case where $X = \mathbb{C}^n$. $\mathbb{C}^n$ can be naturally identified with $\mathbb{R}^{2n}$ as a $\mathcal{C}^\infty$ manifold, so $T_x\mathbb{C}^n = T_x\mathbb{R}^{2n} = \mathbb{R}^{2n} = \mathbb{C}^n$, where the last equality is just the natural identification of $\mathbb{C}^n$ and $\mathbb{R}^{2n}$. Identifying $T_x\mathbb{C}^n$ with $\mathbb{C}^n$ imposes a $\mathbb{C}$ vector space structure on $T_x\mathbb{C}^n$. Let us examine this identification more carefully. We denote the multiplication by i map by $J : T_x\mathbb{C}^n \to T_x\mathbb{C}^n$; it is an $\mathbb{R}$-linear map such that $J^2 = -\operatorname{Id}$. In the standard basis

$$\Big\{\Big(\frac{\partial}{\partial x_1}\Big)_x, \Big(\frac{\partial}{\partial y_1}\Big)_x, \dots, \Big(\frac{\partial}{\partial x_n}\Big)_x, \Big(\frac{\partial}{\partial y_n}\Big)_x\Big\},$$

for $T_x\mathbb{C}^n = T_x\mathbb{R}^{2n}$, we have $J((\frac{\partial}{\partial x_j})_x) = (\frac{\partial}{\partial y_j})_x$ and $J((\frac{\partial}{\partial y_j})_x) = -(\frac{\partial}{\partial x_j})_x$, $1 \leqslant j \leqslant n$. It follows that

$$(a+ib)\nu = a\nu + bJ(\nu).$$

for any $\nu \in T_x\mathbb{C}^n$ and any complex number $a + ib \in \mathbb{C}$

The homogeneous Cauchy–Riemann equations, which encode the $\mathbb{C}$-linearity at x of $(df)_x$ for any holomorphic function germ f, can be written as

$$(df)_x(J\nu) = i(df)_x(\nu) \quad \text{for any } \nu \in T_x\mathbb{C}^n.$$

Remark. This relationship between the complex structure on $T_x\mathbb{C}^n$ and holomorphic functions implies that J is independent of the choice of coordinates on $\mathbb{C}^n$. We can therefore define J on the tangent space of a complex analytic manifold.

Theorem 5.1. *Let X be a complex analytic manifold. For any $x \in X$, there is a unique $\mathbb{R}$-linear map $J = J_x : T_xX \to T_xX$ such that, for any holomorphic function germ at x,*

$$(df)_x(J\nu) = i(df)_x(\nu) \quad \textit{for every } \nu \in T_xX.$$

Moreover, $J^2 = -\operatorname{Id}$ and setting $(a+ib)\nu = a\nu + bJ(\nu)$ for any $a + ib \in \mathbb{C}$ and $\nu \in T_xX$ yields a complex vector space structure on T_xX.

Proof. We start by proving that if J exists then it is unique.

Let $(z_1, \dots, z_n)$ be a system of holomorphic coordinates in a neighbourhood of $x \in X$ and let $(x_1, y_1, x_2, y_2, \dots, x_n, y_n)$ be the underlying real coordinates. Then, for any $j = 1, \dots, n$, $(dz_j)_x = (dx_j)_x + i(dy_j)_x$ or alternatively $(dx_j)_x = \operatorname{Re}(dz_j)_x$ and $(dy_j)_x = \operatorname{Im}(dz_j)_x$.

We now calculate $J(\frac{\partial}{\partial x_k})_x$ for $k = 1, \dots, n$. We have

$$J\Big(\frac{\partial}{\partial x_k}\Big)_x = \sum_{j=1}^{n} \Big[dx_j\Big(J\Big(\frac{\partial}{\partial x_k}\Big)\Big)\Big]\frac{\partial}{\partial x_j} + \Big[dy_j\Big(J\Big(\frac{\partial}{\partial x_k}\Big)\Big)\Big]\frac{\partial}{\partial y_j}.$$

As z_j is holomorphic, $dz_j(\frac{\partial}{\partial x_k})_x = \delta_{jk}$, and

$$dx_j\Big(J\Big(\frac{\partial}{\partial x_k}\Big)_x\Big) = \operatorname{Re} dz_j\Big(J\Big(\frac{\partial}{\partial x_k}\Big)_x\Big) = \operatorname{Re} i dz_j\Big(\frac{\partial}{\partial x_k}\Big)_x = \operatorname{Re} i\delta_{jk} = 0,$$

$$dy_j\Big(J\Big(\frac{\partial}{\partial x_k}\Big)_x\Big) = \operatorname{Im} dz_j\Big(J\Big(\frac{\partial}{\partial x_k}\Big)_x\Big) = \operatorname{Im} i dz_j\Big(\frac{\partial}{\partial x_k}\Big)_x = \delta_{jk}.$$

It follows that $J(\frac{\partial}{\partial x_k})_x = (\frac{\partial}{\partial y_k})_x$ and likewise $J(\frac{\partial}{\partial y_k})_x = -(\frac{\partial}{\partial x_k})_x$ and hence J is unique.

We now define the map J by the above equations with respect to some chosen set of holomorphic coordinates $(z_1, \dots, z_n)$. We saw above for $\mathbb{C}^n$ that J then has the desired properties and the fact that any map having these properties is unique implies that the map J thus defined is independent of our choice of holomorphic coordinates. □

We now consider the link between the complex structures on T_xX and $\mathbb{C}T_xX$. The map $J : T_xX \to T_xX$ is $\mathbb{R}$-linear and $J^2 = -\mathrm{Id}$. This map therefore has no real eigenvalues and if we want to diagonalise it we have to consider the natural extension of J to a $\mathbb{C}$-linear map, also denoted J, from $\mathbb{C}T_xX$ to itself. This extension then has two eigenvalues, i and $-i$. We denote the eigenspace associated to i by $T_x^{1,0}X$ and the eigenspace associated to $-i$ by $T_x^{0,1}X$: we then have

$$\mathbb{C}T_xX = T_x^{1,0}X \oplus T_x^{0,1}X.$$

If $(z_1, \dots, z_n)$ is a system of holomorphic coordinates in a neighbourhood of x then the vectors $((\frac{\partial}{\partial z_1})_x, \dots, (\frac{\partial}{\partial z_n})_x)$ form a basis for $T_x^{1,0}X$ and the vectors $((\frac{\partial}{\partial \bar{z}_1})_x, \dots, (\frac{\partial}{\partial \bar{z}_n})_x)$ form a basis of $T_x^{0,1}X$.

We note that T_xX with the $\mathbb{C}$-vector space structure defined by J is naturally isomorphic to $T_x^{1,0}X$ via the map sending ν to $\frac{1}{2}(\nu - iJ(\nu))$. This map sends the family $((\frac{\partial}{\partial x_1})_x, \dots, (\frac{\partial}{\partial x_n})_x)$ – which is a basis for T_xX as a $\mathbb{C}$-vector space – to the basis $((\frac{\partial}{\partial z_1})_x, \dots, (\frac{\partial}{\partial z_n})_x)$ of $T_x^{1,0}X$ and is therefore an isomorphism.

We define $T^{1,0}(X)$ to be the disjoint union of the spaces $T_x^{1,0}(X)$ for all $x \in X$ and we denote the natural projection from $T^{1,0}(X)$ to X by p.

Definition 5.2. Let X be a complex analytic manifold and let A be an open set in X. A *field of holomorphic vectors* on A is a map $V : A \to T^{1,0}(X)$ such that $p \circ V = \mathrm{Id}$.

6 Differential forms of type (p, q)

Let X be a complex analytic manifold of dimension n and let x be a point in X. The fact that T_xX has a complex vector space structure leads us to give special consideration to those elements in $\mathbb{C}T_xX$ which are $\mathbb{C}$-linear with respect to this structure.

We define the space of differential 1-forms of type $(1, 0)$ at x by

$$\Lambda^{1,0}(T_x^*X) = \{\omega \in \mathbb{C}T_x^*X \mid \omega(J\nu) = i\omega(\nu), \forall\, \nu \in \mathbb{C}T_xX\}.$$

Example. The differentials $(df)_x$ of germs of holomorphic functions at x are of type $(1, 0)$ by definition of J.

If $(z_1, \dots, z_n)$ are local holomorphic coordinates defined in a neighbourhood of x then the family $((dz_1)_x, \dots, (dz_n)_x)$ forms a basis for $\Lambda^{1,0}(T_x^*X)$. The conjugate space $\Lambda^{0,1}T_x^*X = \overline{\Lambda^{1,0}T_x^*X}$, which has a basis given in these coordinates by $((d\overline{z}_1)_x, \dots, (d\overline{z}_n)_x)$, is the space of forms of type $(0,1)$ at x. There is a direct sum decomposition

$$\mathbb{C}T_x^*X = \Lambda^{1,0}T_x^*X \oplus \Lambda^{0,1}T_x^*X. \tag{6.1}$$

We now consider forms of higher degree. If ω is a complex-valued differential form of degree r then it is a linear combination of elements of the form $\omega_1 \wedge \cdots \wedge \omega_r$ where $\omega_j \in \mathbb{C}T_x^*X$. By (3.1), each ω_j, $1 \leqslant j \leqslant r$, can be written in the form $\omega_j' + \omega_j''$, where $\omega_j' \in \Lambda^{1,0}T_x^*X$ and $\omega_j'' \in \Lambda^{0,1}T_x^*X$. It follows that ω is a linear combination of elements of the form $\eta_1 \wedge \cdots \wedge \eta_r$ where each η_j is either of type $(0,1)$ or of type $(1,0)$.

We say that ω is a differential form of type (p,q) or bidegree (p,q) at x if ω is a linear combination of elements of the form $\omega_{i_1} \wedge \cdots \wedge \omega_{i_p} \wedge \overline{\omega}_{j_1} \wedge \cdots \wedge \overline{\omega}_{j_q}$ where all the ω_ν are 1-forms of type $(1,0)$ at x.

We denote by $\mathcal{C}^k_{p,q}(X)$ the subspace of $\mathcal{C}^k_{p+q}(X)$ consisting of $(p+q)$-differential forms which are of type (p,q) at every point. We then have a direct sum decomposition

$$\mathcal{C}^k_r(X) = \bigoplus_{p+q=r} \mathcal{C}^k_{p,q}(X).$$

Note that $\mathcal{C}^k_{p,q}(X) = \{0\}$ if p or $q > n = \dim_{\mathbb{C}} X$.

If $(z_1, \dots, z_n)$ are holomorphic coordinates on a chart domain $U \subset X$ then $dz_j \in \mathcal{C}^\infty_{1,0}(U)$ for any $j = 1, \dots, n$ and any (p,q)-form $\omega \in \mathcal{C}^k_{p,q}(U)$ can then be written uniquely in the form

$$\omega = \sum_{\substack{|I|=p \\ |J|=q}} a_{IJ} dz_I \wedge d\overline{z}_J,$$

where the a_{IJ} are $\mathcal{C}^k$ functions on U and the sum is taken over strictly increasing multi-indices $I = (i_1, \dots, i_p)$ and $J = (j_1, \dots, j_q)$.

7 The $\overline{\partial}$ operator and Dolbeault cohomology

The decomposition of differential 1-forms on a complex analytic manifold into type $(0,1)$ and type $(1,0)$ forms induces a natural decomposition of the exterior differential operator d into a holomorphic differential operator and an antiholomorphic differential operator.

Let X be a complex analytic manifold of dimension n. If f is a $\mathcal{C}^1$ function on X then, for any $x \in X$,

$$(df)_x = \sum_{j=1}^n \frac{\partial f}{\partial z_j}(x)(dz_j)_x + \sum_{j=1}^n \frac{\partial f}{\partial \overline{z}_j}(x)(d\overline{z}_j)_x.$$

We can therefore write $df = \partial f + \overline{\partial} f$ where ∂f is a differential form of type $(1,0)$ on X and $\overline{\partial} f$ is a differential form of type $(0,1)$ on X. We note that the condition $f \in \mathcal{O}(X)$ is then equivalent to $\overline{\partial} f = 0$.

The decomposition $d = \partial + \overline{\partial}$ can be extended to forms of any degree in the following way. Suppose that $\omega \in \mathcal{C}^1_{p,q}(X)$ is given in some local system of holomorphic coordinates $(z_1, \dots, z_n)$ in a neighbourhood of $x \in X$, by $\omega = \sum_{\substack{|I|=p \\ |J|=q}} a_{IJ} dz_I \wedge d\overline{z}_J$. By definition of d,

$$dw = \sum_{\substack{|I|=p \\ |J|=q}} d(a_{IJ}) \wedge dz_I \wedge d\overline{z}_J = \sum_{\substack{|I|=p \\ |J|=q}} (\partial a_{IJ} + \overline{\partial} a_{IJ}) \wedge dz_I \wedge d\overline{z}_J.$$

We then set

$$\partial \omega = \sum_{\substack{|I|=p \\ |J|=q}} \partial(a_{IJ}) dz_I \wedge d\overline{z}_J = \sum_{\substack{|I|=p \\ |J|=q}} \sum_{k=1}^{n} \frac{\partial}{\partial z_k} a_{IJ} dz_k \wedge dz_I \wedge d\overline{z}_J$$

and

$$\overline{\partial} \omega = \sum_{\substack{|I|=p \\ |J|=q}} \overline{\partial}(a_{IJ}) dz_I \wedge d\overline{z}_J = \sum_{\substack{|I|=p \\ |J|=q}} \sum_{k=1}^{n} \frac{\partial a_{IJ}}{\partial \overline{z}_k} d\overline{z}_k \wedge dz_I \wedge d\overline{z}_J.$$

We have therefore defined operators ∂ and $\overline{\partial}$ on $\mathcal{C}^1_{p,q}(X)$ such that $\partial(\mathcal{C}^1_{p,q}(X))$ is contained in $\mathcal{C}_{p+1,q}(X)$ and $\overline{\partial}(\mathcal{C}^1_{p,q}(X))$ is contained in $\mathcal{C}_{p,q+1}(X)$

Proposition 7.1. *The operators ∂ and $\overline{\partial}$ have the following properties:*

a) $d = \partial + \overline{\partial}$ *on* $\mathcal{C}^1_{\bullet}(X)$,
b) $\partial \circ \partial = 0$, $\overline{\partial} \circ \overline{\partial} = 0$ *and* $\partial \circ \overline{\partial} + \overline{\partial} \circ \partial = 0$ *on* $\mathcal{C}^2_{\bullet,\bullet}(X)$,
c) ∂ *and* $\overline{\partial}$ *commute with pullback.*

Proof. Property a) follows immediately from the definitions of ∂ and $\overline{\partial}$.

Consider an element $\omega \in \mathcal{C}^2_{p,q}(X)$. As $d \circ d = 0$ and $d = \partial + \overline{\partial}$,

$$0 = (\partial + \overline{\partial}) \circ (\partial + \overline{\partial})\omega = (\partial \circ \partial)\omega + (\partial \circ \overline{\partial} + \overline{\partial} \circ \partial)\omega + (\overline{\partial} \circ \overline{\partial})\omega.$$

But $(\partial \circ \partial)\omega$ is now of type $(p+2, q)$, $(\partial \circ \overline{\partial} + \overline{\partial} \circ \partial)\omega$ is of type $(p+1, q+1)$ and $(\overline{\partial} \circ \overline{\partial})\omega$ is of type $(p, q+2)$. It follows that each of the terms vanishes since their sum vanishes.

Let $F : X \to Y$ be a holomorphic map: note that if $(\zeta_1, \dots, \zeta_n)$ are holomorphic coordinates in a neighbourhood of a point $y \in Y$ then $F^*\zeta_j = \zeta_j \circ F$ is a holomorphic function in a neighbourhood of $x = F^{-1}(y)$ and hence $F^*(d\zeta_j) = d(\zeta_j \circ F)$ is a $(1,0)$ form and $F^*(d\overline{\zeta}_j) = d(\overline{\zeta_j \circ F})$ is a $(0,1)$ form. Using local coordinates, this implies that $F^*(\mathcal{C}^k_{p,q}(X)) \subset \mathcal{C}^k_{p,q}(X)$ for all

$p, q \geqslant 0$ and $k \geqslant 0$. Since Proposition 8.4 of Appendix A says that $dF^* = F^*d$, we can write

$$\partial(F^*\omega) + \overline{\partial}(F^*\omega) = d(F^*\omega) = F^*(d\omega) = F^*(\partial\omega) + F^*(\overline{\partial}\omega)$$

for any $\omega \in \mathcal{C}_{p,q}(Y)$, where the last equality holds because F^* is linear. Comparing bidegrees, we see that $\partial(F^*\omega) = F^*(\partial\omega)$ and $\overline{\partial}(F^*\omega) = F^*(\overline{\partial}\omega)$. □

The operator $\overline{\partial}$ defined above is called the *Cauchy–Riemann operator.*

We denote the space of (p, q)-forms of class $\mathcal{C}^\infty$ on X by $\mathcal{E}^{p,q}(X)$ for any $0 \leqslant p \leqslant n$ and $0 \leqslant q \leqslant n$. These spaces are equipped with a Fréchet space topology which can be characterised as follows: a sequence $(\omega_j)_{j\in\mathbb{N}}$ of elements in $\mathcal{E}^{p,q}(X)$ converges to 0 if and only if for any chart domain U in X on which ω_j can be written as $\omega_j = \sum_{\substack{|I|=p \\ |J|=q}} \omega^j_{IJ} dz_I \wedge d\overline{z}_J$ the sequences $(\omega^j_{IJ})_{j\in\mathbb{N}}$ converge to zero uniformly on any compact subset of U and so do all their derivatives.

The operator $\overline{\partial} : \mathcal{E}^{p,q}(X) \to \mathcal{E}^{p,q+1}(X)$ is then a continuous linear operator and its kernel, denoted by $Z^{p,q}(X)$, is therefore closed.

Definition 7.2. We define the *Dolbeault cohomology groups* to be the spaces

$$H^{p,q}(X) = Z^{p,q}(X)/\overline{\partial}\mathcal{E}^{p,q-1}(X).$$

These spaces are naturally equipped with the quotient topology which is not generally Hausdorff because the space $\overline{\partial}\mathcal{E}^{p,q-1}(X)$ is not always closed. If $\overline{\partial}\mathcal{E}^{p,q-1}(X)$ is closed then $H^{p,q}(X)$ is a Fréchet space.

These groups encode the obstruction to solving the Cauchy–Riemann equations $\overline{\partial}u = f$ for any $f \in Z^{p,q}(X)$.

We end by defining Dolbeault cohomology groups with support conditions. We denote by c the family of compact subsets of X. For any compact subset K in a manifold M, Φ denotes the family of closed sets in $X = M \setminus K$ whose closure is compact in M and Ψ denotes the family of closed sets in M which do not meet K: for simplicity we let Θ be one of these three families. The space $\mathcal{E}^{p,q}_\Theta(X)$ is then the space of $\mathcal{C}^\infty$ (p, q)-forms on X whose support is contained in the family Θ. If $\theta \in \Theta$, we denote by $\mathcal{E}^{p,q}_\theta(X)$ the subspace of $\mathcal{E}^{p,q}(X)$ consisting of (p, q)-forms supported on θ. Then $\mathcal{E}^{p,q}_\Theta(X) = \bigcup_{\theta\in\Theta} \mathcal{E}^{p,q}_\theta(X)$. We note that if Θ is one of the three families c, Φ or Ψ then X has an exhaustion $(\theta_i)_{i\in\mathbb{N}}$ by elements of Θ (i.e. $X = \bigcup_{i\in\mathbb{N}} \theta_i, \theta_i \subset \mathring{\theta}_{i+1}$). The spaces $\mathcal{E}^{p,q}_\theta(X)$ are closed in $\mathcal{E}^{p,q}(X)$; they are therefore Fréchet spaces and the topology on $\mathcal{E}^{p,q}_\Theta(X)$ is the finest topology for which the inclusions $\mathcal{E}^{p,q}_{\theta_i}(X) \hookrightarrow \mathcal{E}^{p,q}_\Theta(X)$ are all continuous. The operator $\overline{\partial}$ is then a continuous linear operator from $\mathcal{E}^{p,q}_\Theta(X)$ to $\mathcal{E}^{p,q+1}_\Theta(X)$. We set $Z^{p,q}_\Theta(X) = Z^{p,q}(X) \cap \mathcal{E}^{p,q}_\Theta(X)$.

Definition 7.3. The *Dolbeault cohomology groups with support in* Θ are the spaces

$$H^{p,q}_\Theta(X) = Z^{p,q}_\Theta(X)/\overline{\partial}\mathcal{E}^{p,q-1}_\Theta(X).$$

We equip these groups with the quotient topology which is not generally Hausdorff. They encode the obstruction to solving the Cauchy–Riemann equation in the class of forms with support in the family Θ.

8 Complex tangent space to the boundary of a domain

When we come to define CR functions (Chapter IV) and pseudoconvex domains (Chapter VI) we will need the properties of the tangent space to the boundary of a domain with smooth boundary in a complex analytic manifold. The aim of this section is to study the analytic properties of this space: in particular, we will consider its interaction with the complex structure of the surrounding manifold.

We initially only assume that X is a $\mathcal{C}^\infty$ differentiable manifold.

Definition 8.1. Let D be an open set in X. For any $1 \leqslant k \leqslant \infty$ we say that D has $\mathcal{C}^k$ boundary in a neighbourhood of $p \in \partial D$ if there is an open neighbourhood U of p in X and a real-valued $\mathcal{C}^k$ function $r \in \mathcal{C}^k(U)$ such that

$$\begin{cases} U \cap D = \{x \in U \mid r(x) < 0\} \\ dr(x) \neq 0, \quad x \in U. \end{cases} \tag{8.1}$$

We say that ∂D is $\mathcal{C}^k$ if it is $\mathcal{C}^k$ in a neighbourhood of every point. A function $r \in \mathcal{C}^k(U)$ such that (8.1) holds is called a *defining function* for D at p. If U is a neighbourhood of ∂D then r is called a global defining function.

Lemma 8.2. *Let r_1 and r_2 be two defining functions for D which are $\mathcal{C}^k$ on a neighbourhood U of $p \in \partial D$. There is then a strictly positive function $h \in \mathcal{C}^{k-1}(U)$ such that*

$$\begin{cases} r_1 = hr_2 & \text{on } U \\ dr_1(x) = h(x)dr_2(x) & \text{for all } x \in U \cap \partial D. \end{cases} \tag{8.2}$$

Proof. Note that h is unique if it exists because it is continuous on U and equal to r_1/r_2 on $U \setminus \partial D$.

Without loss of generality we can assume that U is contained in a chart domain of X. Consider a point $q \in U \cap \partial D$ and choose coordinates on U such that $q = 0$ and $U \cap \partial D = \{x \in \mathbb{R}^n \mid x_n = 0\}$. We can assume that $r_2(x) = x_n$. For any $x' = (x_1, \dots, x_{n-1})$ close enough to 0, we have $r_1(x', 0) = 0$ and hence

$$r_1(x', x_n) = r_1(x', x_n) - r_1(x', 0) = x_n \int_0^1 \frac{\partial r_1}{\partial x_n}(x', tx_n)dt.$$

We set $h(x) = \int_0^1 \frac{\partial r_1}{\partial x_n}(x', tx_n)dt$; $h(x)$ is then a $\mathcal{C}^{k-1}$ function on U such that $r_1 = hr_2$ on U.

If $k \geqslant 2$ then $dr_1(x) = r_2(x)dh(x) + h(x)dr_2(x) = h(x)dr_2(x)$ for any $x \in U \cap \partial D$.

If $k = 1$, $r_1(x) = h(x)r_2(x) = \big(h(x) - h(x',0)\big)r_2(x) + h(x',0)r_2(x)$ and hence $r_1(x) = h(x',0)dr_2(x',0) + o(x_n)$ as x_n tends to 0 since h is continuous on U and $r_2(x',0) = 0$. It therefore follows that $dr_1(x) = h(x)dr_2(x)$ for any $x \in U \cap \partial D$.

It remains to prove that h is strictly positive on U. As $h = r_1/r_2$ on $U \smallsetminus D$, h is strictly positive on $U \smallsetminus D$ because r_1 and r_2 are defining functions. As $dr_1(x) \neq 0$ on U and $dr_1(x) = h(x)dr_2(x)$ on $U \cap \partial D$, h does not vanish on $U \cap \partial D$. As h is continuous on U it is strictly positive on U. □

If D is a domain with $\mathcal{C}^k$ boundary in a neighbourhood of $p \in \partial D$ then ∂D is a $\mathcal{C}^k$ differentiable manifold in a neighbourhood of p. We can therefore consider the tangent space $T_p(\partial D)$ to ∂D at p.

Proposition 8.3. *If r is a defining function for D at p then*

$$T_p(\partial D) = \{\xi \in T_p(X) \mid dr(p)(\xi) = 0\}. \tag{8.3}$$

If $(x_1, \dots, x_n)$ are local coordinates on X in a neighbourhood of p then $\xi \in T_p(\partial D)$ if and only if

$$\xi = \sum_{j=1}^{n} \xi_j \Big(\frac{\partial}{\partial x_j}\Big)_p \quad \textit{where} \quad \sum_{j=1}^{n} \frac{\partial r}{\partial x_j}(p)\xi_j = 0.$$

Proof. Let U be a neighbourhood of p such that

$$\partial D \cap U = \{x \in U \mid r(x) < 0\} \quad \text{and} \quad dr(x) \neq 0 \text{ for any } x \in U.$$

We denote the inclusion of $\partial D \cap U$ in X by i. This inclusion induces an injective map $di : T_p(\partial D) \hookrightarrow T_p(X)$ such that if α is a curve in ∂D passing through p representing the vector $\nu \in T_p(\partial D)$ then $\xi = di(\nu)$ is the class of $i \circ \alpha$. Since the image of α is contained in ∂D we have $r \circ i \circ \alpha \equiv 0$ and hence

$$dr(p)(\xi) = \frac{d}{dt}(r \circ i \circ \alpha)(0) = 0$$

which proves that on identifying $T_p(\partial D)$ and $di(T_p(\partial D))$

$$T_p(\partial D) \subset \{\xi \in T_p(X) \mid dr(p)(\xi) = 0\}. \tag{8.4}$$

As both sides of (8.4) are vector spaces of dimension $(n-1)$, the inclusion of (8.4) is in fact an equality. □

Remark. Equation (8.3) shows that we can identify $T_p(\partial D)$ with the set of directional derivatives at p which vanish on r.

Assume now that X is a complex analytic manifold of dimension n and D is an open set in X with $\mathcal{C}^k$ boundary in a neighbourhood of $p \in \partial D$. The complex structure on X induces an extra structure on $T_p(\partial D)$.

As we have seen above, we can identify $T_p(\partial D)$ with a real subspace of real dimension $(2n-1)$ in $T_p(X)$. If J is the complex structure on $T_p(X)$ we can consider $JT_p(\partial D)$, which is also a real subspace of real dimension $(2n-1)$ in $T_p(X)$, and hence

$$T_p^{\mathbb{C}}(\partial D) = T_p(\partial D) \cap JT_p(\partial D)$$

is a real subspace of real dimension $(2n-2)$ in $T_p(X)$ which is stable under J. This space is therefore a complex subspace of $T_p(X)$ of complex dimension $(n-1)$. We note that $T_p^{\mathbb{C}}(\partial D) \neq \{0\}$ if and only if $n \geqslant 2$. The space $T_p^{\mathbb{C}}(\partial D)$ is called the *complex tangent space* to ∂D at p. If we identify $T_p(X)$ with the complex structure J with $T_p^{1,0}(X)$ then $T_p^{\mathbb{C}}(\partial D)$ becomes a subspace of $T_p^{1,0}(X)$.

Proposition 8.4. *If r is a defining function for D at $p \in \partial D$ then*

$$T_p^{\mathbb{C}}(\partial D) = \{t \in T_p^{1,0}(X) \mid \partial r(p)(t) = 0\}.$$

If $(z_1, \dots, z_n)$ are holomorphic local coordinates for X in a neighbourhood of p then $t \in T_p^{\mathbb{C}}(\partial D)$ if and only if

$$t = \sum_{j=1}^{n} t_j \Big(\frac{\partial}{\partial z_j}\Big)_p, \quad \textit{where} \sum_{j=1}^{n} \frac{\partial r}{\partial z_j}(p) t_j = 0.$$

Proof. As the function r is real-valued

$$dr(p) = \partial r(p) + \overline{\partial} r(p) = 2 \operatorname{Re} \partial r(p).$$

By definition of $T_p^{\mathbb{C}}(\partial D) = T_p(\partial D) \cap JT_p(\partial D)$ and (8.3),

$$T_p^{\mathbb{C}}(\partial D) = \{t \in T_p^{1,0}(X) \mid dr(p)(t) = dr(p)(Jt) = 0\}.$$

But now $\partial r(p)(Jt) = i\partial r(p)(t)$ since $\partial r(p)$ is a $(1,0)$ differential form at p and hence

$$\operatorname{Re}\big(\partial r(p)(Jt)\big) = -\operatorname{Im} \partial r(p)(t).$$

It follows that

$$\begin{aligned} T_p^{\mathbb{C}}(\partial D) &= \{t \in T_p^{1,0}(X) \mid \operatorname{Re} \partial r(p)(t) = \operatorname{Im} \partial r(p)(t) = 0\} \\ &= \{t \in T_p^{1,0}(X) \mid \partial r(p)(t) = 0\}. \end{aligned}$$ □

Let $\mathbb{C}T_p(\partial D)$ be the complexification of $T_p(\partial D)$. $T_p^{\mathbb{C}}(\partial D)$ is then an $(n-1)$-dimensional subspace of $\mathbb{C}T_p(\partial D)$.

Definition 8.5. The vector space $T_p^{0,1}(\partial D) = \overline{T_p^{\mathbb{C}}(\partial D)}$, the conjugate of $T_p^{\mathbb{C}}(\partial D)$ in $\mathbb{C}T_p(\partial D)$, is called the *space of tangential Cauchy–Riemann operators* at $p \in \partial D$.

Note that if r is a defining function for ∂D at p and $(z_1, \ldots, z_n)$ are local holomorphic coordinates on X in a neighbourhood of p then a vector $\tau \in T_p^{0,1}(\partial D)$ if and only if

$$\tau = \sum_{j=1}^{n} \tau_j \Big(\frac{\partial}{\partial \overline{z}_j}\Big)_p \quad \text{where} \sum_{j=1}^{n} \frac{\partial r}{\partial z_j}(p)\overline{\tau}_j = 0.$$

Example. If $n = 2$ then $T_p^{0,1}(\partial D)$ is a 1-dimensional $\mathbb{C}$-vector space generated by

$$L_p = \frac{\partial r}{\partial \overline{z}_2}(p)\frac{\partial}{\partial \overline{z}_1} - \frac{\partial r}{\partial \overline{z}_1}(p)\frac{\partial}{\partial \overline{z}_2}.$$

Comments

The theory of currents is developed in Schwarz's book [Sc] and de Rham's book [Rh]. De Rham's book [Rh] also contains a discussion of regularisations on manifolds and the Kronecker index and more information on the Kronecker index can be found in [L-T1]. Whilst writing Sections 4 to 8 of this chapter the author relied on the sections dealing with similar material in Section 2 of Chapter III and Section 2 of Chapter II of M. Range's book [Ra]. The interested reader may consult R. Narasimhan's book [Na2] for more details.

III

The Bochner–Martinelli–Koppelman kernel and formula and applications

In this chapter we define one of the fundamental tools of integral representation theory in complex analysis, namely the Bochner–Martinelli–Koppelman kernel. This kernel generalises the Cauchy kernel on $\mathbb{C}$ to $\mathbb{C}^n$. It enables us to prove an integral representation formula, the Bochner–Martinelli–Koppelman formula, which extends Cauchy's formula to (p,q) differential forms in $\mathbb{C}^n$. This formula plays an important role in the study of the operator $\overline{\partial}$: in particular, we prove using this formula our first results on the existence of solutions to the Cauchy–Riemann equation in $\mathbb{C}^n$ by considering the case where the data has compact support. Hartog's phenomenon, a special case of which was studied in Chapter I, follows from the existence of a compactly supported solution to the Cauchy–Riemann equations for $n \geqslant 2$ when the right-hand side of the equation is a compactly supported form of bidegree $(0,1)$. The links between the vanishing of compactly supported Dolbeault cohomology groups in bidegree $(0,1)$ and Hartog's phenomenon will be explored in greater detail in Chapter V. We will also use the Bochner–Martinelli–Koppelman formula to study the regularity of the operator $\overline{\partial}$ by proving a Hölder hypoellipticity theorem.

1 The Bochner–Martinelli–Koppelman kernel and formula

For any $(\xi,\eta) \in \mathbb{C}^n \times \mathbb{C}^n$ we set $\langle \xi,\eta \rangle = \sum_{j=1}^n \xi_j\eta_j$. We consider the set E defined by $E = \{(\xi,\eta) \in \mathbb{C}^n \times \mathbb{C}^n \mid \langle \xi,\eta \rangle \neq 0\}$ and on the open set E we define a differential form μ by $\mu(\xi,\eta) = \langle \xi,\eta \rangle^{-n}\omega'(\xi) \wedge \omega(\eta)$ where

$$\omega(\eta) = d\eta_1 \wedge \cdots \wedge d\eta_n \quad \text{and} \quad \omega'(\xi) = \sum_{j=1}^n (-1)^{j-1}\xi_j d\xi_1 \wedge \cdots \wedge \widehat{d\xi_j} \wedge \cdots \wedge d\xi_n,$$

where $\widehat{d\xi_j}$ indicates the suppression of the term $d\xi_j$.

Lemma 1.1. *The differential form μ has the property that $d\mu = 0$ on E.*

C. Laurent-Thiébaut, *Holomorphic Function Theory in Several Variables: An Introduction*, Universitext, DOI 10.1007/978-0-85729-030-4_3,

Proof. Let us calculate $d\mu$.

$$d\mu = n\langle\xi,\eta\rangle^{-n}\omega(\xi)\wedge\omega(\eta) - n\langle\xi,\eta\rangle^{-n-1}d(\langle\xi,\eta\rangle)\wedge\omega'(\xi)\wedge\omega(\eta).$$

As $d(\langle\xi,\eta\rangle)\wedge\omega'(\xi)\wedge\omega(\eta) = \langle\xi,\eta\rangle\omega(\xi)\wedge\omega(\eta)$ it follows that $d\mu = 0$. □

We denote the diagonal in $\mathbb{C}^n\times\mathbb{C}^n$ by Δ. The image of the map $\widehat{s}$ from $\mathbb{C}^n\times\mathbb{C}^n\setminus\Delta$ to $\mathbb{C}^n\times\mathbb{C}^n$ defined by $\widehat{s}(\zeta,z) = (\overline{\zeta}-\overline{z},\zeta-z)$ is therefore contained in E. We now consider the differential form $\widehat{s}^*\mu$. This form is $\mathcal{C}^\infty$ on $\mathbb{C}^n\times\mathbb{C}^n\setminus\Delta$ and is closed because μ is closed.

Definition 1.2. The *Bochner–Martinelli–Koppelman kernel* is the differential form on $\mathbb{C}^n\times\mathbb{C}^n\setminus\Delta$ defined by $B = \frac{(n-1)!}{(2i\pi)^n}\widehat{s}^*\mu$. We have

$$(1.1)\quad B(z,\zeta) = \frac{(n-1)!}{(2i\pi)^n}\,\frac{\sum_{j=1}^n(-1)^{j-1}(\overline{\zeta}_j-\overline{z}_j)\underset{s\neq j}{\wedge}(d\overline{\zeta}_s-d\overline{z}_s)\wedge\omega(\zeta-z)}{|\zeta-z|^{2n}}.$$

We set

$$B = \sum_{\substack{0\leqslant p\leqslant n\\ 0\leqslant q\leqslant n-1}} B^p_q,$$

where B^p_q is of type (p,q) in z and type $(n-p,n-q-1)$ in ζ. We also set $B^p_{-1} = 0$.

Lemma 1.3. *For any $(z,\zeta)\in\mathbb{C}^n\times\mathbb{C}^n\setminus\Delta$*

$$\overline{\partial}B(z,\zeta) = 0 \quad \textit{and} \quad \overline{\partial}_z B^p_q(z,\zeta) = -\overline{\partial}_\zeta B^p_{q+1}(z,\zeta).$$

In particular, $\overline{\partial}_\zeta B^0_0(z,\zeta) = 0$.

Proof. We denote by $k_{\mathrm{BM}}(x)$ the $\mathcal{C}^\infty$ differential form on $\mathbb{C}^n$ minus 0 defined by $k_{\mathrm{BM}}(x) = \frac{(n-1)!}{(2i\pi)^n}\mu(\overline{x},x)$. This is an $(n,n-1)$-form and hence $\overline{\partial}k_{\mathrm{BM}} = dk_{\mathrm{BM}}$. We note that $B = \tau^*k_{\mathrm{BM}}$ where τ is the holomorphic map from $\mathbb{C}^n\times\mathbb{C}^n$ to $\mathbb{C}^n$ given by $(z,\zeta)\mapsto\zeta-z$: it follows that $\overline{\partial}B = dB$. As $B = \frac{(n-1)!}{(2i\pi)^n}\widehat{s}^*\mu$ and $d\mu = 0$ we have $\overline{\partial}B = dB = 0$. The second formula follows on comparing bidegrees. □

We can give another expression for B using determinants over algebras. If $\mathcal{A}$ is an arbitrary algebra (for example, the algebra of differential forms) and $A = (a_{ij})_{\substack{1\leqslant i\leqslant n\\ 1\leqslant j\leqslant n}}$ is an order n matrix with coefficients in $\mathcal{A}$ then we set

$$\det A = \sum_{\sigma\in\mathfrak{S}_n}\mathcal{E}(\sigma)a_{\sigma(1),1}\cdots a_{\sigma(n),n},$$

where $\mathfrak{S}_n$ is the group of permutations of $\{1,\dots,n\}$ and $\mathcal{E}(\sigma)$ is the signature of the permutation σ. This determinant has the following properties:

i) $\det A$ is multilinear in the columns of A for linear combinations whose coefficients lie in the centre of $\mathcal{A}$.
ii) if there are integers k and ℓ, $1 \leqslant k < \ell \leqslant n$, such that $a_{ik} = b_k c_i$ and $a_{i\ell} = b_\ell c_i$ for any $i \in \{1, \dots, n\}$, where b_k and b_ℓ are arbitrary elements of $\mathcal{A}$ and the c_i are contained in the centre of $\mathcal{A}$, then $\det A = 0$.

We can then write

$$\omega'(\xi) = \frac{1}{(n-1)!} \det \begin{pmatrix} \xi_1 & d\xi_1 \cdots d\xi_1 \\ \vdots & \vdots \quad \vdots \\ \xi_n & \underbrace{d\xi_n \cdots d\xi_n}_{(n-1)} \end{pmatrix}$$

which can also be written as

$$\omega'(\xi) = \frac{1}{(n-1)!} \det(\xi, \underbrace{d\xi}_{n-1}).$$

We deduce from this the following expression for the Bochner–Martinelli–Koppelman kernel

$$B(z,\zeta) = \frac{1}{(2i\pi)^n} \frac{\det(\overline{\zeta} - \overline{z}, \overbrace{d\overline{\zeta} - d\overline{z}}^{n-1}) \wedge \omega(\zeta - z)}{|\zeta - z|^{2n}}.$$

Remark. Note that, for $n = 1$, $B_0^0(z,\zeta) = \frac{1}{2i\pi} d\zeta/(\zeta - z)$. This is the usual Cauchy kernel.

Proposition 1.4. *The Bochner–Martinelli–Koppelman kernel B is a locally integrable differential form on $\mathbb{C}^n \times \mathbb{C}^n$ which defines a current which we also denote by B such that*

$$dB = \overline{\partial} B = [\Delta], \tag{1.2}$$

where $[\Delta]$ is the integration current on Δ, the diagonal in $\mathbb{C}^n \times \mathbb{C}^n$.

Proof. Let $\tau : \mathbb{C}^n \times \mathbb{C}^n \to \mathbb{C}^n$ be the map given by $(z,\zeta) \mapsto \zeta - z$ and let k_{BM} be the $\mathcal{C}^\infty$ differential form on $\mathbb{C}^n \setminus \{0\}$ defined by

$$k_{\mathrm{BM}}(x) = \frac{(n-1)!}{(2i\pi)^n} |x|^{-2n} \sum_{j=1}^n (-1)^{j-1} \overline{x}_j \big(\underset{j \neq s}{\wedge} d\overline{x}_s \big) \wedge \omega(x).$$

The coefficients of k_{BM} are locally integrable on $\mathbb{C}^n$ since they are of order $O(|x|^{-2n+1})$ and it follows that k_{BM} defines a current on $\mathbb{C}^n$. We note that

$$k_{\mathrm{BM}}(x) = \frac{(n-2)!}{(2i\pi)^n} \sum_{j=1}^n (-1)^{j-1} \frac{\partial}{\partial x_j} (-|x|^{-2n+2}) \big(\underset{s \neq j}{\wedge} d\overline{x}_s \big) \wedge \omega(x)$$

and hence

$$dk_{\mathrm{BM}}(x) = \overline{\partial} k_{\mathrm{BM}}(x) = \frac{(n-2)!}{(2i\pi)^n}\Big(\sum_{j=1}^{n} \frac{\partial^2}{\partial x_j \partial \overline{x}_j}(-|x|^{-2n+2})\Big)\omega(\overline{x}) \wedge \omega(x).$$

We recognize the fundamental solution of the Laplacian $E(x) = \frac{(n-2)!}{\pi^n} 1/|x|^{2n-2}$, $x \in \mathbb{C}^n$. It follows that

$$dk_{\mathrm{BM}} = \overline{\partial} k_{\mathrm{BM}} = [0], \tag{1.3}$$

where $[0]$ is the integration current at 0, i.e. the Dirac delta function at 0. We note that $B = \tau^* k_{\mathrm{BM}}$ and B is therefore locally integrable on $\mathbb{C}^n \times \mathbb{C}^n$ and defines a current which we again denote by B. Considering the pullbacks of the currents k_{BM} and $[0]$ under the projection τ, we deduce from (1.3) the following result

$$dB = \overline{\partial} B = \overline{\partial}(\tau^* k_{\mathrm{BM}}) = \tau^*(\overline{\partial} k_{\mathrm{BM}}) = \tau^*[0] = [\tau^{-1}(0)] = [\Delta]. \qquad \square$$

We now prove that if K is a current on $\mathbb{C}^n \times \mathbb{C}^n$ defined by a locally integrable $\mathcal{C}^\infty$ differential form on $\mathbb{C}^n \times \mathbb{C}^n \setminus \Delta$ such that $dK = [\Delta]$ then we can associate an integral representation formula to K. On setting $K = B$ we get the Bochner–Martinelli–Koppelman formula.

Theorem 1.5. *Let K be a current on $\mathbb{C}^n \times \mathbb{C}^n$ defined by a locally integrable $\mathcal{C}^\infty$ differential form on $\mathbb{C}^n \times \mathbb{C}^n \setminus \Delta$ such that $dK = [\Delta]$. Let D be a bounded domain in $\mathbb{C}^n$ with $\mathcal{C}^1$ boundary and let f be a differential form of degree r which is $\mathcal{C}^\infty$ on $\overline{D}$. Then,*

$$\int_{\zeta \in \partial D} f(\zeta) \wedge K(z,\zeta) - \int_{\zeta \in D} d_\zeta f(\zeta) \wedge K(z,\zeta) + d_z \int_{\zeta \in D} f(\zeta) \wedge K(z,\zeta) = \begin{cases} (-1)^r f(z) & \text{on } D \\ 0 & \text{on } \mathbb{C}^n \setminus \overline{D}, \end{cases} \tag{1.4}$$

where d_z denotes differentiation of currents.

Proof. We apply Stokes' formula for the Kronecker index of two currents (Chapter II, Cor. 3.7) to the currents $T = [D \times D]$ and $S = f(\zeta) \wedge K(z,\zeta) \wedge g(z)$ where g is a $(2n-r)$-differential form with compact support in D.

We note that $SS(bT) = (\partial D \times D) \cup (D \times \partial D)$ and $SS(dS)$ is contained in $SS(S) = \Delta \cap \pi_\zeta^{-1}(\operatorname{supp} g)$, where π_ζ is the first projection from $\mathbb{C}^n \times \mathbb{C}^n$ to $\mathbb{C}^n$. As g has compact support in D it follows that $SS(bT)$ and $SS(dS)$ do not meet and hence $\mathcal{K}(bT, S) = \mathcal{K}(T, dS)$. We calculate both sides of this equation: we get

$$\begin{aligned} \mathcal{K}(bT, S) &= \mathcal{K}\big([\partial D \times D] + [D \times \partial D], f(\zeta) \wedge K(z,\zeta) \wedge g(z)\big) \\ &= \int_{\partial D \times D} f(\zeta) \wedge K(z,\zeta) \wedge g(z) \end{aligned}$$

since g is compactly supported in D. Moreover

$$dS = d_\zeta f(\zeta) \wedge K(z,\zeta) \wedge g(z) + (-1)^r f(\zeta) \wedge [\Delta] \wedge g(z) + (-1)^{r+1} f(\zeta) \wedge K(z,\zeta) \wedge d_\zeta g(z)$$

and hence

$$\mathcal{K}(T, dS) = \int_{D\times D} d_\zeta f(\zeta) \wedge K(z,\zeta) \wedge g(z) + (-1)^r \mathcal{K}([D\times D], f(\zeta)\wedge[\Delta]\wedge g(z)) + (-1)^{r+1} \int_{D\times D} f(\zeta)\wedge K(z,\zeta)\wedge d_z g(z)$$

since K is locally integrable on $D \times D$ and $(\Delta \cap \pi_\zeta^{-1}(\operatorname{supp} g))$ does not meet $SS([D \times D])$.

Moreover, as the support of $f(\zeta)\wedge[\Delta]\wedge g(z)$ is relatively compact in $D\times D$,

$$\mathcal{K}\big([D\times D], f(\zeta)\wedge[\Delta]\wedge g(z)\big) = \int_{\mathbb{C}^n} f(z)\wedge g(z).$$

by definition of $[\Delta]$. And finally

$$(1.5)\quad (-1)^r \int_{\mathbb{C}^n} f(z)\wedge g(z) = \int_{D\times\partial D} f(\zeta)\wedge K(z,\zeta)\wedge g(z) - \int_{D\times D} d_\zeta f(\zeta)\wedge K(z,\zeta)\wedge g(z) + \int_{z\in D} (d_z \int_{\zeta\in D} f(\zeta)\wedge K(z,\zeta)\wedge g(z)$$

from which we can deduce the following equality of currents on D

$$(-1)^r f(z) = \int_{\partial D} f(\zeta)\wedge K(z,\zeta) - \int_D d_\zeta f(\zeta)\wedge K(z,\zeta) + d_z \int_D f(\zeta)\wedge K(z,\zeta).$$

If $z \notin \overline{D}$, the differential form $f(\zeta)\wedge K(z,\zeta)$ is $\mathcal{C}^\infty$ on D and $dK(z,\zeta) = d_z K(z,\zeta) + d_\zeta K(z,\zeta) = 0$. The theorem follows on applying the classical Stokes' Theorem (cf. Appendix A, Th. 7.1). □

Remarks 1.6.

1) Note that by Fubini's theorem formula (1.5) can also be written as

$$(1.6)\quad (-1)^r\langle f,g\rangle = \langle [\partial D]\wedge f, \int_{z\in\mathbb{C}^n} K(z,\bullet)\wedge g(z)\rangle - \langle [D]\wedge df, \int_{z\in\mathbb{C}^n} K(z,\bullet)\wedge g(z)\rangle + (-1)^{r+1}\langle [D]\wedge f, \int_{z\in\mathbb{C}^n} K(z,\bullet)\wedge dg(z)\rangle$$

for any $\mathcal{C}^\infty$ differential form g of degree $(2n-r)$ with compact support in D.

2) Theorem 1.5 is still valid for a domain D with piecewise $\mathcal{C}^1$ boundary as in Remark 7.2 2) of Appendix A. We then set $[\partial D] = \sum_{i=1}^{\ell}[S_i^0]$.

3) As the differential form $z \mapsto K(z,\zeta)$ is locally integrable and continuous on $\mathbb{C}^n \setminus \{\zeta\}$ for any $\zeta \in \mathbb{C}^n$ the differential forms

$$\int_{z\in\mathbb{C}^n} K(z,\bullet) \wedge g(z) \quad \text{and} \quad \int_{z\in\mathbb{C}^n} K(z,\bullet) \wedge dg(z)$$

are continuous on $\mathbb{C}^n$, and formula (1.6) is therefore still meaningful for a continuous differential form f on $\overline{D}$ such that df is also continuous on $\overline{D}$. We shall prove the formula in this generality. If f is a continuous differential form on $\overline{D}$ such that df is also continuous on $\overline{D}$ then we construct a family $(f_\varepsilon)_{0<\varepsilon<\varepsilon_0}$ of continuous differential forms with compact support in $\mathbb{C}^n$ such that the forms df_ε are also continuous with compact support in $\mathbb{C}^n$ and the families $(f_\varepsilon)_{0<\varepsilon<\varepsilon_0}$ and $(df_\varepsilon)_{0<\varepsilon<\varepsilon_0}$ converge uniformly on $\overline{D}$ to f and df respectively.

For any $\xi \in \partial D$ we denote the ball of centre ξ and radius r by $B(\xi,r)$. We denote the exterior normal vector to ∂D at ξ by ν_ξ. Choose r and ε_0 small enough that $(\partial D \cap B(\xi,r)) - t\nu_\xi$ is contained in D for any $t \in \,]0,\varepsilon]$. Let $U_1,\dots,U_p$ be a finite subcover of ∂D by such balls and let U_0 be a relatively compact open set in D such that $(U_0,U_1,\dots,U_p)$ is a cover of $\overline{D}$. Let $(\chi_j)_{0\leqslant j\leqslant p}$ be a family of positive $\mathcal{C}^\infty$ functions with compact support in $\mathbb{C}^n$ such that $\operatorname{supp}\chi_j \subset U_j$ and $\sum_{j=0}^{p}\chi_j = 1$ on $\overline{D}$. We define f_ε by

$$f_\varepsilon(z) = \chi_0(z)f(z) + \sum_{j=1}^{p} \chi_j(z) f(z-\varepsilon\nu_j).$$

If $\varepsilon \in \,]0,\varepsilon_0[$ then f_ε is a continuous form with compact support in a neighbourhood of $\overline{D}$ and by the uniform continuity of f on $\overline{D}$ the family $(f_\varepsilon)_{0<\varepsilon<\varepsilon_0}$ converges uniformly to f on $\overline{D}$. Moreover,

$$\begin{aligned} df_\varepsilon(z) = \chi_0(z)df(z) + \sum_{j-1}^{p} \chi_j(z)df(z-\varepsilon\nu_j) + d\chi_0(z)\wedge f(z) \\ + \sum_{j=1}^{p} d\chi_j(z)\wedge f(z-\varepsilon\nu_j). \end{aligned}$$

The differential forms df_ε are continuous and have compact support in a neighbourhood of $\overline{D}$ for any $\varepsilon \in \,]0,\varepsilon_0[$. By the uniform continuity of f and df on $\overline{D}$ the family $(df_\varepsilon)_{0<\varepsilon<\varepsilon_0}$ converges uniformly to df on $\overline{D}$.

Regularising the differential forms f_ε by convolution, we get a family $(\widetilde{f}_\varepsilon)_{0<\varepsilon<\varepsilon_0}$ of $\mathcal{C}^\infty$ differential forms on $\mathbb{C}^n$ such that $(\widetilde{f}_\varepsilon)_{0<\varepsilon<\varepsilon_0}$ converges uniformly to f on $\overline{D}$ and $(d\widetilde{f}_\varepsilon)_{0<\varepsilon<\varepsilon_0}$ converges uniformly to df on $\overline{D}$ as ε tends to 0. The currents $([\partial D]\wedge\widetilde{f}_\varepsilon)_{0<\varepsilon<\varepsilon_0}$, $([D]\wedge d\widetilde{f}_\varepsilon)_{0<\varepsilon<\varepsilon_0}$ and $([D]\wedge f_\varepsilon)_{0<\varepsilon<\varepsilon_0}$ then converge to $[\partial D]\wedge f$, $[D]\wedge df$ and $[D]\wedge f$ respectively in the weak topology on zero-order currents. It follows that (1.6) and hence (1.4) can be extended to the case where f and df are only assumed continuous on $\overline{D}$.

Theorem 1.7 (The Bochner–Martinelli–Koppelman formula). *Let D be a bounded domain in $\mathbb{C}^n$ with $\mathcal{C}^1$ boundary and let f be a continuous (p,q)-differential form on $\overline{D}$ such that $\overline{\partial} f$ is also continuous on $\overline{D}$ $(0 \leqslant p \leqslant n$, $0 \leqslant q \leqslant n)$. Then*

$$\int_{\zeta\in\partial D} f(\zeta)\wedge B_q^p(z,\zeta) - \int_{\zeta\in D} \overline{\partial}_\zeta f(\zeta)\wedge B_q^p(z,\zeta)$$
$$+\overline{\partial}_z \int_{\zeta\in D} f(\zeta)\wedge B_{q-1}^p(z,\zeta) = \begin{cases} (-1)^{p+q} f(z) & \text{on } D \\ 0 & \text{on } \mathbb{C}^n \setminus \overline{D}, \end{cases}$$

where $\overline{\partial}_z$ denotes differentiation of currents.

Proof. This follows immediately from Theorem 1.5 and Remark 1.6 on replacing K by the Bochner–Martinelli–Koppelman kernel B. Bidegree considerations allow us to replace the operator d by $\overline{\partial}$ and the kernel B by a B_q^p. □

2 Existence of solutions to the $\overline{\partial}$ equation for compactly supported data

Let D be an open set in $\mathbb{R}^n$. For any real number α, $0<\alpha<1$, and any function $f{:}D\to\mathbb{C}$ we define the Hölder norm of order α of f on D by

$$|f|_{\alpha,D} = \sup_{x\in D}|f(x)| + \sup_{\substack{x,x'\in D\\ x\neq x'}} \frac{|f(x)-f(x')|}{|x-x'|^\alpha}.$$

The set $\Lambda^\alpha(D) = \{f : D\to\mathbb{C} \mid |f|_{\alpha,D} < +\infty\}$ is the vector space of Hölder continuous functions of order α on D. It is a Banach space and any function f in $\Lambda^\alpha(D)$ is bounded and uniformly continuous on D. We denote the space of continuous functions on D such that $|f|_{\alpha,K} < +\infty$ for any compact set K in D by $\mathcal{C}^\alpha(D)$ and for any integer k we denote the space of $\mathcal{C}^k$ functions on D all of whose kth-order partial derivatives are contained in $\mathcal{C}^\alpha(D)$ by $\mathcal{C}^{k+\alpha}(D)$.

If D is an open set in $\mathbb{C}^n$ then we denote the space of (p,q) differential forms whose coefficients are in $\mathcal{C}^{k+\alpha}(D)$ by $\mathcal{C}^{k+\alpha}_{p,q}(D)$. If $f\in\mathcal{C}^\alpha_{p,q}(D)$ can be written as $f=\sum_{I,J} f_{IJ}dz_I\wedge d\overline{z}_J$ where $I=(i_1,\ldots,i_p)$ with $i_1<\cdots<i_p$ and $J=(j_1,\ldots,j_q)$ with $j_1<\cdots<j_q$, then we define the Hölder norm of order α of f on D by $|f|_{\alpha,D} = \sup_{I,J}|f_{I,J}|_{\alpha,D}$.

Let B be the Bochner–Martinelli–Koppelman kernel defined in Section 1 and let D be a bounded domain with $\mathcal{C}^1$ boundary. If f is a bounded differential form defined on ∂D then for any $z\in D$ we set

$$\widetilde{B}_{\partial D} f(z) = \int_{\zeta\in\partial D} f(\zeta)\wedge B(z,\zeta). \tag{2.1}$$

The differential form $\widetilde{B}_{\partial D} f$ is $\mathcal{C}^\infty$ on D.

If f is a bounded differential form defined on D then for any $z \in D$ we set

(2.2) $$\widetilde{B}_D f(z) = \int_{\zeta \in D} f(\zeta) \wedge B(z,\zeta).$$

As B has an integrable singularity along the locus $\zeta = z$ the differential form $\widetilde{B}_D f$ is well defined on D. Note that if f is of type (p,q) then

(2.3) $$\widetilde{B}_D f = \int_{\zeta \in D} f(\zeta) \wedge B^p_{q-1}(\cdot,\zeta)$$

and in particular $\widetilde{B}_D f = 0$ if $q = 0$.

Proposition 2.1. *Let D be a bounded open set in $\mathbb{C}^n$. Then,*

1) *for any bounded differential form f on D the differential form $\widetilde{B}_D f$ is $\mathcal{C}^\alpha$ on D for any α such that $0 < \alpha < 1$.*
Moreover, for any α such that $0 < \alpha < 1$ there is a constant C_α such that, for any bounded differential form f on D,

$$|\widetilde{B}_D f|_{\alpha,D} \leqslant C_\alpha |f|_{0,D}$$

2) *if f is a $\mathcal{C}^k$ bounded differential form then $\widetilde{B}_D f$ is $\mathcal{C}^{k+\alpha}$ on D for any α such that $0 < \alpha < 1$.*

Proof. By definition of the kernel B (cf. formula (1.1)) there is a constant C such that

$$|\widetilde{B}_D f(z) - \widetilde{B}_D f(\xi)| \leqslant C|f|_{0,D} \sum_{j=1}^n \int_D \left| \frac{\overline{\zeta}_j - \overline{z}_j}{|\zeta_j - z_j|^{2n}} - \frac{\overline{\zeta}_j - \overline{\xi}_j}{|\zeta - \xi|^{2n}} \right| d\sigma_{2n},$$

where σ_{2n} is the Lebesgue measure on $\mathbb{R}^{2n}$.

Lemma 2.2. *Let $n \geqslant 1$ be an integer and let R be a strictly positive real number. There is then a constant C such that, for any s and $t \in \mathbb{R}^n$ such that $|s| \leqslant R$ and $|t| \leqslant R$,*

$$\int_{\substack{x \in \mathbb{R}^n \\ |x| < R}} \left| \frac{x_1 - t_1}{|x - t|^n} - \frac{x_1 - s_1}{|x - s|^n} \right| d\sigma_n \leqslant C|t - s| \cdot \big| \log|t - s| \big|.$$

Proof (of the Lemma). Given s and $t \in \mathbb{R}^n$ such that $|s| \leqslant R$ and $|t| \leqslant R$ consider the following sets:

$$W_1 = \Big\{ x \in \mathbb{R}^n \mid |x - t| \leqslant \frac{|t - s|}{2} \Big\},$$
$$W_2 = \Big\{ x \in \mathbb{R}^n \mid |x - s| \leqslant \frac{|t - s|}{2} \Big\},$$
$$W_3 = \Big\{ x \in \mathbb{R}^n \mid |x| < R,\ |x - t| \geqslant \frac{|t - s|}{2},\ |x - s| \geqslant \frac{|t - s|}{2} \Big\},$$

Then $\{x \in \mathbb{R}^n \mid |x| < R\} \subset W_1 \cup W_2 \cup W_3$.

Note that $|x-s| \geqslant \big||x-t| - |t-s|\big| \geqslant \frac{1}{2}|t-s|$ for any $x \in W_1$ and hence

$$\int_{W_1} \left| \frac{x_1-t_1}{|x-t|^n} - \frac{x_1-s_1}{|x-s|^n} \right| d\sigma_n \leqslant \int_{W_1} |x-t|^{1-n} d\sigma_n + C|t-s|^{1-n} \int_{W_1} d\sigma_n$$
$$\leqslant C'|t-s|$$

by integrating in spherical coordinates. Similarly,

$$\int_{W_2} \left| \frac{x_1-t_1}{|x-t|^n} - \frac{x_1-s_1}{|x-s|^n} \right| d\sigma_n \leqslant C'|t-s|.$$

Moreover,

$$\left| \frac{x_1-t_1}{|x-t|^n} - \frac{x_1-s_1}{|x-s|^n} \right|$$
$$\leqslant \left| \frac{(x_1-t_1)(|x-s|-|x-t|)\sum_{\nu=0}^{n-1}|x-s|^{\nu}|x-t|^{n-\nu-1} - |t_1-s_1|\,|x-t|^n}{|x-s|^n|x-t|^n} \right|$$
$$\leqslant \big||x-s|-|x-t|\big| \left| \sum_{\nu=1}^{n-1} \frac{|x-s|^{\nu}|x-t|^{n-\nu}}{|x-s|^n|x-t|^n} \right| + \frac{|t-s|}{|x-s|^n}$$
$$\leqslant n|t-s| \left(\max\left\{ \frac{1}{|x-s|^n}, \frac{1}{|x-t|^n} \right\} + \frac{1}{|x-s|^n} \right).$$

Since $|t| \leqslant R$ and $|s| \leqslant R$ it follows that

$$\int_{W_3} \left| \frac{x_1-t_1}{|x-t|^n} - \frac{x_1-s_1}{|x-s|^n} \right| d\sigma_n \leqslant C''|t-s| \int_{\{|t-s|/2 \leqslant |y| \leqslant 2R\}} \frac{d\sigma_n}{|y|^n}$$
$$\leqslant C'''|t-s| \int_{|t-s|/2}^{2R} \frac{dr}{r} \leqslant C|t-s| \cdot \big|\log|t-s|\big|. \qquad \square$$

End of the proof of the proposition. By the above lemma,

$$|\widetilde{B}_D f(z) - \widetilde{B}_D f(\xi)| \leqslant C|f|_{0,D}|z-\xi|\big|\log|z-\xi|\big|.$$

As $\sup_{z,\xi \in D} |z-\xi|^{1-\alpha}\big|\log|z-\xi|\big| < +\infty$ for any $\alpha \in]0,1[$ claim 1) follows.

Assume that f is $\mathcal{C}^k$ and bounded on D. Given $z \in D$, let us prove that $\widetilde{B}_D f$ is $\mathcal{C}^{k+\alpha}$ in a neighbourhood of z. Consider a function $\chi \in \mathcal{D}(D)$ such that $\chi \equiv 1$ on a neighbourhood U_z of z: $\widetilde{B}_D(1-\chi)f$ is then $\mathcal{C}^\infty$ on U_z because B is $\mathcal{C}^\infty$ on $\mathbb{C}^n \times \mathbb{C}^n \setminus \Delta$. It is therefore enough to prove that $\widetilde{B}_D \chi f$ is $\mathcal{C}^{k+\alpha}$ on U_z. Suppose that f is a (p,q)-form. Then $\widetilde{B}_D \chi f(\xi) = \int_{\zeta \in D} \chi(\zeta) f(\zeta) \wedge B^p_{q-1}(\xi,\zeta)$. Set $f(\zeta) = \sum_{\substack{|I|=p\\|J|=q}} f_{IJ}(\zeta) d\zeta_I \wedge d\overline{\zeta}_J$. Then

$$B^p_{q-1}(\xi,\zeta) = \sum_{\substack{|K|=n-q-2\\|L|=q-1\\|M|=n-p\\|N|=p}} B_{KLMN} d\overline{\zeta}_K \wedge d\zeta_M \wedge d\overline{\xi}_L \wedge d\xi_N.$$

and $\widetilde{B}_D\chi f(\xi) = \sum_{L,N} h_{L,N} d\overline{\xi}_L \wedge d\xi_N$, where

$$h_{L,N}(\xi) = \sum_{I,J,K,M} \int_{\zeta\in D} \chi(\zeta) f_{IJ}(\zeta) B_{KLMN}(\xi,\zeta) d\zeta_I \wedge d\overline{\zeta}_J \wedge d\overline{\zeta}_K \wedge d\zeta_M.$$

As χ has compact support in D,

$$h_{L,N}(\xi) = \sum_{I,J,K,M} \int_{\zeta\in\mathbb{C}^n} \chi(\zeta+\xi) f_{IJ}(\zeta+\xi) B_{KLMN}(\zeta,\xi+\zeta) d\zeta_I \wedge d\overline{\zeta}_J \wedge d\overline{\zeta}_K \wedge d\zeta_M.$$

As $B = \tau^* k_{\mathrm{BM}}$ the functions $B_{KLMN}(\xi,\xi+\zeta)$ are independent of ξ and are locally integrable with respect to ζ. Since f is $\mathcal{C}^k$ we can differentiate k times under the integral sign, which proves that $\widetilde{B}_D\chi f$ is $\mathcal{C}^k$ and if $|\nu| = k$ then

$$\begin{aligned} D^\nu h_{L,N}(\xi) &= \\ &= \sum_{I,J,K,M} \int_{\zeta\in\mathbb{C}^n} D^\nu(\chi f_{IJ})(\zeta+\xi) B_{KLMN}(\xi,\xi+\zeta) d\zeta_I \wedge d\overline{\zeta}_J \wedge d\overline{\zeta}_K \wedge d\zeta_M \\ &= \sum_{I,J,KM} \int_{\zeta\in\mathbb{C}^n} D^\nu(\chi f_{IJ})(\zeta) B_{KLMN}(\xi,\zeta) d\zeta_I \wedge d\overline{\zeta}_J \wedge d\overline{\zeta}_K \wedge d\zeta_M. \end{aligned}$$

It the follows from the definition of B_{KLMN} that there is a constant C such that, for any $\xi,\omega \in D$,

$$\begin{aligned} |D^\nu h_{LN}(\xi) - D^\nu h_{LN}(\omega)| &< C \sup_{\zeta\in D} \Big|\sum_{I,J} D^\nu(\chi f_{IJ})(\zeta)\Big| \\ &\quad \times \sum_{j=1}^n \int_{\zeta\in D} \left| \frac{\overline{\zeta}_j - \overline{\xi}_j}{|\zeta-\xi|^n} - \frac{\overline{\zeta}_j - \overline{\omega}_j}{|\zeta-\omega|^n} \right| d\sigma_{2n} \end{aligned}$$

and we complete the proof of the theorem as in the proof of 1). □

Theorem 2.3. *Let D be a bounded domain in $\mathbb{C}^n$ with $\mathcal{C}^1$ boundary and let f be a continuous (p,q)-differential form on $\overline{D}$ such that $\overline{\partial} f$ is also continuous on $\overline{D}$, $0 \leqslant p \leqslant n$, $0 \leqslant q \leqslant n$. The differential forms $\widetilde{B}_{\partial D} f$, $\widetilde{B}_D \overline{\partial} f$, $\widetilde{B}_D f$ and $\overline{\partial}\widetilde{B}_D f$ are then continuous on D and*

$$(2.4) \qquad (-1)^{p+q} f = \widetilde{B}_{\partial D} f - \widetilde{B}_D \overline{\partial} f + \overline{\partial}\widetilde{B}_D f \quad \text{on } D.$$

Proof. The regularity of the differential forms $\widetilde{B}_{\partial D} f$, $\widetilde{B}_D \overline{\partial} f$ and $\widetilde{B}_D f$ follows from Proposition 2.1. The equation (2.4) holds for currents by Theorem 1.7, but as the differential forms $\widetilde{B}_{\partial D} f, \widetilde{B}_D \overline{\partial} f$ and f are continuous $\overline{\partial}\widetilde{B}_D f$ is continuous. □

Corollary 2.4 (Bochner–Martinelli formula). *Let D be a bounded domain with $\mathcal{C}^1$ boundary in $\mathbb{C}^n$ and let f be a $\mathcal{C}^1$ function on $\overline{D}$. Then,*

$$\int_{\zeta\in\partial D} f(\zeta)\wedge B_0^0(z,\zeta) - \int_{\zeta\in D} \overline{\partial}_\zeta f(\zeta)\wedge B_0^0(z,\zeta) = f(z) \quad \textit{for any } z\in D.$$

If f is holomorphic on D and continuous on $\overline{D}$ then

$$\int_{\zeta\in\partial D} f(\zeta)\wedge B_0^0(z,\zeta) = f(z) \quad \textit{for any } z\in D.$$

Note that if $n=1$ then Corollary 2.4 is simply the Cauchy–Green formula and Cauchy's formula in $\mathbb{C}$.

Corollary 2.5.

1) *Let q be an integer such that $1\leqslant q\leqslant n-1$ and let $f\in \mathcal{C}^k_{p,q}(\mathbb{C}^n)$ be a (p,q)-differential form which is $\mathcal{C}^k$ for some $k\geqslant 0$, $\overline{\partial}$-closed and compactly supported in $\mathbb{C}^n$. There is then a $(p,q-1)$-differential form u which is $\mathcal{C}^{k+\alpha}$ for any $\alpha\in\,]0,1[$ such that $\overline{\partial}u=f$ in $\mathbb{C}^n$.*
2) *If D is a bounded domain with $\mathcal{C}^1$ boundary in $\mathbb{C}^n$ then for any (p,n)-form f in $\mathcal{C}^k_{p,n}(\overline{D})$ there is a $(p,n-1)$-differential form u which is $\mathcal{C}^{k+\alpha}$ on D for any $\alpha\in\,]0,1[$ such that $\overline{\partial}u=f$ in D.*

Proof. We start by proving 1). Let D be a bounded domain with $\mathcal{C}^1$ boundary in $\mathbb{C}^n$ such that D contains the support of f. Applying Theorem 2.3 we get

$$(-1)^{p+q}f(z) = \overline{\partial}\widetilde{B}_D f(z) \quad \text{for any } z\in D$$

since $B_{\partial D}f = \widetilde{B}_D\overline{\partial}f = 0$ because f is $\overline{\partial}$-closed with compact support in D. Setting $u(z)=\widetilde{B}_D f(z) = \int_{\zeta\in\mathbb{C}^n} f(\zeta)\wedge B^p_{q-1}(z,\zeta)$ we see that u is a $\mathcal{C}^{k+\alpha}$ form for any $\alpha\in\,]0,1[$ by Proposition 2.1 and $\overline{\partial}u=f$ on $\mathbb{C}^n$.

We now prove 2). Note that $B^p_n=0$ for any $p\in\{0,\dots,n\}$ and it follows from the Bochner–Martinelli–Koppelman formula that

$$(-1)^{p+n}f(z) = \overline{\partial}_z\int_{\zeta\in D} f(\zeta)\wedge B^p_{n-1}(z,\zeta) \quad \text{for any } z\in D,$$

where this equation is to be understood as an equality of currents if f is only continuous. Setting $u(z)=\int_{\zeta\in D} f(\zeta)\wedge B^p_{n-1}(z,\zeta)$ the function u is then the solution we seek and $u\in\mathcal{C}^{k+\alpha}$ for any $\alpha\in\,]0,1[$ by Proposition 2.1. □

Example. Assume that $n=q=1$. If D is a bounded domain with $\mathcal{C}^1$ boundary in $\mathbb{C}$ and f is a $\mathcal{C}^k$ function for some $k\geqslant 0$ on $\overline{D}$ then the partial differential equation

$$(*) \qquad \frac{\partial}{\partial\overline{z}}u = f$$

has a solution u_0, which is $\mathcal{C}^{k+\alpha}$ for any $\alpha \in]0,1[$, given by

$$u_0(z) = \frac{1}{2i\pi} \int_{\zeta \in D} f(\zeta) \frac{d\zeta \wedge d\overline{\zeta}}{\zeta - z}.$$

To prove this, we simply apply statement 2) of Corollary 2.5 to the $(0,1)$-differential form $f(z)d\overline{z}$. The general solution to $(*)$ is then of the form $u = u_0 + h$ where h is a holomorphic function on D.

3 Regularity of $\overline{\partial}$

We now extend the Bochner–Martinelli–Koppelman formula to compactly supported currents and deduce a Hölder hypoellipticity theorem for the operator $\overline{\partial}$ from it.

We continue to denote the Bochner–Martinelli–Koppelman kernel on $\mathbb{C}^n$ by $B(z,\zeta)$. If f is a continuous compactly supported differential form on $\mathbb{C}^n$ then for any $z \in \mathbb{C}^n$ we set

$$\widetilde{B}_{\mathbb{C}^n} f(z) = \int_{z \in \mathbb{C}^n} f(\zeta) \wedge B(z,\zeta).$$

Lemma 3.1. *The map $\widetilde{B}_{\mathbb{C}^n}$ is a continuous linear map from $\mathcal{D}_\bullet(\mathbb{C}^n)$ to $\mathcal{C}^\infty_\bullet(\mathbb{C}^n)$. Moreover, if f is a (p,q)-differential form then $\widetilde{B}_{\mathbb{C}^n} f$ has bidegree $(p,q-1)$.*

Proof. It follows easily from Proposition 2.1 that if f is $\mathcal{C}^\infty$ then $\widetilde{B}_{\mathbb{C}^n} f$ is a $\mathcal{C}^\infty$ differential form. Note that if f is of type (p,q) then $\widetilde{B}_{\mathbb{C}^n} f(z) = \int_{\zeta \in \mathbb{C}^n} f(\zeta) \wedge B^p_{q-1}(z,\zeta)$ and hence $\widetilde{B}_{\mathbb{C}^n} f$ is of bidegree $(p,q-1)$.

Consider an element $f \in \mathcal{D}_\bullet(\mathbb{C}^n)$. We denote by $D^{\alpha\overline{\beta}} f$ the differential form obtained by replacing the coefficients f_{IJ} of f by their derivatives $D^{\alpha\overline{\beta}} f_{IJ}$. Repeating the proof of Proposition 2.1 it is easy to show that $D^{\alpha\overline{\beta}}(\widetilde{B}_{\mathbb{C}^n} f) = \widetilde{B}_{\mathbb{C}^n}(D^{\alpha\overline{\beta}} f)$ – as moreover $|\widetilde{B}_{\mathbb{C}^n} f|_{0,\mathbb{C}^n} \leqslant C(\operatorname{supp} f)|f|_{0,\mathbb{C}^n}$ this proves that $\widetilde{B}_{\mathbb{C}^n}$ is continuous because a sequence of differential forms with compact support converges to 0 in $\mathcal{D}_\bullet(\mathbb{C}^n)$ if all the terms in the sequence are supported on a fixed compact set and converge uniformly to 0 along with all their derivatives. □

If T is a compactly supported current on $\mathbb{C}^n$ then we can define $\widetilde{B}'T$ by setting

$$\langle \widetilde{B}'T, \varphi \rangle = \langle T, \widetilde{B}_{\mathbb{C}^n} \varphi \rangle = \left\langle T, \int_{\zeta \in \mathbb{C}^n} \varphi(\zeta) \wedge B(\cdot,\zeta) \right\rangle \tag{3.1}$$

for any $\varphi \in \mathcal{D}_\bullet(\mathbb{C}^n)$. This defines a current on $\mathbb{C}^n$, and we note that if T is of bidegree (p,q) then $\widetilde{B}'T$ is of bidegree $(p,q-1)$.

Proposition 3.2. *Let f be a continuous (p,q)-differential form with compact support in $\mathbb{C}^n$ and denote by T_f the current defined by f. Then,*

$$\widetilde{B}'T_f = (-1)^{p+q-1}\widetilde{B}_{\mathbb{C}^n} f.$$

Proof. Consider an element $\varphi \in \mathcal{D}_{n-p,n-q+1}(\mathbb{C}^n)$. By definition of $\widetilde{B}'$,

$$\begin{aligned}
\langle \widetilde{B}'T_f, \varphi\rangle &= \langle T_f, \widetilde{B}_{\mathbb{C}^n}\varphi\rangle \\
&= \int_{z\in\mathbb{C}^n} f(z) \wedge \Big(\int_{\zeta\in\mathbb{C}^n} \varphi(\zeta) \wedge B(z,\zeta)\Big) \\
&= (-1)^{p+q-1} \int_{\zeta\in\mathbb{C}^n} \Big(\int_{z\in\mathbb{C}^n} f(z) \wedge B(z,\zeta)\Big) \wedge \varphi(\zeta).
\end{aligned}$$

As $B(z,\zeta) = B(\zeta,z)$ it follows that $\langle \widetilde{B}'T_f, \varphi\rangle = (-1)^{p+q-1}\langle \widetilde{B}_{\mathbb{C}^n} f, \varphi\rangle$. □

Proposition 3.3. *For any current T with compact support in $\mathbb{C}^n$ the current $\widetilde{B}'T$ is $\mathcal{C}^\infty$ on $\mathbb{C}^n \setminus \operatorname{supp} T$. More precisely, there is a differential form f with $\mathcal{C}^\infty$ coefficients on $\mathbb{C}^n \setminus \operatorname{supp} T$ such that*

$$\widetilde{B}'T\big|_{\mathbb{C}^n \setminus \operatorname{supp} T} = T_f.$$

Proof. It will be enough to find such a form f for any relatively compact open set $U \subset \mathbb{C}^n \setminus \operatorname{supp} T$. Fix such a U and let χ be a $\mathcal{C}^\infty$ function on $\mathbb{C}^n$ such that $\chi = 1$ in a neighbourhood of $\operatorname{supp} T$ and $\chi = 0$ in a neighbourhood of $\overline{U}$. If $\varphi \in \mathcal{D}_\bullet(U)$ then, by definition of $\widetilde{B}'T$,

$$\langle \widetilde{B}'T, \varphi\rangle = \Big\langle T, \int_{\zeta\in\mathbb{C}^n} \varphi(\zeta) \wedge B(\cdot,\zeta)\Big\rangle.$$

As $\chi = 0$ in a neighbourhood of the support of φ and $\chi = 1$ in a neighbourhood of the support of T, we have

$$\begin{aligned}
\langle \widetilde{B}'T, \varphi\rangle &= \Big\langle \chi(z)T, \int_{\zeta\in\mathbb{C}^n} (1-\chi)(\zeta)\varphi(\zeta) \wedge B(z,\zeta)\Big\rangle \\
&= \Big\langle T, \int_{\zeta\in\mathbb{C}^n} \chi(z)(1-\chi(\zeta))\varphi(\zeta) \wedge B(z,\zeta)\Big\rangle \\
&= (-1)^{d^o\varphi+1}\langle(\langle T, \chi(z)(1-\chi(\zeta))B(z,\zeta)\rangle), \varphi(\zeta)\rangle
\end{aligned}$$

since $\chi(z)(1-\chi(\zeta))B(z,\zeta)$ is a $\mathcal{C}^\infty$ differential form on a neighbourhood of $\operatorname{supp} T \times \mathbb{C}^n$. It follows that $\widetilde{B}'T$ is equal to the $\mathcal{C}^\infty$ differential form $\zeta \mapsto \langle T, \chi(z)(1-\chi(\zeta))B(z,\zeta)\rangle$ on U. □

We now prove a Bochner–Martinelli–Koppelman formula for compactly supported currents using the operator $\widetilde{B}'$.

Theorem 3.4. *The following representation formula holds for any current T with compact support in $\mathbb{C}^n$*

$$T = -(\widetilde{B}'\overline{\partial}T + \overline{\partial}\widetilde{B}'T). \tag{3.2}$$

Proof. Assume that T has bidegree (p,q) and consider a form $\varphi \in \mathcal{D}_{n-p,n-q}(\mathbb{C}^n)$. By formula (2.4) we have

$$(-1)^{p+q}\varphi = \overline{\partial}\widetilde{B}_{\mathbb{C}^n}\varphi - \widetilde{B}_{\mathbb{C}^n}\overline{\partial}\varphi \tag{3.3}$$

because φ has compact support. Applying T to both sides of (3.3) we get

$$\begin{aligned}(-1)^{p+q}\langle T,\varphi\rangle &= \langle T, \overline{\partial}\widetilde{B}_{\mathbb{C}^n}\varphi\rangle - \langle T, \widetilde{B}_{\mathbb{C}^n}\overline{\partial}\varphi\rangle \\ &= (-1)^{p+q-1}\big(\langle \widetilde{B}'\overline{\partial}T,\varphi\rangle + \langle \overline{\partial}\widetilde{B}'T,\varphi\rangle\big)\end{aligned}$$

and the theorem follows. □

We can deduce a regularity result for the Cauchy–Riemann operator from this formula.

Theorem 3.5. *Let X be a complex analytic manifold and let T be a degree 0 current on X. If $\overline{\partial}T = T_f$ and f is a $(0,1)$-differential form which is $\mathcal{C}^k$ on X, $k = 0,1,\dots,\infty$, then*

$$T = T_g, \quad \textit{for some } g \in \bigcap_{0<\alpha<1} \mathcal{C}^{k+\alpha}(X).$$

In particular, if $\overline{\partial}T = 0$ then $T = T_h$ for some holomorphic function h on X.

Proof. Since the statement is local we can assume that X is an open set in $\mathbb{C}^n$. It is then enough to prove that for any open set $U \Subset X$ there is a function $g \in \bigcap_{0<\alpha<1} \mathcal{C}^{k+\alpha}(U)$ such that $T = T_g$ on U. Let χ be a $\mathcal{C}^\infty$ function with compact support in X such that $\chi = 1$ in a neighbourhood of $\overline{U}$. Applying formula (3.2) to χT, we get

$$-\chi T = \widetilde{B}'(\overline{\partial}(\chi T)) = \widetilde{B}'(T\overline{\partial}\chi) + \widetilde{B}'(\chi T_f) \tag{3.4}$$

since T is of degree 0. As $\operatorname{supp}(T\overline{\partial}\chi) \cap U = \varnothing$ it follows from Proposition 3.3 that $\widetilde{B}'(T\overline{\partial}\chi) = T_{g_1}$ where $g_1 \in \mathcal{C}^\infty(U)$. Moreover $\widetilde{B}'(\chi T_f)$ is equal to $\widetilde{B}_{\mathbb{C}^n}(\chi f)$ by Proposition 3.2 and it follows that $\widetilde{B}'(\chi T_f) \in \bigcap_{0<\alpha<1} \mathcal{C}^{k+\alpha}(\mathbb{C}^n)$ by Proposition 2.1. Setting $g = g_1 + \widetilde{B}'(\chi T_f)\big|_U$, it then follows from (3.4) that $T\big|_U = T_g$. □

Remarks 3.6.

1) Theorem 3.5 remains valid for a current T of bidegree $(p,0)$ because any $\overline{\partial}$-closed $(p,0)$-differential form has holomorphic coefficients and is therefore $\mathcal{C}^\infty$.

2) The following result, dealing with the situation where we replace the degree 0 current T in Theorem 3.5 by a current of bidegree (p,q), $q \geqslant 1$, will be proved in Chapter V, § 4: there is a $g \in \bigcap_{0<\alpha<1} \mathcal{C}^{k+\alpha}_{p,q}(X)$ such that $\overline{\partial} T = \overline{\partial} T_g$, i.e. if the equation $\overline{\partial} u = f$ for some $f \in \mathcal{C}^k_{p,q+1}(X)$, has a current solution then it has a solution which is $\mathcal{C}^{k+\alpha}$ for any $\alpha \in]0,1[$.

4 Hartogs' phenomenon

The aim of this section is to show that if D is an open set in $\mathbb{C}^n$, $n \geqslant 2$, and K is a compact subset of D such that $D \smallsetminus K$ is connected then any holomorphic function on $D \smallsetminus K$ can be extended to a holomorphic function on the whole of D. In Chapter I we proved this theorem for a polydisc D in $\mathbb{C}^n$ when $n \geqslant 2$.

We start by giving a more precise version of Corollary 2.5 for a $(0,1)$-form f.

Theorem 4.1. *Let $f \in \mathcal{C}^k_{0,1}(\mathbb{C}^n)$ be a $\overline{\partial}$-closed $\mathcal{C}^k$ differential $(0,1)$-form $(k \geqslant 0)$ with compact support in $\mathbb{C}^n$ where $n \geqslant 2$. There is then a compactly supported function u which is $\mathcal{C}^{k+\alpha}$ for any $\alpha \in]0,1[$ such that $\overline{\partial} u = f$ on $\mathbb{C}^n$. Moreover, u is given by the integral representation formula*

$$u(z) = \int_{\zeta \in \mathbb{C}^n} f(\zeta) \wedge B^0_0(z,\zeta). \tag{4.1}$$

Proof. We proved in Corollary 2.5 that the function u given by (4.1) is a $\mathcal{C}^{k+\alpha}$ solution of the equation $\overline{\partial} u = f$ on $\mathbb{C}^n$. It remains to show that u has compact support in $\mathbb{C}^n$. Note that u is holomorphic outside the support of f because $\overline{\partial} u = f$. Moreover, if u is given by (4.1) then

$$|u(z)| \leqslant \frac{C}{\operatorname{dist}(z, \operatorname{supp} f)} |f|_{0,\mathbb{C}^n}$$

which implies that $u(z) \to 0$ when $|z|$ tends to ∞.

For any $z \in \mathbb{C}^n$, $n \geqslant 2$, we set $z = (z', z_n)$ where $z' \in \mathbb{C}^{n-1}$ and $z_n \in \mathbb{C}$. As the support of f is compact there is a number $R > 0$ such that if $|z'| > R$ then the set $\{z'\} \times \mathbb{C}$ does not meet the support of f. Fix such a z'. The function $z_n \mapsto u(z', z_n)$ is then holomorphic and tends to 0 when $|z_n|$ tends to infinity. By Liouville's theorem this function is therefore identically zero. This shows that u vanishes on the open set $\{z' \in \mathbb{C}^{n-1} \mid |z'| > R\} \times \mathbb{C}$ in the complement of $\operatorname{supp} f$. As u is holomorphic on $\mathbb{C}^n \smallsetminus \operatorname{supp} f$ it is therefore identically zero on the non-bounded connected component of $\mathbb{C}^n \smallsetminus \operatorname{supp} f$, which implies that u has compact support. □

Remark. Theorem 4.1 does not hold for $n = 1$. This is implied by the following more general fact: if f is a continuous $\overline{\partial}$-closed differential $(0,n)$-form with compact support in $\mathbb{C}^n$ then the equation $\overline{\partial} u = f$ does not have a compactly supported solution in general. Indeed, if u_0 is a solution of the equation $\overline{\partial} u = f$ with compact support in $\mathbb{C}^n$ and D is an open set with $\mathcal{C}^1$ boundary in $\mathbb{C}^n$ containing the support of u_0 then

$$\int_{\partial D} u_0 \wedge dz_1 \wedge \cdots \wedge dz_n = 0.$$

But applying Stokes' formula we see that

$$\begin{aligned} \int_{\partial D} u_0 \wedge dz_1 \wedge \cdots \wedge dz_n &= \int_D d(u_0 \wedge dz_1 \wedge \cdots \wedge dz_n) \\ &= \int_{\mathbb{C}^n} f \wedge dz_1 \wedge \cdots \wedge dz_n \end{aligned}$$

and this last term is not generally zero.

Theorem 4.2. *Let D be a bounded open set in $\mathbb{C}^n$ such that $\mathbb{C}^n \setminus D$ is connected and assume that $n > 1$. For any holomorphic function f on a neighbourhood of ∂D there is then a holomorphic function F on a neighbourhood of $\overline{D}$ which is equal to f on ∂D.*

Proof. We denote the neighbourhood of ∂D on which f is defined and holomorphic by $U_{\partial D}$ and choose a $\mathcal{C}^\infty$ function χ supported on $U_{\partial D}$ which is equal to 1 in a neighbourhood of ∂D. The function $\widetilde{f} = \chi f$ is then defined on $\mathbb{C}^n$ and coincides with f on a neighbourhood of ∂D. We set

$$g = \begin{cases} \overline{\partial} \widetilde{f} & \text{on } D \\ 0 & \text{on } \mathbb{C}^n \setminus D. \end{cases}$$

This is a $\mathcal{C}^\infty$ differential $(0,1)$-form on $\mathbb{C}^n$ which is supported on D since $\overline{\partial} \widetilde{f} = \overline{\partial} f = 0$ in a neighbourhood of ∂D. Moreover, g is $\overline{\partial}$-closed in $\mathbb{C}^n$, and we can therefore consider the equation $\overline{\partial} u = g$. By Theorem 4.1, this equation has a solution u_0 which has compact support and is therefore zero on some open set in $\mathbb{C}^n \setminus D$ since D is bounded. As $\overline{\partial} u_0 = g$, u_0 is holomorphic on $\mathbb{C}^n \setminus D$ and since $\mathbb{C}^n \setminus D$ is connected by hypothesis, u_0 vanishes identically on $\mathbb{C}^n \setminus D$. It follows that $F = \widetilde{f} - u_0$ is the function we seek. □

Corollary 4.3. *Let D be a domain in $\mathbb{C}^n$ and let K be a compact subset of D such that $D \setminus K$ is connected. Any holomorphic function on $D \setminus K$ can then be extended to a holomorphic function on D.*

Proof. We may assume without loss of generality that K is connected. Let D' be a bounded domain in $\mathbb{C}^n$ such that $\mathbb{C}^n \setminus D'$ is connected, D' is contained in D and D' contains K. Such an open set exists because $D \setminus K$ is connected, so if $D \neq \mathbb{C}^n$ we can set $D' = \{z \in D \mid \operatorname{dist}(z, K) < \varepsilon\}$ for small enough ε and if $D = \mathbb{C}^n$ we can set $D' = B(0, R)$ for large enough R. If f is holomorphic on $D \setminus K$ then it is holomorphic in a neighbourhood of $\partial D'$ and by Theorem 4.2

there is therefore a holomorphic function F defined on a neighbourhood of $\overline{D}'$ such that $f\big|_{\partial D'} = F\big|_{\partial D'}$. As $\partial D'$ is a real hypersurface in $\mathbb{C}^n$ and f and F are holomorphic in a a neighbourhood of $\partial D'$ and are equal on $\partial D'$, they are equal on the connected component of their common domain of definition containing $\partial D'$. The function $\widetilde{F} = \begin{cases} F & \text{on } D' \\ f & \text{on } D \smallsetminus K \end{cases}$ is then the extension we seek. □

Comments

The formula known as the Bochner–Martinelli formula extending the Cauchy formula to a bounded domain with smooth boundary in $\mathbb{C}^n$ was discovered independently by E. Martinelli [Ma1] in 1938 and S. Bochner [Bo] in 1940. These two authors used it to produce a rigorous proof of Hartogs' phenomenon ([Ma2] and [Bo]). It was generalised to differential forms by W. Koppelman [Ko] in 1967.

Proofs of the Bochner–Martinelli–Koppelman formula different from the one given here are presented in [He/Le1] and [Ra]. The regularity properties given in Section 2 are also discussed in these two books. The regularity theorem for $\overline{\partial}$ is proved in Section 1 of the first chapter of Henkin and Leiterer's book [He/Le2] and Range gives a rather different style of proof of Hartogs' phenomenon in Section 2 of Chapter IV of [Ra]. In [Ho2], Hörmander gives a different proof of Theorem 4.1, based on the one-dimensional Cauchy–Green formula, and deduces Hartogs' phenomenon from it.

IV

Extensions of CR functions

Whilst studying Hartogs' phenomenon in Chapter III we proved that if D is a simply connected bounded domain in $\mathbb{C}^n$, $n \geqslant 2$, then any holomorphic function defined on a neighbourhood of the boundary of D can be extended to a holomorphic function on D. It follows that the restriction to ∂D of a holomorphic function defined in a neighbourhood of ∂D is the boundary value of a holomorphic function on D which is continuous on $\overline{D}$. We now try to characterise the boundary values of holomorphic functions on a bounded domain $D \subset \mathbb{C}^n$ which are continuous on $\overline{D}$. The main result of this chapter is Bochner's extension theorem for CR functions defined on the boundary of a domain. Its proof uses the Bochner–Martinelli transform which is studied in Section 1. We also prove our first generalisation of Bochner's theorem to CR functions which are only defined on part of the boundary of the domain. This generalisation is also based on the properties of the Bochner–Martinelli transform but it requires two extra ingredients: Stokes' formula for CR functions and the integrals of the Bochner–Martinelli kernel.

1 The Bochner–Martinelli transform

Let U be an open set in $\mathbb{C}^n$ and let V be a real smooth $\mathcal{C}^1$ hypersurface in U (i.e. V is a real $\mathcal{C}^1$ submanifold of real dimension $(2n-1)$ contained in the open set U in $\mathbb{C}^n \simeq \mathbb{R}^{2n}$) such that $U \setminus V$ has exactly two connected components U^+ and U^-. We assume that V is oriented and its orientation is that of ∂U^+.

Definition 1.1. Let f be a continuous function with compact support defined on V. We define the *Bochner–Martinelli transform* of f by

$$F(z) = \int_{\zeta \in V} f(\zeta) B_0^0(z,\zeta) \quad \text{for any } z \in U \setminus V.$$

We note that F is a $\mathcal{C}^\infty$ (and even real analytic) function on $U \setminus \operatorname{supp} f$ because $B_0^0(z,\zeta)$ is a differential form with real analytic coefficients on $\mathbb{C}^n \times \mathbb{C}^n \setminus \Delta$.

C. Laurent-Thiébaut, *Holomorphic Function Theory in Several Variables: An Introduction*,
Universitext, DOI 10.1007/978-0-85729-030-4_4,

The aim of this section is to study F in a neighbourhood of a point contained in $\operatorname{supp} f$.

Remark 1.2. Let $D \Subset U$ be a domain with piecewise $\mathcal{C}^1$ boundary such that $D \subset U^+$ and $\partial D \cap \dot{V} \supset \operatorname{supp} f$. If $z \in U \smallsetminus (V \cap \partial D)$ then

$$F(z) = \int_{\zeta \in V \cap \partial D} f(\zeta) B_0^0(z,\zeta) = \int_{\zeta \in \partial D} f(\zeta) B_0^0(z,\zeta).$$

It is easy to construct such a domain D. We can always assume that V is defined by $V = \{z \in U \mid r(z) = 0\}$, where r is a $\mathcal{C}^1$ function on U such that $dr(z) \neq 0$ for any $z \in V$. Let D' be a domain with $\mathcal{C}^1$ boundary in V containing the support of f. There is then a $\mathcal{C}^1$ function, ρ, defined in a neighbourhood of D' such that

$$D' = \{z \in V \cap U_{D'} \mid \rho(z) < 0\},$$
$$\partial D' = \{z \in U_{D'} \mid \rho(z) = r(z) = 0\}$$

and $$d\rho(z) \wedge dr(z) \neq 0 \quad \text{for any } z \in \partial D'.$$

For any $\varepsilon > 0$ we set $V_\varepsilon = \{z \in U \mid r(z) = \varepsilon\}$ if $U^+ = \{z \in U \mid r(z) > 0\}$ and $V_\varepsilon = \{z \in U \mid r(z) = -\varepsilon\}$ if $U^+ = \{z \in U \mid r(z) < 0\}$. For any small enough ε $dr(z) \neq 0$ for any $z \in V_\varepsilon \cap \{\rho \leqslant 0\}$ and $d\rho(z) \wedge dr(z) \neq 0$ for any $z \in V_\varepsilon \cap \{\rho = 0\}$. Assume r is chosen such that $U^+ = \{z \in U \mid r(z) < 0\}$ we set

$$D = \{z \in U \mid -\varepsilon < r(z) < 0\} \cap \{z \in U_{D'} \mid \rho(z) < 0\},$$

and D is then a domain with piecewise $\mathcal{C}^1$ boundary satisfying the conditions of Remark 1.2.

Theorem 1.3. *Let f be a $\mathcal{C}^\alpha$ function $(0 < \alpha \leqslant 1)$ which has compact support in V. The functions $F|_{U^+}$ and $F|_{U^-}$ then have continuous extensions F^+ and F^- to $U^+ \cup V$ and $U^- \cup V$ and on V these extensions satisfy the Plemelj formula*

$$F^+|_V - F^-|_V = f.$$

Proof. Let D be a domain with piecewise $\mathcal{C}^1$ boundary satisfying the conditions of Remark 1.2. Then $F(z) = \int_{\zeta \in \partial D} f(\zeta) B_0^0(z,\zeta)$.

Let $\widetilde{f}$ be a $\mathcal{C}^\alpha$ extension of f to U. We can construct $\widetilde{f}$ as follows: let $(U_i)_{i\in I}$ be an open cover of V by open subsets such that for any $i \in I$ there is a $\mathcal{C}^1$ map $h_i : U_i \to \mathbb{R}^{2n}$ such that $h_i(V) = \{x \in \mathbb{R}^{2n} \mid x_1 = 0\}$ and if $U_i \cap \operatorname{supp} f \neq \varnothing$ then $U_i \cap (\partial D \smallsetminus V) = \varnothing$. We define $\widetilde{f}_i$ by $\widetilde{f}_i(z) = f \circ h_i^{-1}(0, (h_i)_2(z), \dots, (h_i)_{2n}(z))$ for any $z \in U_i$; the function $\widetilde{f}_i$ is then $\mathcal{C}^\alpha$ on U_i and satisfies $\widetilde{f}_i|_{U_i \cap V} = f$. We set $U_0 = \mathbb{C}^n \smallsetminus V$ and let $(\chi_i)_{i \in I \cup \{0\}}$ be a $\mathcal{C}^\infty$ partition of unity subordinate to the open cover $(U_i)_{i\in I\cup\{0\}}$. The function $\widetilde{f} = \sum_{i\in I} \chi_i \widetilde{f}_i$ is then an extension with the required properties.

We set $F_0(z) = F(z) - \widetilde{f}(z)$ for any $z \in D$ and $F_0(z) = F(z)$ for any $z \notin \overline{D}$. By the Bochner–Martinelli formula (Chapter III, Corollary 2.4 and Remark 1.6 2))

$$F_0(z) = \int_{\zeta \in \partial D} \big(f(\zeta) - \widetilde{f}(z)\big) B_0^0(z, \zeta).$$

Since $\widetilde{f}$ is $\mathcal{C}^\alpha$ and $\widetilde{f}\big|_{\partial D} = f$ we have

$$\big(f(\zeta) - \widetilde{f}(z)\big) B_0^0(z, \zeta) = O\big(1/|\zeta - z|^{2n-1-\alpha}\big).$$

The differential form $(f(\zeta) - \widetilde{f}(z)) B_0^0(z, \zeta)$ therefore has locally integrable coefficients on ∂D, and it follows that F_0 is continuous on U. The existence of functions F^+ and F^- such that $F^+\big|_V - F^-\big|_V = f$ then follows easily from the existence of F_0. □

Remark. Under the hypotheses of Theorem 1.3 it is in fact possible to prove that F^+ and F^- are $\mathcal{C}^\alpha$ on $U^+ \cup V$ and $U^- \cup V$ respectively (cf. [Ci]).

Theorem 1.4. *Let $z_0 \in V$ be given. We set $\Delta_r = V \cap B(z_0, r)$ for any $r > 0$ and we denote by n_z the unit normal vector to V at z in the direction of U^+. For any small enough r and continuous function f with compact support in V the quantity $F(z + tn_z) - F(z - tn_z)$ converges to $f(z)$ uniformly with respect to z in Δ_r as t tends to 0 along the positive real axis.*

Proof. Choose $r > 0$ such that $B(z_0, 2r) \subset U$ and decompose f as $f = f_0 + f_1$ where $\operatorname{supp} f_1 \subset B(z_0, 2r)$ and $f_0 \equiv 0$ on $B(z_0, r) \cap V$. The Bochner–Martinelli transform F_0 of f_0 is obviously continuous on $B(z_0, r)$ and it is therefore enough to study the Bochner–Martinelli transform F_1 of f_1.

Without loss of generality we can assume that $z_0 = 0$ and r is small enough that:

1) $\partial B(0, 2r)$ is transverse to V. We then let D be a domain with $\mathcal{C}^1$ boundary contained in $B(0, 2r) \cap U^+$ such that $\partial D \cap V$ contains the support of f_1.
2) if n_z is the normal unit vector to V at z in the direction of U^+ then there is a constant $C < 1$ such that $|(\zeta - z, n_z)| \leqslant C|\zeta - z|$ for any $z, \zeta \in B(0, 2r) \cap V$.

For any $z \in \Delta_r$,

$$\int_{\zeta \in \partial D} f_1(z) B_0^0(z + tn_z, \zeta) = \begin{cases} f(z) & \text{for small enough } t > 0 \\ 0 & \text{for } t < 0. \end{cases}$$

by the Bochner–Martinelli formula. It follows that if $|t| \leqslant t_0$ is small enough then

$$\begin{aligned} (1.1) \quad G(z, t) &= F_1(z + tn_z) - F_1(z - tn_z) - f(z) \\ &= \int_{\zeta \in \partial D} [f_1(\zeta) - f_1(z)]\big(B_0^0(z + tn_z, \zeta) - B_0^0(z - tn_z, \zeta)\big). \end{aligned}$$

Lemma 1.5. *There is a constant C such that*

$$\int_{\partial D} \left|B_0^0(z+tn_z,\zeta) - B_0^0(z-tn_z,\zeta)\right| d\sigma(\zeta) \leqslant C, \quad \textit{for any } z \in \Delta_r,\ 0 < t \leqslant t_0.$$

Proof of Lemma 1.5. We set

$$A(t,\zeta,) = |B_0^0(z+tn_z,\zeta) - B_0^0(z-tn_z,\zeta)|.$$

By calculations similar to those given in the proof of Lemma 2.2 of Chapter III, we get the following estimation

$$A(t,\zeta) \leqslant C_1|t| \Big(\max\Big(\frac{1}{|\zeta-(z-tn_z)|^{2n}}, \frac{1}{|\zeta-(z+tn_z)|^{2n}}\Big) + \frac{1}{|\zeta-(z-tn_z)|^{2n}}\Big).$$

But now $|\zeta - z \pm tn_z|^2 \geqslant (1-c)[|\zeta-z|^2+t^2]$ for any $z,\zeta \in B(0,2r)\cap V$ since r is chosen such that $|(\zeta - z \mid n_z)| \leqslant c|\zeta - z|$. It follows that

$$A(t,\zeta) \leqslant C_2 \frac{|t|}{(|\zeta-z|^2+t^2)^n}. \tag{1.2}$$

For any $z \in \Delta_r$ and γ such that $0 < \gamma < r$,

$$\int_{\partial D} A(t,\zeta)d\sigma(\zeta) = \int_{\partial D \setminus B(z,\gamma)} A(t,\zeta)d\sigma(\zeta) + \int_{\partial D \cap B(z,\gamma)} A(t,\zeta)d\sigma(\zeta).$$

The first integral on the right-hand side is clearly uniformly bounded with respect to z in Δ_r. After a choice of parameterisation of $V \cap B(z,\gamma)$, the second integral can be bounded above by

$$I(t) = \int_{\substack{x\in\mathbb{R}^{2n-1}\\ |x|\leqslant R}} \frac{t}{(|x|^2+t^2)^n} dV(x).$$

By a change of variable $y = tx$ we get

$$I(t) = \int_{\substack{y\in\mathbb{R}^{2n-1}\\ |y|\leqslant R/t}} \frac{t^{2n}dV(y)}{(t^2|y|^2+t^2)^n} \leqslant \int_{\mathbb{R}^{2n-1}} \frac{dV(y)}{(|y|^2+1)^n} < +\infty,$$

which completes the proof of the lemma. □

End of the proof of Theorem 1.4. Fix $\varepsilon > 0$ and choose $\eta > 0$ such that $|f(\zeta) - f(z)| < \varepsilon/C$ whenever $|\zeta - z| < \eta$. We then cut ∂D into two parts, $\partial D \cap B(z,\eta)$ and $\partial D \setminus B(z,\eta)$. We get the following estimation from (1.1) and Lemma 1.5:

$$|G(z,t)| \leqslant \varepsilon + 2|f|_{\partial D} \int_{\partial D \setminus B(z,\eta)} \left|B_0^0(z+tn_z,\zeta) - B_0^0(z-tn_z,\zeta)\right| d\sigma(\zeta). \tag{1.3}$$

On $\partial D \setminus B(z,\eta)$ the function $A(t,\zeta) = \left|B_0^0(z+tn_z,\zeta) - B_0^0(z-tn_z,\zeta)\right|$ depends continuously on (t,ζ) and by (1.2) we have

$$A(t,\zeta) \leqslant C_2 \frac{|t|}{(\eta^2+t^2)^n}.$$

It follows that the integral on the right-hand side of (1.3) tends to 0 as t tends to 0 and, moreover, this convergence is uniform with respect to z in Δ_r since $B_0^0(z,\zeta)$ only depends on $z-\zeta$. There is therefore a t_0' such that if $0<t<t_0'$ then, for any $z \in \Delta_r$, $|G(z,t)| < 2\varepsilon$. □

Corollary 1.6. *If f is continuous and compactly supported in V and $F\big|_{U^-}$ has a continuous extension F^- to $U^- \cup V$ then $F\big|_{U^+}$ has a continuous extension F^+ to $U^+ \cup V$ and*

$$F^+\big|_V - F^-\big|_V = f.$$

Proof. Let $z_0 \in V$ be fixed and let $w \in U^+$ be close enough to z_0. We denote the orthogonal projection of w to V by z. Then $w = z + tn_z$, where n_z is the unit normal vector to V at z. We will prove that $F(w)$ tends to $F^-(z_0)+f(z_0)$ as w tends to z_0, which will prove the corollary

$$(1.4) \quad \begin{aligned} |F(w) - F^-(z_0) - f(z_0)| &\leqslant |F(z+tn_z) - F(z-tn_z) - f(z)| \\ &\quad + |f(z) - f(z_0)| + |F(z+tn_z) - F^-(z_0)|. \end{aligned}$$

Choose $\varepsilon > 0$. If w is close enough to z_0 then it follows from Theorem 1.4 that the first term in the right-hand side of (1.4) is bounded above by $\varepsilon/3$. Moreover, if w is close enough to z_0 then z is close to z_0 and by continuity of f the second term is also bounded above by $\varepsilon/3$. And finally, if w tends to z_0 then $z - tn_z$ also tends to z_0 and the continuity of F^- on $U^- \cup V$ implies that the third term is also bounded above by $\varepsilon/3$. □

Example. Let D be a bounded domain with $\mathcal{C}^1$ boundary and consider a function $f \in \mathcal{C}(\partial D)$. Assume that $\int_{\partial D} f(\zeta) B_0^0(z,\zeta) = 0$ for any $z \notin \overline{D}$: the function $F(z) = \int_{\partial D} f(\zeta) B_0^0(z,\zeta)$, which is defined on D, then has a continuous extension to $\overline{D}$ and $F(z) = f(z)$ for any $z \in \partial D$.

2 CR functions on a real hypersurface

Let V be a real oriented $\mathcal{C}^1$ hypersurface in $\mathbb{C}^n$. The aim of this section is to define a class of functions on V which contains the restrictions of holomorphic functions defined in a neighbourhood of V and which is equal to the set of restrictions of holomorphic functions on D when V is the boundary of the domain D.

Definition 2.1. A continuous function f on V is said to be a *Cauchy–Riemann* (CR) function if for any $\mathcal{C}^\infty$ differential form λ of bidegree $(n, n-2)$ defined on a neighbourhood of V such that $\operatorname{supp}\lambda \cap V$ is compact we have

$$\int_V f\overline{\partial}\lambda = 0.$$

Remark. In terms of currents, f is CR on V means exactly that $f[V]^{0,1}$ is $\overline{\partial}$-closed, where $[V]^{0,1}$ is the bidegree $(0,1)$ part of the integration current on V.

Example. If F is a holomorphic function in a neighbourhood of V then $f = F|_V$ is CR on V. Indeed, let λ be a $\mathcal{C}^\infty$ differential $(n, n-2)$-form in a neighbourhood of V such that $\operatorname{supp}\lambda \cap V$ is compact and let D be a bounded domain with piecewise $\mathcal{C}^1$ boundary contained in the domains of definition of F and λ such that $\partial D \cap V \supset \operatorname{supp}\lambda \cap V$ and the orientation on V is the same as the orientation on ∂D. We then have

$$\begin{aligned}
\int_V f\overline{\partial}\lambda &= \int_{\partial D} F\overline{\partial}\lambda && \text{by definition of } D \text{ and because } f = F|_V \\
&= \int_{\partial D} \overline{\partial}(F\lambda) && \text{because } F \text{ is holomorphic} \\
&= \int_{\partial D} d(F\lambda) = 0 && \text{by Stoke's formula since } \lambda \text{ is of type } (n, n-2).
\end{aligned}$$

Definition 2.2. A function $f \in \mathcal{C}^1(V)$ is said to be *Cauchy–Riemann* (CR) on V if $\nu(f) = 0$ for any $p \in V$ and any $\nu \in T_p^{0,1}(V)$.

Remark. If V is defined by $\{z \in U \mid r(z) = 0\}$, where U is an open set in $\mathbb{C}^n$ and r is a real-valued $\mathcal{C}^1$ function on U such that $dr(z) \neq 0$ for all $z \in U$, then a function $f \in \mathcal{C}^1(V)$ is CR if and only if for any $p \in V$

$$\sum_{j=1}^n t_j \frac{\partial f}{\partial \overline{z}_j}(p) = 0$$

for any $t \in \mathbb{C}^n$ such that $\sum_{j=1}^n \frac{\partial r}{\partial z_j}(p)\overline{t}_j = 0$.

We will now prove that if f is a $\mathcal{C}^1$ function on V then the above two definitions coincide.

Lemma 2.3. *Assume that V is $\mathcal{C}^k$ for some k, $1 \leqslant k \leqslant \infty$. If $f \in \mathcal{C}^k(V)$ is a* CR *function in the sense of Definition 2.2 then there is a neighbourhood U of V and an extension $\widetilde{f} \in \mathcal{C}^{k-1}(U)$ of f such that*

i) $\widetilde{f}|_V = f$.
ii) *All the $k-1$st order derivatives of $\widetilde{f}$ are differentiable at any point $z \in V$ and $D^\alpha \widetilde{f}$ is continuous on V for any α such that $|\alpha| = k$.*

iii) $\overline{\partial}\widetilde{f}(z) = 0$ *for any* $z \in V$.

Proof. For any $p \in V$ let $L_1(p), \dots, L_{n-1}(p)$ be a basis for $T_p^{1,0}(V)$ and let $L_n(p) \neq 0$ be a vector such that $T_p^{1,0}(\mathbb{C}^n) = T_p^{1,0}(V) \oplus \mathbb{C}L_n(p)$. By Definition 2.2, f is CR if and only if $\overline{L_j(p)}f = 0$, $j = 1, \dots, n-1$, for any $p \in V$.

We seek an extension $\widetilde{f}$ of f such that $\partial\widetilde{f}/\partial\overline{z}_j(p) = 0$ for any $p \in V$ and hence $\overline{L}_n(p)\widetilde{f} = 0$ for any $p \in V$. Let us check that this condition determines the Taylor series of $\widetilde{f}$ up to order 1 at any point in V. If V is defined by $\{z \in U \mid r(z) = 0\}$, where U is an open subset of $\mathbb{C}^n$ and r is a real-valued $\mathcal{C}^k$ function on U such that $dr(z) \neq 0$ for any $z \in U$, then we can choose

$$L_n(z) = \sum_{j=1}^{n} \frac{\partial r}{\partial \overline{z}_j}(z)\Big(\frac{\partial}{\partial z_j}\Big)_z, \qquad z \in U.$$

Then $(L_n - \overline{L}_n)(r) = 0$ and hence $L_n(z) - \overline{L}_n(z)$ is an element of $\mathbb{C}T_z(V)$ for any $z \in V$. It follows that if $\overline{L}_n(z)\widetilde{f} = 0$ for any $z \in V$ then

$$L_n(z)\widetilde{f} = [L_n(z) - \overline{L}_n(z)]\widetilde{f} = [L_n(z) - \overline{L}_n(z)]f,$$

for any $z \in V$, since $\widetilde{f}\big|_V = f$. Moreover, $L_j(z)\widetilde{f} = L_j(z)f$ and $\overline{L}_j(z)\widetilde{f} = 0$ for any $z \in V$ and $1 \leqslant j \leqslant n-1$. As $\mathbb{C}T_z\mathbb{C}^n$ is generated by $(L_i(z), \overline{L}_i(z))_{1\leqslant i\leqslant n}$ for any $z \in V$, $\nu\widetilde{f}$ is therefore entirely determined for any $\nu \in \mathbb{C}T_z\mathbb{C}^n$ and $z \in V$.

We can assume that $\overline{L}_n r(z) = 1$ for any $z \in V$ and f is the restriction of a $\mathcal{C}^k$ function on U which we again denote by f. We then set $\widetilde{f}(z) = f(z) - r(z)(\overline{L}_n f)(z)$ for any $z \in U$. It is clear that $\widetilde{f}$ is $\mathcal{C}^{k-1}$ and $\widetilde{f}\big|_V = f$. Since $r\big|_{\partial D} \equiv 0$ and r is of class $\mathcal{C}^k$ it is easy to prove ii).

By definition, $\overline{\partial}\widetilde{f}(z) = 0$ for any $z \in V$ if and only if $\overline{L}_j(z)\widetilde{f} = 0$ for any $z \in V$ and $j = 1, \dots, n$. But $\overline{L}_j(z)\widetilde{f} = \overline{L}_j(z)f = 0$ for any $z \in V$ because f is CR and $\overline{L}_n(z)\widetilde{f} = \overline{L}_n(z)f - (\overline{L}_n(z)r)(\overline{L}_n(z)f) = 0$ because $\overline{L}_n(z)r = 1$ for any $z \in V$. □

We now show that the two definitions are the same when f is $\mathcal{C}^1$ on V.

Suppose that $f \in \mathcal{C}^1(V)$ is CR in the sense of Definition 2.2 and let $\widetilde{f}$ be an extension of f to a neighbourhood of V for which the conclusions of Lemma 2.3 hold. If λ is a $\mathcal{C}^\infty$ differential $(n, n-2)$-form on a neighbourhood of V such that $\operatorname{supp}\lambda \cap V$ is compact then

$$\begin{aligned}
\int_V f\overline{\partial}\lambda &= \int_V \widetilde{f}\overline{\partial}\lambda && \text{since } \widetilde{f}\big|_V = f \\
&= \int_V \overline{\partial}(\widetilde{f}\lambda) && \text{since } \overline{\partial}\widetilde{f} = 0 \text{ on } V \\
&= \int_V d(\widetilde{f}\lambda) = 0 && \text{by Stokes' formula since } \lambda \text{ is of type } (n, n-2).
\end{aligned}$$

Conversely, if $f \in \mathcal{C}^1(V)$ is CR in the sense of Definition 2.1 and $\widetilde{f}$ is a $\mathcal{C}^1$ extension of f to a neighbourhood of V then $\overline{\partial}(\widetilde{f}[V]^{0,1}) = 0$ which implies that $(\overline{\partial}\widetilde{f}) \wedge [V]^{0,1} = 0$ since $[V]$ is a closed current. If r is a defining function for V then $[V]^{0,1} = \delta\overline{\partial}r$ and hence $\overline{\partial}\widetilde{f} \wedge [V]^{0,1} = 0$ is equivalent to

$$\Big(\frac{\partial r}{\partial \overline{z}_k}\frac{\partial f}{\partial \overline{z}_j} - \frac{\partial r}{\partial \overline{z}_j}\frac{\partial f}{\partial \overline{z}_k}\Big) = 0 \quad \text{whenever } 1 \leqslant j < k \leqslant n.$$

For any $p \in \partial D$ the vectors

$$\Big(\frac{\partial r}{\partial \overline{z}_k}(p)\Big(\frac{\partial}{\partial \overline{z}_j}\Big)_p - \frac{\partial r}{\partial \overline{z}_j}(p)\Big(\frac{\partial}{\partial \overline{z}_k}\Big)_p\Big), \quad 1 \leqslant j < k \leqslant n,$$

are a generating family for $T_p^{0,1}(V)$ and hence $\nu(f) = 0$ for any $\nu \in T_p^{0,1}(V)$, $p \in V$.

3 Bochner's theorem

Let D be a bounded domain with $\mathcal{C}^1$ boundary in $\mathbb{C}^n$. We will study the following problem: given a continuous function f on ∂D, what are the conditions on f and D under which f can be extended to a function F which is continuous on $\overline{D}$ and holomorphic on D.

We know by Hartogs' extension theorem (Chap. III, Th. 4.2) that the problem can be solved for $n \geqslant 2$ if the boundary of D is connected and f is the restriction of a holomorphic function defined on a neighbourhood of ∂D.

Let us start by finding some necessary conditions for the problem to have a solution.

1) The function f must be CR.
Indeed, let α be a $\mathcal{C}^\infty$ differential form of bidegree $(n, n-2)$ which is defined in a neighbourhood of ∂D. Assume that f can be extended to a continuous function F on $\overline{D}$ which is holomorphic on D and let χ be a $\mathcal{C}^\infty$ function with compact support in the domain of definition of α which is identically 1 on a neighbourhood of ∂D. Then,

$$\begin{aligned}
\int_{\partial D} f\overline{\partial}\alpha &= \int_{\partial D} F\overline{\partial}(\chi\alpha) && \text{by definition of } F \text{ and } \chi \\
&= \int_D d(F\overline{\partial}(\chi\alpha)) && \text{by Stokes' formula} \\
&= \int_D \overline{\partial}(F\overline{\partial}(\chi\alpha)) && \text{since } \alpha \text{ is of bidegree } (n, n-2) \\
&= \int_D \overline{\partial}F \wedge \overline{\partial}(\chi\alpha) = 0 && \text{since } F \text{ is holomorphic on } D.
\end{aligned}$$

2) It must be the case that ∂D is connected or $\int_{\partial D} f(\zeta)B_0^0(z,\zeta) = 0$ for any $z \notin \overline{D}$.
   Suppose that ∂D has two connected components Γ_1 and Γ_2. Let Γ_1 be the component which bounds the unbounded component of $\mathbb{C}^n \setminus D$ and denote the restriction of f to Γ_i by f_i, $i = 1, 2$. If the problem can be solved then there is a continuous function F on $\overline{D}$ which is holomorphic on D such that $F\big|_{\Gamma_i} = f_i$ for $i = 1, 2$. If K is a bounded connected component of $\mathbb{C}^n \setminus \overline{D}$ then by Hartogs' theorem F can be extended to a holomorphic function $\widetilde{F}$ on $D \cup K$. By the Bochner–Martinelli formula we then have

$$\widetilde{F}(z) = \int_{\zeta \in \Gamma_1} f(\zeta) \wedge B_0^0(z,\zeta) = -\int_{\zeta \in \Gamma_2} f(\zeta) \wedge B_0^0(z,\zeta)$$

   for any $z \in \overset{\circ}{K}$ and hence

$$\int_{\partial D} f(\zeta)B_0^0(z,\zeta) = 0 \quad \text{for any } z \notin \overline{D}$$

   since this integral clearly vanishes if z is in the unbounded component of $\mathbb{C}^n \setminus \overline{D}$.

Theorem 3.1 (Bochner's theorem). *Let D be a relatively compact domain with $\mathcal{C}^1$ boundary in $\mathbb{C}^n$, $n \geqslant 2$, such that $\mathbb{C}^n \setminus \overline{D}$ is connected. If f is a continuous* CR *function on ∂D then there is a continuous function F on $\overline{D}$ which is holomorphic on D such that $F\big|_{\partial D} = f$. Moreover, $F(z) = \int_{\zeta \in \partial D} f(\zeta)B_0^0(z,\zeta)$ for any $z \in D$.*

Proof. We start by proving that F is unique if it exists. If F_1 and F_2 are two holomorphic extensions of f to D then $F_1 - F_2$ is a holomorphic function on D which is continuous on $\overline{D}$ and has the property that $F_1 - F_2\big|_{\partial D} = 0$. By the maximum principle it follows that $F_1 - F_2 \equiv 0$ on $\overline{D}$.

Moreover, if F exists then by the Bochner–Martinelli formula

$$F(z) = \int_{\zeta \in \partial D} F(\zeta)B_0^0(z,\zeta) = \int_{\zeta \in \partial D} f(\zeta)B_0^0(z,\zeta) \quad \text{for any } z \in D$$

since $F\big|_{\partial D} = f$. (This gives a second proof of the uniqueness of F.)

We now prove that F exists. Consider the function

$$F(z) = \int_{\partial D} f(\zeta)B_0^0(z,\zeta).$$

This function is $\mathcal{C}^\infty$ on $\mathbb{C}^n \setminus \partial D$ and by our previous results on the Bochner–Martinelli transform (cf. Cor. 1.6) it is enough to show that F is holomorphic on $\mathbb{C}^n \setminus \partial D$ and vanishes identically on $\mathbb{C}^n \setminus \overline{D}$. Let us now calculate $\overline{\partial} F$ on $\mathbb{C}^n \setminus \overline{D}$. We have $\overline{\partial} F(z) = \int_{\zeta \in \partial D} f(\zeta)\overline{\partial}_z B_0^0(z,\zeta)$ but $\overline{\partial}_z B_0^0(z,\zeta) =$

$-\overline{\partial}_\zeta B_1^0(z,\zeta)$ on $\mathbb{C}^n \times \mathbb{C}^n \setminus \Delta$ and hence $\overline{\partial} F(z) = -\int_{\zeta\in\partial D} f(\zeta)\overline{\partial}_\zeta B_1^0(z,\zeta) = 0$ for any given $z \in \mathbb{C}^n \setminus \partial D$ since f is a CR function on the compact set ∂D. The function F is therefore holomorphic on $\mathbb{C}^n \setminus \partial D$. Note that $F(z)$ tends to 0 as $|z|$ tends to infinity since $|F(z)| \leqslant C/\operatorname{dist}(z,\partial D)$. As F is holomorphic on $\mathbb{C}^n \setminus \overline{D}$, $n \geqslant 2$ and $\mathbb{C}^n \setminus \overline{D}$ is connected, F vanishes identically on $\mathbb{C}^n \setminus \overline{D}$ (cf. the proof of Theorem 4.1 in Chapter IV). □

Theorem 3.2. *Let D be a relatively compact domain with $\mathcal{C}^1$ boundary in $\mathbb{C}^n$ and let f be a continuous function on ∂D. A necessary and sufficient condition for the existence of a continuous function F on $\overline{D}$, holomorphic on D, such that $F|_{\partial D} = f$, is that for any $\overline{\partial}$-closed $\mathcal{C}^\infty$ differential $(n, n-1)$-form α defined on a neighbourhood of $\overline{D}$,*

$$\int_{\partial D} f\alpha = 0.$$

Proof.

1) *Necessity.* Assume there is a continuous F on $\overline{D}$, holomorphic on D, such that $F|_{\partial D} = f$ and consider a $\overline{\partial}$-closed $\mathcal{C}^\infty$ differential $(n, n-1)$-form α in a neighbourhood of $\overline{D}$. Then,

$$\begin{aligned} \int_{\partial D} f\alpha &= \int_{\partial D} F\alpha && \text{by definition of } F \\ &= \int_D d(F\alpha) && \text{by Stokes' formula} \\ &= \int_D \overline{\partial} F \wedge \alpha + F\overline{\partial}\alpha = 0 && \text{since } F \text{ and } \alpha \text{ are } \overline{\partial}\text{-closed.} \end{aligned}$$

2) *Sufficiency.* We start by proving that f is CR. Let λ be a $\mathcal{C}^\infty$ differential form of bidegree $(n, n-2)$ defined in a neighbourhood of ∂D and let χ be a $\mathcal{C}^\infty$ function with compact support in the domain of definition of λ which is equal to 1 in a neighbourhood of ∂D. Then,

$$\int_{\partial D} f\overline{\partial}\lambda = \int_{\partial D} f\overline{\partial}(\chi\lambda) = 0$$

since $\overline{\partial}(\chi\lambda)$ is a $\overline{\partial}$-closed $\mathcal{C}^\infty$ differential $(n, n-1)$-form defined in a neighbourhood of $\overline{D}$.
Consider the function $F(z) = \int_{\zeta\in\partial D} f(\zeta)B_0^0(z,\zeta)$. This function is holomorphic on $\mathbb{C}^n\setminus\partial D$ by the proof of the above theorem. If $z \in \mathbb{C}^n \setminus \overline{D}$ then $B_0^0(z,\cdot)$ is a $\overline{\partial}$-closed $\mathcal{C}^\infty$ differential $(n, n-1)$-form on $\mathbb{C}^n \setminus \{z\}$, which is a neighbourhood of $\overline{D}$. The hypothesis on f then implies that

$$F(z) = \int_{\zeta\in\partial D} f(\zeta)B_0^0(z,\zeta) = 0 \quad \text{for any } z \in \mathbb{C}^n \setminus \overline{D}.$$

It then follows from Corollary 1.6 that f is the desired extension.

We end this section by studying the regularity of the extension F when f is $\mathcal{C}^k$.

Theorem 3.3. *Let D be a bounded domain in $\mathbb{C}^n$ with $\mathcal{C}^m$ boundary and let f be a* CR *function which is $\mathcal{C}^k$ for some $1 \leqslant k \leqslant m$ on ∂D. Assume that either*

- *$\mathbb{C}^n \setminus \overline{D}$ is connected and $n \geqslant 2$ or*
- *$\int_{\partial D} f\alpha = 0$ for any $\overline{\partial}$-closed $\mathcal{C}^\infty$ differential $(n, n-1)$-form α in a neighbourhood of $\overline{D}$.*

There is then a $\mathcal{C}^k$ function F on $\overline{D}$, holomorphic on D, such that

$$F\big|_{\partial D} = f.$$

Proof. The existence and uniqueness of F follow from Theorems 3.1 and 3.2. It only remains to study its regularity. We recall that

$$F(z) = \int_{\zeta \in \partial D} f(\zeta) B_0^0(z, \zeta) \quad \text{for any } z \in \mathbb{C}^n \setminus \partial D.$$

We start by showing that F is $\mathcal{C}^1$ on $\overline{D}$. Differentiating under the integral sign we see that

$$\frac{\partial F}{\partial z_j}(z) = \int_{\zeta \in \partial D} f(\zeta) \frac{\partial}{\partial z_j} B_0^0(z, \zeta) \quad \text{for any } z \in \mathbb{C}^n \setminus \partial D.$$

As B_0^0 is of type $(n, n-1)$ with respect to ζ and $\overline{\partial}_\zeta$-closed for any $j \in \{1, \ldots, n\}$ we can write

$$B_0^0 = d\zeta_j \wedge B_j,$$

where B_j is a $(n-1, n-1)$-differential form such that $\overline{\partial}_\zeta B_j = 0$. Moreover, the coefficients of B_j are functions in $\zeta - z$. It follows that

$$\frac{\partial B_0^0}{\partial z_j} = -d\zeta_j \wedge \frac{\partial B_j}{\partial \zeta_j} = -\partial_\zeta B_j = -d_\zeta B_j.$$

Let $\widetilde{f}$ be an extension of f satisfying the conclusions of Lemma 2.3. For any $z \in \mathbb{C}^n \setminus \partial D$ we then have

$$\begin{aligned}
\frac{\partial F}{\partial z_j}(z) &= -\int_{\zeta \in \partial D} \widetilde{f}(\zeta) d_\zeta B_j(z, \zeta) \\
&= \int_{\zeta \in \partial D} d_\zeta \widetilde{f}(z) \wedge B_j(z, \zeta) && \text{by Stokes' formula} \\
&= \int_{\zeta \in \partial D} \frac{\partial \widetilde{f}}{\partial \zeta_j}(\zeta) d\zeta_j \wedge B_j(z, \zeta) && \text{since } \overline{\partial} \widetilde{f}\big|_{\partial D} = 0 \\
&= \int_{\zeta \in \partial D} \frac{\partial \widetilde{f}}{\partial \zeta_j}(\zeta) B_0^0(z, \zeta) && \text{by definition of } B_j.
\end{aligned}$$

As $F \equiv 0$ on $\mathbb{C}^n \setminus \overline{D}$ the same is true of $\partial F/\partial z_j$ and it therefore follows from Corollary 1.6 that $\partial F/\partial z_j\big|_D$ can be extended continuously to $\overline{D}$.

The general case is proved by induction on k. Assume the result holds for some $k \geqslant 1$. Let $f \in \mathcal{C}^{k+1}(\partial D)$ be a CR function on ∂D and let $\widetilde{f}$ be an extension of f satisfying the conclusions of Lemma 2.3. We have just proved that if $F(z) = \int_{\zeta\in\partial D} f(\zeta)B_0^0(z,\zeta)$ for any $z \in \mathbb{C}^n \setminus \partial D$ then for any $j = 1,\dots,n$

$$\frac{\partial F}{\partial z_j}(z) = \int_{\zeta\in\partial D} \frac{\partial \widetilde{f}}{\partial \zeta_j}(\zeta)B_0^0(z,\zeta) \quad \text{for any } z \in \mathbb{C}^n \setminus \partial D.$$

Since $\partial F/\partial z_j$ is holomorphic on D and continuous on $\overline{D}$ and $F\big|_{\partial D} = f$, $\partial F/\partial z_j\big|_{\partial D}$ is CR and $\partial F/\partial z_j\big|_{\partial D} = \partial \widetilde{f}/\partial \zeta_j\big|_{\partial D} \in \mathcal{C}^k(\partial D)$. It follows from the induction hypothesis that $\partial F/\partial z_j \in \mathcal{C}^k(\overline{D})$ for any $j = 1,\dots,n$, or in other words, $F \in \mathcal{C}^{k+1}(\overline{D})$. □

4 Stokes' formula for CR functions

Let V be a real $\mathcal{C}^1$ hypersurface in $\mathbb{C}^n$, $n \geqslant 2$.

Let D be a relatively compact domain with $\mathcal{C}^1$ boundary in V. If f is a $\mathcal{C}^1$ CR function on V then it follows from Stokes' formula that, for any differential $(n, n-2)$-form α which is $\mathcal{C}^1$ in a neighbourhood of V,

$$\int_D f\overline{\partial}\alpha = \int_{\partial D} f\alpha. \tag{4.1}$$

The aim of this paragraph is to extend formula (4.1) to the case where the function f is only assumed to be continuous.

Lemma 4.1. *If D is a relatively compact domain with $\mathcal{C}^1$ boundary in V then there is a bounded domain $\widetilde{D}$ in $\mathbb{C}^n$ with $\mathcal{C}^1$ boundary in a neighbourhood of V such that $\widetilde{D} \cap V = D$ and V cuts ∂D transversally.*

Proof. Since ∂D is a $\mathcal{C}^1$ submanifold of codimension 2 in $\mathbb{C}^n$ there are real-valued functions r and ρ such that

a) r is defined and $\mathcal{C}^1$ on a neighbourhood U_V of V in $\mathbb{C}^n$, $V = \{z \in U_V \mid r(z) = 0\}$ and $dr(z) \neq 0$ for every $z \in U_V$.
b) ρ is defined and $\mathcal{C}^1$ on a neighbourhood $U_{\partial D}$ of ∂D in $\mathbb{C}^n$, $\partial D = \{z \in V \cap U_{\partial D} \mid \rho(z) = 0\}$, $d\rho(z) \neq 0$ for every $z \in U_{\partial D}$ and $D \cap U_{\partial D} = \{z \in V \cap U_{\partial D} \mid \rho(z) < 0\}$.
c) $dr(z) \wedge d\rho(z) \neq 0$ for every $z \in U_V \cap U_{\partial D}$.

We then define $\widetilde{D}$ as follows: $\widetilde{D} \subset \{z \in U_V \mid -\varepsilon < r(z) < \varepsilon\}$, $\widetilde{D} \cap V = D$ and $\widetilde{D} \cap U_{\partial D} = \{z \in U_{\partial D} \mid \rho(z) < 0\}$, where ε is chosen small enough that $U_{\partial D} \cap \{z \in U_V \mid -\varepsilon < r(z) < \varepsilon\} \neq \varnothing$. □

Lemma 4.2. *Let D be a relatively compact domain with $\mathcal{C}^1$ boundary in V and let $\widetilde{D}$ be a domain associated to D by Lemma 4.1. If φ is a continuous differential $(2n-1)$-form on V and $\widetilde{\varphi}$ is a continuous extension of φ to $\mathbb{C}^n$ then*

$$\int_D \varphi = \mathcal{K}([\widetilde{D}], [V] \wedge \widetilde{\varphi}),$$

where $\mathcal{K}(\cdot, \cdot)$ is the Kronecker index.

Proof. We will prove that $\mathcal{K}([\widetilde{D}], [V]^{0,1} \wedge \widetilde{\varphi})$ exists and is equal to $\int_{\widetilde{D}\cap V} \widetilde{\varphi}$ which is equal to $\int_D \varphi$ by definition of $\widetilde{D}$ and $\widetilde{\varphi}$. Using a partition of unity we can assume that the support of $\widetilde{\varphi}$ is small enough that we are in the following situation: up to change of $\mathcal{C}^1$ coordinates, V is defined by the equation $x_1 = 0$ and $\widetilde{D}$ is defined by the equation $x_2 < 0$ in a neighbourhood of the support of $\widetilde{\varphi}$.

Let $(\theta_\varepsilon)_{\varepsilon>0}$ and $(\alpha_{\varepsilon'})_{\varepsilon'>0}$ be two families of regularising functions. The following then holds:

$$\begin{aligned}&\mathcal{K}([\widetilde{D}], [V] \wedge \widetilde{\varphi})\\ &= \lim_{\varepsilon,\varepsilon'\to 0} \int_{\mathbb{R}^{2n}} \Big(\int_{\mathbb{R}^{2n-1}} \varphi(0,z')\theta_\varepsilon(x-(0,z'))dz'\Big)\Big(\int_{\mathbb{R}^{2n}} \chi_2^+(y)\alpha_{\varepsilon'}(x-y)dy\Big)dx\end{aligned}$$

where $z' = (z_2, \ldots, z_n)$, $\widetilde{\varphi} = \varphi dx_2 \wedge \cdots \wedge dx_n + dx_1 \wedge \psi$ and χ_2^+ is the characteristic function of the set $\{x \in \mathbb{R}^{2n} \mid x_2 > 0\}$. Applying Fubini's theorem and linear changes of coordinates we get

$$\begin{aligned}&\mathcal{K}([\widetilde{D}], [V] \wedge \widetilde{\varphi})\\ &= \lim_{\varepsilon,\varepsilon'\to 0} \int_{\mathbb{R}^{2n-1}} \varphi(0,z') \int_{\mathbb{R}^{2n}} \theta_\varepsilon(x) \int_{\mathbb{R}^{2n}} \chi_2^+(y)\alpha_{\varepsilon'}(x+(0,z')-y)dydxdz'\\ &= \lim_{\varepsilon,\varepsilon'\to 0} \int_{\mathbb{R}^{2n-1}} \varphi(0,z') \int_{\mathbb{R}^{2n}} \theta_\varepsilon(x) \int_{\mathbb{R}^{2n}} \chi_2^+(x+(0,z')-y)\alpha_{\varepsilon'}(y)dydxdz'\\ &= \lim_{\varepsilon,\varepsilon'\to 0} \int_{\mathbb{R}^{2n}} \theta_\varepsilon(x) \int_{\mathbb{R}^{2n}} \alpha_{\varepsilon'}(y) \int_{\mathbb{R}^{2n-1}} \varphi(0,z')\chi_2^+(x+(0,z')-y)dz'dydx.\end{aligned}$$

Note that the function χ_2^+ is independent of the variables $x_1, x_3, \ldots, x_{2n}$ and set $z'' = (z_3, \ldots, z_{2n})$. Then,

$$\begin{aligned}&\mathcal{K}([\widetilde{D}], [V] \wedge \widetilde{\varphi}) =\\ &\lim_{\varepsilon,\varepsilon'\to 0} \int_{\mathbb{R}^{2n}} \theta_\varepsilon(x) \int_{\mathbb{R}^{2n}} \alpha_{\varepsilon'}(y) \int_{\mathbb{R}^{2n-2}} \int_{\mathbb{R}} \varphi(0,u-x_2+y_2,z'')\chi_{\{u>0\}}dudz''dydx.\end{aligned}$$

It follows that, since $\int_{\mathbb{R}^{2n}} \theta_\varepsilon(x)dx = \int_{\mathbb{R}^{2n}} \alpha_{\varepsilon'}(y)dy = 1$,

$$\begin{aligned}&\mathcal{K}([\widetilde{D}], [V] \wedge \widetilde{\varphi}) - \int_{\widetilde{D}\cap V} \widetilde{\varphi} = \lim_{\varepsilon,\varepsilon'\to 0} \int_{\mathbb{R}^{2n}} \theta_\varepsilon(x)\Big(\int_{\mathbb{R}^{2n}} \alpha_{\varepsilon'}(y)\\ &\Big[\int_{\mathbb{R}^{2n-2}} \Big(\int_{\mathbb{R}} (\varphi(0,u-x_2+y_2,z'') - \varphi(0,u,z''))\chi_{\{u>0\}}du\Big)dz''\Big]dy\Big)dx.\end{aligned}$$

As the function φ is continuous with compact support it is uniformly continuous and hence $\varphi(0, u - x_2 + y_2, z'')$ tends to $\varphi(0, u, z'')$ uniformly with respect to (u, z'') for any $x \in \operatorname{supp}\theta_\varepsilon$ and $y \in \operatorname{supp}\alpha_{\varepsilon'}$, as ε and ε'' tend to 0. It follows that

$$\mathcal{K}([\widetilde{D}], [V] \wedge \widetilde{\varphi}) = \int_{\widetilde{D} \cap V} \widetilde{\varphi}. \qquad \square$$

Theorem 4.3. *Let D be a relatively compact domain with $\mathcal{C}^1$ boundary in V. For any continuous* CR *function f on V and any differential $(n, n-2)$-form α which is $\mathcal{C}^1$ in a neighbourhood of V we have*

$$\int_D f\overline{\partial}\alpha = \int_{\partial D} f\alpha.$$

Proof. Let $\widetilde{D}$ be an open set in $\mathbb{C}^n$ associated to D by Lemma 4.1, let $\widetilde{f}$ be a continuous extension of f to $\mathbb{C}^n$ and let $\widetilde{\alpha}$ be a $\mathcal{C}^1$ differential $(n, n-2)$-form on $\mathbb{C}^n$ equal to α in a neighbourhood of V. It follows from Lemma 4.2 that

$$\int_D f\overline{\partial}\alpha = \mathcal{K}([\widetilde{D}], [V]^{0,1} \wedge \widetilde{f\overline{\partial}\widetilde{\alpha}})$$

since α has bidegree $(n, n-2)$. But the function f is CR and the current $[V]^{0,1} \wedge \widetilde{f}\widetilde{\alpha}$ has bidegree $(n, n-1)$, so it follows that

$$d([V]^{0,1} \wedge \widetilde{f}\widetilde{\alpha}) = \overline{\partial}([V]^{0,1} \wedge \widetilde{f}\widetilde{\alpha}) = [V]^{0,1} \wedge \widetilde{f\overline{\partial}\widetilde{\alpha}}.$$

Finally, we get

$$\int_D f\overline{\partial}\alpha = \mathcal{K}([\widetilde{D}], d([V]^{0,1} \wedge \widetilde{f}\widetilde{\alpha}))$$

and by Stokes' formula for the Kronecker index (Chap. II, Corollary 3.7),

$$\int_D f\overline{\partial}\alpha = \mathcal{K}(b[\widetilde{D}], [V]^{0,1} \wedge \widetilde{f}\widetilde{\alpha}) = \int_{\partial\widetilde{D} \cap V} \widetilde{f}\widetilde{\alpha} = \int_{\partial D} f\alpha. \qquad \square$$

5 Integral of the Bochner–Martinelli kernel

We recall that the Bochner–Martinelli kernel

$$B_0^0(z, \zeta) = \frac{(n-1)!}{(2i\pi)^n} \frac{\sum_{j=1}^n (-1)^{j-1}(\overline{\zeta}_j - \overline{z}_j) \bigwedge_{s \neq j} d\overline{\zeta}_s}{|\zeta - z|^{2n}} \wedge d\zeta_1 \wedge \cdots \wedge d\zeta_n$$

is a $\mathcal{C}^\infty$ differential form on $\mathbb{C}^n \times \mathbb{C}^n \setminus \Delta$ such that $\overline{\partial}_\zeta B_0^0(z, \zeta) = 0$ for any $z \in \mathbb{C}^n$ and $\zeta \in \mathbb{C}^n \setminus \{z\}$.

In this section we will find explicit differential forms which are solutions of the equation $\overline{\partial}u = B_0^0(z, \cdot)$ on certain open sets in $\mathbb{C}^n \setminus \{z\}$.

Given $z \in \mathbb{C}^n$ we set for any $k = 1, \ldots, n$

$$(5.1)\quad \Omega_k(z,\zeta) = \frac{(-1)^k (n-2)!}{(2i\pi)^n} \frac{1}{(\zeta_k - z_k)|\zeta - z|^{2n-2}}$$

$$\cdot \Big[\sum_{j=1}^{k-1} (-1)^j (\overline{\zeta}_j - \overline{z}_j) \bigwedge_{s \neq j,k} d\overline{\zeta}_s + \sum_{j=k+1}^{n} (-1)^{j-1} (\overline{\zeta}_j - \overline{z}_j) \bigwedge_{s \neq j,k} d\overline{\zeta}_s\Big] \wedge d\zeta_1 \wedge \cdots \wedge d\zeta_n.$$

The differential form $\Omega_k(z,\cdot)$ is $\mathcal{C}^\infty$ on $\mathbb{C}^n \setminus \{\zeta \in \mathbb{C}^n \mid \zeta_k = z_k\}$.

Lemma 5.1. *The differential form $\Omega_k(z,\cdot)$ has the property that*

$$d\Omega_k(z,\cdot) = \overline{\partial}\Omega_k(z,\cdot) = B_0^0(z,\cdot) \quad \text{on } \mathbb{C}^n \setminus \{\zeta \in \mathbb{C}^n \mid \zeta_k = z_k\}.$$

Proof. As the function $1/(\zeta_k - z_k)$ is holomorphic on $\mathbb{C}^n \setminus \{\zeta \in \mathbb{C}^n \mid \zeta_k = z_k\}$ we have

$$\overline{\partial}_\zeta \Big(\frac{1}{(\zeta_k - z_k)|\zeta - z|^{2n-2}} \Big) = \frac{-(n-1)}{(\zeta_k - z_k)} \frac{\sum_{p=1}^n (\zeta_p - z_p) d\overline{\zeta}_p}{|\zeta - z|^{2n}}.$$

On the other hand

$$\overline{\partial}_\zeta \Big(\sum_{j=1}^{k-1} (-1)^j (\overline{\zeta}_j - \overline{z}_j) \bigwedge_{s \neq j,k} d\overline{\zeta}_s + \sum_{j=k+1}^{n} (-1)^{j-1} (\overline{\zeta}_j - \overline{z}_j) \bigwedge_{s \neq j,k} d\overline{\zeta}_s \Big)$$
$$= -(n-1) \bigwedge_{s \neq k} d\overline{\zeta}_s.$$

A straightforward calculation then gives us that

$$\overline{\partial}_\zeta \Omega_k(z,\zeta) = B_0^0(z,\zeta).$$

The form $\Omega_k(z,\cdot)$ has bidegree $(n, n-2)$, so $d\Omega_k(z,\cdot) = \overline{\partial}\Omega_k(z,\cdot)$. □

Given a holomorphic function φ on $\mathbb{C}^n$ and a point $z \in \mathbb{C}^n$ we set $N_z = \{\zeta \in \mathbb{C}^n \mid \varphi(\zeta) = \varphi(z)\}$. We will find integrals for the Bochner–Martinelli kernel on open sets of the form $U_z = \mathbb{C}^n \setminus N_z$. These integrals will be useful for proving the extension theorem in Section 6. Lemma 5.1 solves this problem for the coordinate map $\varphi(z) = z_k$.

Lemma 5.2. *If φ is a holomorphic function on $\mathbb{C}^n$ then there are n holomorphic functions $(h_1, \ldots, h_n)$ on $\mathbb{C}^n \times \mathbb{C}^n$ such that, for any $(z,\zeta) \in \mathbb{C}^n \times \mathbb{C}^n$,*

$$(5.2)\quad \varphi(\zeta) - \varphi(z) = \sum_{k=1}^{n} h_k(z,\zeta)(\zeta_k - z_k).$$

Proof. For $k = 1, \dots, n$ we set

$$h_k(z,\zeta) = \frac{1}{(\zeta_k - z_k)}\Big[\varphi(z_1,\dots,z_{k-1},\zeta_k,\dots,\zeta_n) - \varphi(z_1,\dots,z_k,\zeta_{k+1},\dots,\zeta_n)\Big].$$

The function h_k is holomorphic on $\mathbb{C}^n \times \mathbb{C}^n \setminus \{(z,\zeta) \mid z_k = \zeta_k\}$ and can be extended by continuity to $\mathbb{C}^n \times \mathbb{C}^n$ because the function φ is holomorphic with respect to the variable z_k. Riemann's theorem (Chap. I, Th. 6.4) then implies that h_k is holomorphic on $\mathbb{C}^n \times \mathbb{C}^n$. Equation (5.2) follows immediately from the definition of the functions h_k. □

Remark. Lemma 5.2 no longer holds if φ is only defined on an open set Ω in $\mathbb{C}^n$. We will see some conditions on Ω under which Lemma 5.2 still holds in Chapter VIII.

We can now define

$$\Phi(z,\zeta) = \frac{1}{\varphi(\zeta) - \varphi(z)} \sum_{k=1}^{n} h_k(z,\zeta)(\zeta_k - z_k)\Omega_k(z,\zeta). \tag{5.3}$$

The differential form $\Phi(z,\cdot)$ is $\mathcal{C}^\infty$ on $\mathbb{C}^n \setminus N_z$.

Proposition 5.3. *The differential form $\Phi(z,\cdot)$ has the property that*

$$d\Phi(z,\cdot) = \overline{\partial}\Phi(z,\cdot) = B_0^0(z,\cdot) \quad on\ \mathbb{C}^n \setminus N_z.$$

Proof. The differential form $\Phi(z,\cdot)$ is of type $(n, n-2)$ and it follows that $d\Phi(z,\cdot) = \overline{\partial}\Phi(z,\cdot)$. Moreover, since the functions φ, h_k, $k = 1,\dots,n$ are holomorphic and $(\zeta_k - z_k)\Omega_k(z,\zeta)$ is defined on $\mathbb{C}^n \setminus \{z\}$ we have

$$\overline{\partial}_\zeta\Phi(z,\zeta) = \frac{1}{\varphi(\zeta) - \varphi(z)} \sum_{k=1}^{n} h_k(z,\zeta)\overline{\partial}_\zeta\big((\zeta_k - z_k)\Omega_k(z,\zeta)\big)$$

for any $\zeta \in \mathbb{C}^n \setminus N_z$. It then follows from Lemma 5.1 and (5.2) that $\overline{\partial}_\zeta\Phi(z,\cdot) = B_0^0(z,\cdot)$ on $\mathbb{C}^n \setminus N_z$ by extending by continuity. □

6 An extension theorem for CR functions

Consider the following geometric situation: let V be a real closed oriented $\mathcal{C}^1$ hypersurface in $\mathbb{C}^n$, $n \geqslant 2$, and let Γ be a domain with $\mathcal{C}^1$ boundary in V. We assume that:

i) $\partial\Gamma$ is contained in $M = \{z \in \mathbb{C}^n \mid \operatorname{Re}\varphi(z) = 0\}$, where φ is some holomorphic function on $\mathbb{C}^n$.
ii) $\Gamma \subset \{z \in \mathbb{C}^n \mid \operatorname{Re}\varphi(z) > 0\}$.

iii) $\partial\Gamma$ is the boundary of a bounded domain A in M. We denote by D the bounded domain in $\mathbb{C}^n$ whose boundary is $\overline{\Gamma} \cup \overline{A}$. We suppose that the orientation on V is equal to that on ∂D.

In this section we will prove that if f is a continuous CR function on Γ then there is a unique continuous function F on $D \cup \Gamma$ which is holomorphic on D such that $F\big|_\Gamma = f$.

This result is a special case of an extension problem for CR functions defined on part of the boundary of a domain which can be formulated as follows. Let D be a bounded domain in $\mathbb{C}^n$ and let K be a compact set in ∂D such that $\partial D \setminus K$ is a real connected $\mathcal{C}^1$ hypersurface in $\mathbb{C}^n \setminus K$. Under what conditions on K does any continuous CR function on $\partial D \setminus K$ extend to a holomorphic function on D which is continuous on $\overline{D} \setminus K$? If $K = \varnothing$ then the problem is solved by Bochner's theorem. In this section we solve the case where K is contained in the zero set of the real part of a holomorphic function on $\mathbb{C}^n$. We will give alternative cohomological and geometric conditions on K which imply that extension is possible at the ends of Chapters V and VII and in Chapter VIII.

Theorem 6.1. *Let f be a continuous* CR *function on Γ. There is then a unique holomorphic function F on D, continuous on $D\cup\Gamma$, such that $F\big|_\Gamma = f$.*

Proof. We start by proving that F is unique if it exists. For any given $z_0 \in D$ choose $\varepsilon > 0$ small enough that $z_0 \in D_\varepsilon = D \cap \{\zeta \in \mathbb{C}^n \mid \operatorname{Re}\varphi(\zeta) > \varepsilon\}$. Set $\Gamma_\varepsilon = \Gamma \cap \{\zeta \in \mathbb{C}^n \mid \operatorname{Re}\varphi(\zeta) > \varepsilon\}$ and $A_\varepsilon = D \cap \{\zeta \in \mathbb{C}^n \mid \operatorname{Re}\varphi(\zeta) = \varepsilon\}$. Then, $\partial D_\varepsilon = \overline{\Gamma}_\varepsilon \cup \overline{A}_\varepsilon$. Assume that F exists: it is then holomorphic on D_ε and continuous on $\overline{D}_\varepsilon$, so we can apply the Bochner–Martinelli formula:

$$\begin{aligned} F(z_0) &= \int_{\partial D_\varepsilon} F(\zeta) B_0^0(z_0,\zeta) \quad \text{since } z_0 \in D_\varepsilon \\ &= \int_{\Gamma_\varepsilon} f(\zeta) B_0^0(z_0,\zeta) + \int_{A_\varepsilon} F(\zeta) B_0^0(z_0,\zeta) \quad \text{since } F\big|_{\Gamma_\varepsilon} = f. \end{aligned}$$

Set $N_z = \{\zeta \in \mathbb{C}^n \mid \varphi(\zeta) = \varphi(z)\}$ for any $z \in \mathbb{C}^n$ and note that $N_{z_0} \cap \{\zeta \in \mathbb{C}^n \mid \operatorname{Re}\varphi(\zeta) = \varepsilon\} = \varnothing$ since $z_0 \in D_\varepsilon$. There is therefore a neighbourhood of A_ε which does not meet N_{z_0}, and if Φ is the differential form associated to φ defined in (5.3), we have $\overline{\partial}_\zeta \Phi(z_0,\zeta) = B_0^0(z_0,\zeta)$ on this neighbourhood. As F is holomorphic and $\Phi(z_0,\cdot)$ has bidegree $(n, n-2)$ on D, applying Stokes' formula gives us that

$$F(z_0) = \int_{\Gamma_\varepsilon} f(\zeta) B_0^0(z_0,\zeta) - \int_{\partial\Gamma_\varepsilon} f(\zeta)\Phi(z_0,\zeta) \tag{6.1}$$

since $F\big|_\Gamma = f$ and $\partial\Gamma_\varepsilon \subset \Gamma$, which proves that F is unique.

Let us now prove that F exists. We start by assuming that the data f is continuous on $\overline{\Gamma}$. Letting ε tend to 0 in (6.1) we see that if F exists then

$$F(z) = \int_\Gamma f(\zeta) B_0^0(z,\zeta) - \int_{\partial\Gamma} f(\zeta)\Phi(z,\zeta) \tag{6.2}$$

for every $z \in D$. Consider the function F defined by formula (6.2). This is a $\mathcal{C}^\infty$ function on $\mathbb{C}^n \setminus (M \cup \Gamma)$ because $B_0^0(z,\cdot)$ is defined and $\mathcal{C}^\infty$ on $\mathbb{C}^n \setminus \{z\}$ and $\Phi(z,\cdot)$ is $\mathcal{C}^\infty$ on $\mathbb{C}^n \setminus N_z$, where $N_z = \{\zeta \in \mathbb{C}^n \mid \varphi(z) = \varphi(\zeta)\}$. We will show that F is the extension we seek and for this we will need the following theorem which will be proved in Chapter VII. Let φ be a holomorphic function on $\mathbb{C}^n$, $n > 2$, and set $U_\varepsilon = \{z \in \mathbb{C}^n \mid -\varepsilon < \operatorname{Re}\varphi(z) < \varepsilon\}$ for any $\varepsilon > 0$. The open set U_ε is then a domain of holomorphy (cf. Chap. VI, Prop. 1.16) and every differential $(n, n-2)$-form which is $\mathcal{C}^\infty$ on U_ε is $\overline{\partial}$-exact on U_ε, i.e. $H^{n,n-2}(U_\varepsilon) = 0$ (cf. Chap. VII, Th. 7.4).

First step: F is holomorphic on $\mathbb{C}^n \setminus (M \cup \Gamma)$. Differentiating under the integral sign in (6.2) we get, for any $z \in \mathbb{C}^n \setminus (\Gamma \cup M)$,

$$\overline{\partial} F(z) = \int_\Gamma f(\zeta)\overline{\partial}_z B_0^0(z,\zeta) - \int_{\partial\Gamma} f(\zeta)\overline{\partial}_z \Phi(z,\zeta)$$

and since $\overline{\partial}_z B_0^0 = -\overline{\partial}_\zeta B_1^0$ on $\mathbb{C}^n \times \mathbb{C}^n \setminus \Delta$ it follows that

$$\overline{\partial} F(z) = -\int_\Gamma f(\zeta)\overline{\partial}_\zeta B_1^0(z,\zeta) - \int_{\partial\Gamma} f(\zeta)\overline{\partial}_z \Phi(z,\zeta).$$

If $n = 2$ then a straightforward calculation proves that $\overline{\partial} F(z) = 0$ for any $z \in D$ since $B_1^0(z,\cdot)$ is $\mathcal{C}^\infty$ in a neighbourhood of $\overline{\Gamma}$ for any $z \in D$. If $n > 2$ then Proposition 5.3 implies that

$$\overline{\partial}_\zeta(\overline{\partial}_z \Phi + B_1^0) = \overline{\partial}_z(\overline{\partial}_\zeta \Phi) + \overline{\partial}_\zeta B_1^0 = \overline{\partial}_z B_0^0 + \overline{\partial}_\zeta B_1^0 = 0$$

for any $(z,\zeta) \in \mathbb{C}^n \times \mathbb{C}^n$ such that $\zeta \notin N_z$. Fix $z \in D$ and choose $\varepsilon > 0$ small enough that $z \notin U_\varepsilon = \{\zeta \in \mathbb{C}^n \mid -\varepsilon < \operatorname{Re}\varphi(\zeta) < \varepsilon\}$. The differential form $\overline{\partial}_z \Phi(z,\cdot) + B_1^0(z,\cdot)$ is then a $\overline{\partial}$-closed $(n, n-2)$-form of class $\mathcal{C}^\infty$ on U_ε since $N_z \cap U_\varepsilon = \varnothing$. As $H^{n,n-2}(U_\varepsilon) = 0$, there is a differential $(n, n-3)$-form θ_ε which is $\mathcal{C}^\infty$ on U_ε such that $\overline{\partial}\theta_\varepsilon = \overline{\partial}_z \Phi(z,\cdot) + B_1^0(z,\cdot)$ on U_ε. Let $\widetilde{\theta}_\varepsilon$ be a $\mathcal{C}^\infty$ differential $(n, n-3)$-form on $\mathbb{C}^n$ equal to θ_ε in a neighbourhood of $\partial\Gamma$. Then,

$$\overline{\partial} F(z) = \int_\Gamma f(\zeta)\overline{\partial}_\zeta(\overline{\partial}_\zeta \widetilde{\theta}_\varepsilon(\zeta) - B_1^0(z,\zeta)) - \int_{\partial\Gamma} f(\zeta)(\overline{\partial}_\zeta \widetilde{\theta}_\varepsilon(\zeta) - B_1^0(z,\zeta)).$$

Applying Theorem 4.3, we get $\overline{\partial} F(z) = 0$.

Second step: F vanishes identically on $\mathbb{C}^n \setminus (\overline{D} \cup \Gamma)$. Consider the open set $U = \{z \in \mathbb{C}^n \mid \operatorname{Re}\varphi(z) \notin [0, \sup_{\overline{D}} \operatorname{Re}\varphi]\}$. For any $z \in U$ the set N_z does not meet $\overline{D}$ and it follows that $\Phi(z,\cdot)$ is $\mathcal{C}^\infty$ in a neighbourhood of $\overline{\Gamma}$. As $\overline{\partial}_\zeta \Phi(z,\cdot) = B_0^0(z,\cdot)$ on $\mathbb{C}^n \setminus N_z$, formula (6.2) and Theorem 4.3 imply that $F(z) = 0$ for any $z \in U$. Moreover, the open set U meets all the connected components of $\mathbb{C}^n \setminus (\overline{D} \cup M)$ and F is holomorphic on $\mathbb{C}^n \setminus (\overline{D} \cup M)$, so by the analytic continuation principle F vanishes identically on $\mathbb{C}^n \setminus (\overline{D} \cup M)$.

Third step: $F\big|_{\Gamma} = f$. Set $F_1(z) = \int_{\Gamma} f(\zeta) B_0^0(z,\zeta)$ and $F_2(z) = \int_{\partial\Gamma} f(\zeta)\Phi(z,\zeta)$. By definition of F we have $F = F_1 - F_2$. The function F_2 is $\mathcal{C}^\infty$ on $\mathbb{C}^n \setminus M$ because the set N_z of singularities of $\Phi(z,\cdot)$ does not meet $\partial\Gamma$ for any $z \notin M$. The function F_1 is the Bochner–Martinelli transform of f on Γ and it is $\mathcal{C}^\infty$ on $\mathbb{C}^n \setminus \overline{\Gamma}$.

We set $F_j^+ = F_j\big|_D$ and $F_j^- = F_j\big|_{\mathbb{C}^n \setminus (\overline{D} \cup M)}$ for $j = 1, 2$. Note that F_1^- can be extended continuously to $\widetilde{F}_1^-$ on $(\mathbb{C}^n \setminus (\overline{D} \cup M)) \cup \Gamma$. Indeed, $F_1^- = F\big|_{\mathbb{C}^n \setminus (\overline{D} \cup M)} + F_2^-$ and by the second step F vanishes on $\mathbb{C}^n \setminus (\overline{D} \cup M)$. The continuity of F_2 on $\mathbb{C}^n \setminus M$ then completes the proof. Corollary 1.6 says that F_1^+ can then be continuously extended to $\widetilde{F}_1^+$ on $D \cup \Gamma$ and $\widetilde{F}_1^+\big|_{\Gamma} - \widetilde{F}_1^-\big|_{\Gamma} = f$, so $F\big|_D$ can be continuously extended to $\widetilde{F}$ on $D \cup \Gamma$ and $\widetilde{F}\big|_{\Gamma} = f$ because F_2 is continuous on $\mathbb{C}^n \setminus M$.

This completes the proof of the theorem when f is continuous on $\overline{\Gamma}$. If f is only continuous on Γ then f is continuous on $\overline{\Gamma}_\varepsilon$ for any $\varepsilon > 0$ and we can apply the above to the set D_ε. We thus obtain a family of holomorphic functions $(F_\varepsilon)_{\varepsilon > 0}$ such that $F_\varepsilon = F_{\varepsilon'}\big|_{D_\varepsilon \cup \Gamma_\varepsilon}$ whenever $\varepsilon' > \varepsilon$ by uniqueness. The function F defined by $F\big|_{D_\varepsilon \cup \Gamma_\varepsilon} = F_\varepsilon$ is then the extension we seek. □

Remark. Although it involves using a result which will not be proved until Chapter VII, we prefer to include Theorem 6.1 in this chapter rather than in Chapter VIII – where it would also have fitted in – because our method of proof is similar to the proof of Bochner's theorem.

Comments

On reading carefully the proof of Hartogs' extension theorem in Bochner's article [Bo] one notes that the extension exists even if the data is just a CR function on the boundary of the domain. The proof of Bochner's extension theorem given here is due to Harvey and Lawson [Ha/La] and Čirka [Ci]. The case where the boundary of the domain is not assumed connected is due to Weinstock [We]. The extension theorem of Section 6 was proved by Lupacciolu and Tomassini [Lu/To] for locally Lipschitz data. The proof of Stokes' formula for continuous CR functions which enables us to extend Lupacciolu and Tomassini's result to continuous CR functions can be found in [L-T2]. Most of these results are gathered together in Kytmanov's book [Ky].

V

Extensions of holomorphic and CR functions on manifolds

The aim of this chapter is to study the Hartogs–Bochner phenomenon on complex analytic manifolds. We start by studying the relationship between Hartogs' phenomenon and the vanishing of the Dolbeault cohomology group with compact support in bidegree $(0, 1)$. We then give some cohomological conditions which enable us to extend a CR function of class $\mathcal{C}^\infty$ defined on a subset of the boundary of a domain to a holomorphic function on the whole domain. This work generalises the geometric situation studied at the end of Chapter IV. Proving similar results for CR functions of class $\mathcal{C}^k$ requires two extra elements: a theorem on local resolutions of $\overline{\partial}$ and an isomorphism theorem between the various cohomology groups $H^{p,q}_\alpha(X)$. This isomorphism theorem follows from the local resolution and some sheaf-theoretic results which are given in Appendix B. The existence of the resolution is proved by solving $\overline{\partial}$ in convex domains with $\mathcal{C}^2$ boundary using a new integral formula, the Cauchy–Fantappié formula.

1 Cohomology with compact support and Hartogs' phenomenon

In this section we will study the link between Hartogs' extension phenomenon on a complex analytic manifold and the vanishing of certain Dolbeault cohomology groups on this manifold.

Let X be a complex analytic manifold of dimension n. We will say that Hartogs' phenomenon holds on X if for any relatively compact domain D in X such that $X \setminus D$ is connected and any holomorphic function f on a neighbourhood $U_{\partial D}$ of the boundary of D there is a holomorphic function F defined on a neighbourhood of $\overline{D}$ which is equal to f on a neighbourhood of ∂D.

In Theorem 4.2 of Chapter III we proved that Hartogs' phenomenon holds on $\mathbb{C}^n$ for any $n \geqslant 2$.

Theorem 1.1. *Let X be a non-compact complex analytic manifold. Suppose that for any $\mathcal{C}^\infty$ differential $(0, 1)$-form v with compact support in X, $\overline{\partial}$-exact*

C. Laurent-Thiébaut, *Holomorphic Function Theory in Several Variables: An Introduction*, Universitext, DOI 10.1007/978-0-85729-030-4_5, © Springer-Verlag London Limited 2011

in a neighbourhood of its support there is a $\mathcal{C}^\infty$ function u with compact support in X such that $\overline{\partial}u = v$ on X. Hartogs' phenomenon then holds on X.

Proof. Let D be a relatively compact domain in X such that $X \setminus D$ is connected and let f be a holomorphic function on a neighbourhood $U_{\partial D}$ of the boundary of D. Let $\chi \in \mathcal{D}(X)$ be a function such that $\operatorname{supp}\chi \Subset U_{\partial D}$ and $\chi \equiv 1$ on a neighbourhood $V_{\partial D} \subset U_{\partial D}$ of ∂D. Set $\widetilde{f} = \chi f$: the function $\widetilde{f}$ is then $\mathcal{C}^\infty$ on X and holomorphic on $V_{\partial D}$. Set $v = \overline{\partial}\widetilde{f}$ on $D \cup V_{\partial D}$ and $v = 0$ on $X \setminus D$. Since $\widetilde{f}$ is holomorphic on $V_{\partial D}$, v is $\mathcal{C}^\infty$ on X and its support is contained in D. Moreover, $v = \overline{\partial}\widetilde{f}$ on $D \cup V_{\partial D}$, and v is therefore $\overline{\partial}$-exact in a neighbourhood of its support. By hypothesis there is a $\mathcal{C}^\infty$ function u with compact support in X such that $\overline{\partial}u = v$ in X. The function u is therefore holomorphic on $X \setminus \operatorname{supp} v$, and in particular u is holomorphic on a neighbourhood of $X \setminus D$. Moreover, as u has compact support it vanishes on an open set in $X \setminus \overline{D}$ and by analytic continuation it therefore vanishes on a neighbourhood of $X \setminus D$ because $X \setminus D$ is connected. We can then simply set $F = \widetilde{f} - u$. Indeed, we then have $F = \widetilde{f} = f$ on some neighbourhood of ∂D and $\overline{\partial}F = \overline{\partial}\widetilde{f} - \overline{\partial}u = 0$ on $D \cup V_{\partial D}$. □

Definition 1.2. Let X be a differentiable $\mathcal{C}^k$ manifold, where $0 \leqslant k \leqslant \infty$. We define the number $e(X)$ of *ends* of X to be the maximum number of non-relatively compact connected components of $X \setminus K$ for a compact subset K in X.

We will consider *one-ended manifolds*, i.e. manifolds such that $e(X) = 1$. It follows from the definition that if X is one-ended then for any compact subset K in X we can find a compact subset L in X containing K such that $X \setminus L$ is connected.

Examples.

1) An open set U in $\mathbb{C}$ is one-ended if and only if it is simply connected.
2) X is a compact manifold if and only if $e(X) = 0$.
3) A manifold X is one-ended if X has an exhaustion function, i.e. a function $\varphi : X \to \mathbb{R}$ such that $\{x \in X \mid \varphi(x) \leqslant c\} \Subset X$ for every $c \in \mathbb{R}$, which satisfies the following extra condition; there is a constant c_0 such that $\{x \in X \mid \varphi(x) = c\}$ is connected for any $c \geqslant c_0$.

Theorem 1.3. *Let X be a one-ended complex analytic manifold. If Hartogs' phenomenon holds on X then for any $\overline{\partial}$-exact $\mathcal{C}^\infty$ differential $(0,1)$-form v with compact support in X there is a $\mathcal{C}^\infty$ function u with compact support in X such that $\overline{\partial}u = v$ in X.*

Proof. Let v be a $\mathcal{C}^\infty$ differential $(0,1)$-form with compact support in X such that $v = \overline{\partial}w$ on a neighbourhood U of the support of v. The function w is then holomorphic on $U \setminus \operatorname{supp} v$. By Hartogs' phenomenon, w can be extended holomorphically to each of the relatively compact connected components of

$U \setminus \operatorname{supp} v$ to a function which we again denote by w. As X is one-ended, there is a relatively compact domain D containing $\operatorname{supp} v$ and contained in the union of U and the relatively compact connected components of $X \setminus \operatorname{supp} v$ such that $X \setminus D$ is connected. The function w is then defined in a neighbourhood of $\overline{D}$. It is holomorphic on a neighbourhood of ∂D and by Hartogs' phenomenon its restriction to this neighbourhood can be extended to a holomorphic function $\widetilde{w}$ on a neighbourhood of $\overline{D}$. We set $u = w - \widetilde{w}$ in a neighbourhood of $\overline{D}$ and $u = 0$ on $X \setminus D$. The function u is then $\mathcal{C}^\infty$ on X because $w - \widetilde{w} = 0$ in a neighbourhood of ∂D: moreover $\overline{\partial} u = \overline{\partial} w = v$ in a neighbourhood of $\overline{D}$ because $\widetilde{w}$ is holomorphic, and $\overline{\partial} u = 0 = v$ on $X \setminus D$, because $\operatorname{supp} v \subset D$. □

Remark. The hypothesis that X is one-ended is indispensable in Theorem 1.3. Indeed, consider the case where $X = \mathbb{C}^n \setminus \{0\}$, $n \geqslant 2$. Hartogs' phenomenon holds in X, since X is an open set in $\mathbb{C}^n$, $n \geqslant 2$. Consider a function $\chi \in \mathcal{D}(X)$ such that $\chi(z) = 1$ whenever $3/4 < |z| < 5/4$, $\chi(z) = 2$ whenever $7/4 < |z| < 9/4$ and the support of χ does not meet

$$\{z \in \mathbb{C}^n \mid \ |z| < 2/3\} \cup \{z \in \mathbb{C}^n \mid 4/3 < |z| < 5/3\} \cup \{z \in \mathbb{C}^n \mid \ |z| > 7/3\}.$$

Set $v = \overline{\partial}\chi$ on $3/4 < |z| < 9/4$ and $v = 0$ otherwise. This is a $\mathcal{C}^\infty$ differential $(0,1)$-form on X with compact support which is $\overline{\partial}$-exact in a neighbourhood of its support but there is no compactly supported function u such that $\overline{\partial} u = v$. If u existed then u would be holomorphic on $X \setminus \{z \in \mathbb{C}^n \mid 5/4 < |z| < 7/4\}$, and would vanish on $|z| < r$ for small enough r and on $|z| > R$ for large enough R, and would hence be zero on $X \setminus \{z \in \mathbb{C}^n \mid 5/4 < |z| < 7/4\}$ by analytic continuation. The function $\chi - u$ would then be holomorphic on $\{z \in \mathbb{C}^n \mid 3/4 < |z| < 9/4\}$, constant and equal to 1 in a neighbourhood of $|z| = 1$ and constant and equal to 2 in a neighbourhood of $|z| = 2$, which is impossible.

The following result linking Dolbeault cohomology with compact support on X and Hartogs' phenomenon is an immediate consequence of Theorem 1.1.

Corollary 1.4. *Let X be a non-compact complex analytic manifold. If*

$$H_c^{0,1}(X) = 0$$

then Hartogs' phenomenon holds on X.

We end this section by stating a partial converse to Corollary 1.4 which follows from Theorem 1.3.

Corollary 1.5. *Let X be a non-compact complex analytic manifold. Assume that X is one-ended and every $\overline{\partial}$-closed $\mathcal{C}^\infty$ differential $(0,1)$-form with compact support in X is $\overline{\partial}$-exact on X. If Hartogs' phenomenon holds on X then*

$$H_c^{0,1}(X) = 0.$$

Remarks.

1) If $X = \mathbb{C}^n$ then the conditions of Corollary 1.5 hold and Hartogs' phenomenon is therefore equivalent to the vanishing of the Dolbeault cohomology group $H_c^{0,1}(\mathbb{C}^n)$.
2) If X is an open connected set in $\mathbb{C}^n$ whose complement has no bounded connected components and $H^{0,1}(X) = 0$ then $H_c^{0,1}(X) = 0$ since Hartogs' phenomenon holds in all open sets in $\mathbb{C}^n$.

2 Extension of $\mathcal{C}^\infty$ CR functions

Throughout this section X will be a non-compact complex analytic manifold.

Definition 2.1. Let V be a $\mathcal{C}^\infty$ submanifold of X. A $\mathcal{C}^\infty$ function f on V is said to be a $\mathcal{C}^\infty$ CR function if f has a $\mathcal{C}^\infty$ extension $\widetilde{f}$ to a neighbourhood of V such that $\overline{\partial}\widetilde{f}$ vanishes to all orders along V.

Definition 2.2. Let X be a non-compact complex analytic manifold and let K be a compact set in X. We say that the pair (X, K) *has the* $\mathcal{C}^\infty$ *Hartogs–Bochner property* if for any relatively compact domain D in X such that

1) $\partial D \setminus K$ is a $\mathcal{C}^\infty$ submanifold of $X \setminus K$,
2) $D \setminus K = \mathrm{Int}(\overline{D \setminus K})$,
3) $X \setminus (\overline{D} \cup K)$ is connected

and every $\mathcal{C}^\infty$ CR function f on $\partial D \setminus K$ there is a $\mathcal{C}^\infty$ function F on $\overline{D} \setminus K$, holomorphic on $D \setminus K$, such that $F\big|_{\partial D \setminus K} = f$.

As in Chapter III, we denote the family of closed sets in $X \setminus K$ whose closure in X is compact by Φ.

Theorem 2.3. *Let X be a non-compact complex analytic manifold and let K be a compact set in X. We assume that*

$$H_\Phi^{0,1}(X \setminus K) = 0.$$

The pair (X, K) then has the $\mathcal{C}^\infty$ Hartogs–Bochner property.

Proof. We consider a relatively compact domain D in X satisfying conditions 1), 2) and 3) of Definition 2.2 and a $\mathcal{C}^\infty$ CR function f on $\partial D \setminus K$. By definition of a $\mathcal{C}^\infty$ CR function, f can be extended to a $\mathcal{C}^\infty$ function $\widetilde{f}$ on $X \setminus K$ such that $\overline{\partial}\widetilde{f}$ vanishes to all orders on $\partial D \setminus K$. We set $g = (\chi_{\overline{D} \setminus K})(\overline{\partial}\widetilde{f})$, where $\chi_{\overline{D} \setminus K}$ is the characteristic function of $\overline{D} \setminus K$ and g is a $\mathcal{C}^\infty$ differential $(0,1)$-form on $X \setminus K$ supported on $\overline{D} \setminus K$. Since $\overline{D} \setminus K$ is contained in Φ and we have assumed that $H_\Phi^{0,1}(M \setminus K) = 0$ there is a $\mathcal{C}^\infty$ function h on $X \setminus K$ with support in Φ such that $\overline{\partial}h = g$. The function h is holomorphic on $X \setminus (\overline{D} \cup K)$ and vanishes on an open set in $X \setminus (\overline{D} \cup K)$ because X is not compact. As $X \setminus (\overline{D} \cup K)$ is connected, h vanishes on $X \setminus (\overline{D} \cup K)$ by the principle of analytic continuation and h therefore vanishes on $X \setminus (D \cup K)$ by continuity. The function $F = \widetilde{f} - h$ is then the extension we seek. Indeed, $\overline{\partial} F = \overline{\partial} \widetilde{f} - \overline{\partial} h = 0$ on $D \setminus K$ and $F\big|_{\partial D \setminus K} = f$ because h vanishes on $X \setminus (D \cup K)$. □

Corollary 2.4. *Let X be a non-compact complex analytic manifold such that $H_c^{0,1}(X) = 0$. The pair $(X, \varnothing)$ then has the $\mathcal{C}^\infty$ Hartogs–Bochner property. In other words, for any relatively compact domain D in X with $\mathcal{C}^\infty$ boundary such that $X \setminus \overline{D}$ is connected and any $\mathcal{C}^\infty$ CR function f on ∂D there is a $\mathcal{C}^\infty$ function F on $\overline{D}$, holomorphic on D, such that $F\big|_{\partial D} = f$.*

We end this section by giving conditions on K and X which imply the vanishing of the cohomology group $H_\Phi^{0,1}(X \setminus K)$.

Theorem 2.5. *Let X be a complex analytic manifold of dimension n, $n \geqslant 2$, and let K be a compact set in X such that $X \setminus K$ does not have a relatively compact connected component. We assume that*

i) $H_c^{0,1}(X) = 0$,
ii) *the compact set K has a decreasing sequence of neighbourhoods $(U_n)_{n \in \mathbb{N}}$ such that $\bigcap_{n \in \mathbb{N}} U_n = K$ and for every $n \in \mathbb{N}$ the map $H_c^{0,2}(U_n) \to H_c^{0,2}(X)$ induced by inclusion is injective.*

We then have $H_\Phi^{0,1}(X \setminus K) = 0$.

Lemma 2.6. *Let X be a complex analytic manifold, let K be a compact set in X and let U be a relatively compact neighbourhood of K. We assume that*

i) $H_c^{0,1}(X) = 0$,
ii) *the map $H_c^{0,2}(U) \to H_c^{0,2}(X)$ induced by inclusion is injective.*

For any $\overline{\partial}$-closed differential form $f \in \mathcal{C}_{0,1}^\infty(X \setminus K)$ vanishing outside of a compact set in X there is then a function $g \in \mathcal{C}^\infty(X \setminus \overline{U})$ vanishing outside of a compact set in X such that $\overline{\partial} g = f$ in $X \setminus \overline{U}$.

Proof. Let χ be a $\mathcal{C}^\infty$ function on X such that $\chi \equiv 0$ on a neighbourhood of K and $\chi \equiv 1$ on a neighbourhood of $X \setminus U$. The differential form $\overline{\partial}(\chi f)$ is then $\mathcal{C}^\infty$ and $\overline{\partial}$-closed with compact support in U, so by ii) there is a differential form $h \in \mathcal{C}_{0,1}^\infty(X)$ with compact support in U such that $\overline{\partial} h = \overline{\partial}(\chi f)$ on X. Consider the differential form $\chi f - h$. This form is $\overline{\partial}$-closed, $\mathcal{C}^\infty$ and compactly supported in X. Hypothesis i) then implies the existence of a function $g \in \mathcal{D}(X)$ such that $\chi f - h = \overline{\partial} g$. The function $g\big|_{X \setminus \overline{U}}$ is then the function we seek because $\chi f - h = f$ on $X \setminus \overline{U}$. □

Proof of Theorem 2.5. Let $f \in \mathcal{C}^\infty_{0,1}(X \setminus K)$ be a $\overline{\partial}$-closed differential form vanishing outside a compact set in X. Let g_n be the solution of the equation $\overline{\partial} g = f$ in $X \setminus \overline{U}_n$ which exists by Lemma 2.6. The function $g_{n+1} - g_n$ is then holomorphic on $X \setminus \overline{U}_n$ and vanishes outside a compact set in X. It therefore vanishes on all the non-relatively compact components of $X \setminus U_n$ by analytic continuation. The function g defined by $g = g_n$ on the non-relatively compact connected components of $X \setminus U_n$ is then the function we seek because the union of the non-relatively compact components of the sets $X \setminus U_n$ is $X \setminus K$. □

Theorem 2.7. *Let X be a complex analytic manifold and let K be a compact set in X. Suppose that $H^{p,q}_c(X) = 0$ for some integer $p \geqslant 0$ and some integer $q \geqslant 1$. The natural map $i : H^{p,q}_\Phi(X \setminus K) \to H^{p,q}(X \setminus K)$ is then injective. In particular, if $H^{p,q}(X \setminus K) = 0$ then $H^{p,q}_\Phi(X \setminus K) = 0$.*

Proof. We have to prove that if $f \in \mathcal{C}^\infty_{p,q}(X \setminus K)$ is a $\overline{\partial}$-closed differential form which vanishes outside a compact set in X and $f = \overline{\partial} g$ on $X \setminus K$ for some $g \in \mathcal{C}^\infty_{p,q-1}(X \setminus K)$ then we can find a $g_0 \in \mathcal{C}^\infty_{p,q-1}(X \setminus K)$ which vanishes outside a compact set in X such that $f = \overline{\partial} g_0$ in $X \setminus K$. Consider a function $\chi \in \mathcal{D}(X)$ such that $\chi \equiv 1$ on a neighbourhood of $K \cup \operatorname{supp} f$. We set $\widetilde{g} = \chi g$. Then $\overline{\partial}\widetilde{g} = \overline{\partial}\chi \wedge g + \chi \overline{\partial} g = \overline{\partial}\chi \wedge g + f$ and the differential form $\overline{\partial}\chi \wedge g$ can be extended by 0 to an $\overline{\partial}$-closed (p,q)-form with compact support on X. Since $H^{p,q}_c(X) = 0$, there is an $h \in \mathcal{D}^{p,q-1}(X)$ such that $\overline{\partial} h = \overline{\partial}\chi \wedge g$ on X and it follows that $\overline{\partial}\widetilde{g} = \overline{\partial} h + f$. The differential form $g_0 = \widetilde{g} - h$ is then the form we seek because h has compact support on X. □

3 The Cauchy–Fantappié formula and Dolbeault's lemma

The aim of this section is to construct new integral formulae which will enable us to solve $\overline{\partial}$ for non-compactly supported data on domains in $\mathbb{C}^n$.

Definition 3.1. Let D be a bounded domain in $\mathbb{C}^n$. A $\mathcal{C}^1$ map $w(z,\zeta) = (w_1(z,\zeta), \ldots, w_n(z,\zeta))$ defined for any $z \in D$ and any ζ in some neighbourhood $U_{\partial D}$ of ∂D with values in $\mathbb{C}^n$ is a *Leray section* for D if

$$\langle w(z,\zeta), \zeta - z\rangle \neq 0 \quad \text{for any } (z,\zeta) \in D \times \partial D.$$

Example. The map $w(z,\zeta) = \overline{\zeta} - \overline{z}$ is a Leray section for any bounded domain D in $\mathbb{C}^n$. Indeed,

$$\langle w(z,\zeta), \zeta - z\rangle = \langle \overline{\zeta} - \overline{z}, \zeta - z\rangle = |\zeta - z|^2 \neq 0 \quad \text{for any } z \neq \zeta.$$

Throughout the following D will be a bounded domain with $\mathcal{C}^1$ boundary in $\mathbb{C}^n$ and $w(z,\zeta)$ will be a Leray section for D. For any $\lambda \in [0,1]$ and any $z \in D$, we then set

$$\eta^w(z,\zeta,\lambda) = (1-\lambda)\frac{w(z,\zeta)}{\langle w(z,\zeta), \zeta - z\rangle} + \lambda\frac{\overline{\zeta} - \overline{z}}{|\zeta - z|^2},$$

$$\overline{\omega}'_{z,\zeta,\lambda}(\eta^w(z,\zeta,\lambda)) = \sum_{j=1}^{n}(-1)^{j-1}\eta_j^w(z,\zeta,\lambda)(\overline{\partial}_{z,\zeta} + d_\lambda)\eta_1^w(z,\zeta,\lambda) \wedge \cdots$$
$$\wedge \widehat{(\overline{\partial}_{z,\zeta} + d_\lambda)\eta_j^w(z,\zeta,\lambda)} \wedge \cdots \wedge (\overline{\partial}_{z,\zeta} + d_\lambda)\eta_n^w(z,\zeta,\lambda).$$

We also set

$$K^{\eta^w}(z,\zeta,\lambda) = \frac{(n-1)!}{(2i\pi)^n}\overline{\omega}'_{z,\zeta,\lambda}(\eta^w(z,\zeta,\lambda) \wedge \omega(\zeta - z)$$

for any $z \in D$ and $\zeta \in U_{\partial D}$ such that $\langle w(z,\zeta), \zeta - z\rangle \neq 0$. We also write

$$\overline{\omega}'_{z,\zeta}(w(z,\zeta)) = \sum_{j=1}^{n}(-1)^{j-1}w_j(z,\zeta)\overline{\partial}_{z,\zeta}w_1(z,\zeta) \wedge \cdots$$
$$\wedge \widehat{\overline{\partial}_{z,\zeta}w_j(z,\zeta)} \wedge \cdots \wedge \overline{\partial}_{z,\zeta}w_n(z,\zeta)$$

and, for any $z \in D$ and $\zeta \in U_{\partial D}$ such that $\langle w(z,\zeta), \zeta - z\rangle \neq 0$,

$$K^w(z,\zeta) = \frac{(n-1)!}{(2i\pi)^n}\frac{\overline{\omega}'_{z,\zeta}(w(z,\zeta)) \wedge \omega(\zeta - z)}{\langle w(z,\zeta), \zeta - z\rangle^n}.$$

We note that $K^{\overline{\zeta}-\overline{z}}$ is the Bochner–Martinelli kernel B defined in Chapter III. The kernel $K^{\eta^w}(z,\zeta,\lambda)$ is a continuous differential form of degree $(2n-1)$ on the set $\{(z,\zeta,\lambda) \in D \times U_{\partial D} \times [0,1] \mid \langle w(z,\zeta), \zeta - z\rangle \neq 0\}$. The kernel $K^w(z,\zeta)$ is a continuous differential form of bidegree $(n, n-1)$ on the set $\{(z,\zeta) \in D \times U_{\partial D} \mid \langle w(z,\zeta), \zeta - z\rangle \neq 0\}$.

Lemma 3.2.

i) *We have* $(\overline{\partial}_{z,\zeta} + d_\lambda)K^{\eta^w}(z,\zeta,\lambda) = 0$ *as currents for any* $z \in D, \zeta \in \partial D$ *and* $\lambda \in [0,1]$,

ii) $K^{\eta^w}(z,\zeta,\lambda)\big|_{\lambda=0} = K^w(z,\zeta)$ *and* $K^{\eta^w}(z,\zeta,\lambda)\big|_{\lambda=1} = K^{\overline{\zeta}-\overline{z}}(z,\zeta) = B(z,\zeta)$.

Proof. As the function $(\zeta - z)$ is holomorphic in (z,ζ) and independent of λ, we have $(\overline{\partial}_{z,\zeta} + d_\lambda)\omega(\zeta - z) = 0$ and it follows that

$$(\overline{\partial}_{z,\zeta} + d_\lambda)K^{\eta^w}(z,\zeta,\lambda)$$
$$= \frac{n!}{(2i\pi)^n}(\overline{\partial}_{z,\zeta} + d_\lambda)\eta_1^w(z,\zeta,\lambda) \wedge \cdots \wedge (\overline{\partial}_{z,\zeta} + d_\lambda)\eta_n^w(z,\zeta,\lambda) \wedge \omega(\zeta - z).$$

But $\langle \eta^w(z,\zeta,\lambda), \zeta - z\rangle = 1$ on $D \times U_{\partial D} \times [0,1]$, which implies that $\sum_{j=1}^n (\zeta_j - z_j)(\overline{\partial}_{z,\zeta} + d_\lambda)\eta_j^w(z,\zeta,\lambda) = 0$. It follows that, for any $z \in D$, $\zeta \in U_{\partial D} \setminus \{z\}$ and $\lambda \in [0,1]$,

$$(\overline{\partial}_{z,\zeta} + d_\lambda)\eta_1^w(z,\zeta,\lambda) \wedge \cdots \wedge (\overline{\partial}_{z,\zeta} + d_\lambda)\eta_n^w(z,\zeta,\lambda) = 0$$

which proves that $(\overline{\partial}_{z,\zeta} + d_\lambda)K^{\eta^w}(z,\zeta,\lambda) = 0$.

The claim ii) follows from the definitions of the differential forms K^{η^w}, K^w and $B = K^{\overline{\zeta}-\overline{z}}$ and from the following general result

$$\mu(\xi,\eta) = \langle \xi,\eta\rangle^{-n}\omega'(\xi) \wedge \omega(\eta) = \omega'\Big(\frac{\xi}{\langle \xi,\eta\rangle^n}\Big) \wedge \omega(\eta)$$

which can be obtained by a straightforward calculation. (See Chapter III, § 1 for the definition of μ.) □

Theorem 3.3. *Let D be a bounded domain in $\mathbb{C}^n$ with $\mathcal{C}^1$ boundary and let $w(z,\zeta)$ be a Leray section for D. Assume that w is $\mathcal{C}^2$ with respect to z and all the partial derivatives of w of order at most 2 with respect to z are continuous on $D \times U_{\partial D}$. We then define the operator T by*

$$Tf(z) = \Big(\int_{(\zeta,\lambda)\in\partial D\times[0,1]} f(\zeta) \wedge K^{\eta^w}(z,\zeta,\lambda) + \int_{\zeta\in D} f(\zeta) \wedge B(z,\zeta)\Big)$$

for any continuous differential form f on $\overline{D}$.

i) *$Tf \in \mathcal{C}^k_{p,q-1}(D)$ if $f \in \mathcal{C}^k_{p,q}(D) \cap \mathcal{C}_{p,q}(\overline{D})$, w is $\mathcal{C}^{k+1}$ with respect to z and all the partial derivatives of w of order at most $(k+1)$ with respect to z are continuous on $D \times U_{\partial D}$.*
ii) *If f is a continuous differential (p,q)-form on $\overline{D}$ such that $\overline{\partial} f$ is also continuous on $\overline{D}$ for some $0 \leqslant p,q \leqslant n$ then*

$$(3.1) \qquad (-1)^{p+q} f = \int_{\zeta\in\partial D} f(\zeta) \wedge K^w(\cdot,\zeta) - T\overline{\partial} f + \overline{\partial} T f \quad \text{on } D.$$

Proof. If w satisfies the hypotheses of i) then the kernel $K^{\eta^w}(z,\zeta,\lambda)$ is $\mathcal{C}^k$ with respect to z. It follows by differentiating under the integral sign that $\int_{(\zeta,\lambda)\in\partial D\times[0,1]} f(\zeta) \wedge K^{\eta^w}(z,\zeta,\lambda)$ is $\mathcal{C}^k$ on D whenever f is continuous on ∂D. Moreover, Proposition 2.1 of Chapter III says that if $f \in \mathcal{C}^k_{p,q}(D) \cap \mathcal{C}_{p,q}(\overline{D})$ then the differential form $\widetilde{B}_D f = \int_{\zeta\in D} f(\zeta) \wedge B(\cdot,\zeta)$ is $\mathcal{C}^{k+\alpha}$ on D for any $\alpha \in\,]0,1[$. It follows that Tf is $\mathcal{C}^k$. On $D \times \partial D \times [0,1]$ the only non-zero contribution to the wedge product $f(\zeta) \wedge K^{\eta^w}(z,\zeta,\lambda)$ comes from the bidegree $(p,q-1)$ part of $K^{\eta^w}(z,\zeta,\lambda)$ with respect to z and as $\widetilde{B}_D f$ has bidegree $(p,q-1)$ the form Tf also has bidegree $(p,q-1)$.

By the Bochner–Martinelli–Koppelman formula (Chap. III, Th. 2.3), to prove (3.1) it is enough to prove that

$$(3.2)\quad \overline{\partial}_z \int_{(\zeta,\lambda)\in\partial D\times[0,1]} f(\zeta)\wedge K^{\eta^w}(z,\zeta,\lambda) = \int_{\zeta\in\partial D} f(\zeta)\wedge B(z,\zeta) - \int_{\zeta\in\partial D} f(\zeta)\wedge K^w(z,\zeta) + \int_{(\zeta,\lambda)\in\partial D\times[0,1]} \overline{\partial} f(\zeta)\wedge K^{\eta^w}(z,\zeta,\lambda).$$

Fix $z\in D$ and apply Stokes' formula to the differential form $f(\zeta)\wedge K^{\eta^w}(z,\zeta,\lambda)$ on $\partial D\times[0,1]$. We get

$$(3.3)\quad \int_{(\zeta,\lambda)\in\partial D\times[0,1]} (d_\zeta+d_\lambda)(f(\zeta)\wedge K^{\eta^w}(z,\zeta,\lambda)) = \int_{(\zeta,\lambda)\in\partial(\partial D\times[0,1])} f(\zeta)\wedge K^{\eta^w}(z,\zeta,\lambda).$$

As $\partial(\partial D\times[0,1]) = \partial D\times\{0\} - \partial D\times\{1\}$ it follows from Lemma 3.2, ii) that

$$(3.4)\quad \int_{(\zeta,\lambda)\in\partial(\partial D\times[0,1])} f(\zeta)\wedge K^{\eta^w}(z,\zeta,\lambda) = \int_{\zeta\in\partial D} f(\zeta)\wedge K^w(z,\zeta) - \int_{\zeta\in\partial D} f(\zeta)\wedge B(z,\zeta).$$

For bidegree reasons $(f(\zeta)\wedge K^{\eta^w}(z,\zeta,\lambda))$ is saturated in $d\zeta$, so

$$\begin{aligned}(d_\zeta+d_\lambda)(f(\zeta)\wedge K^{\eta^w}(z,\zeta,\lambda)) &= (\overline{\partial}_\zeta+d_\lambda)\big(f(\zeta)\wedge K^{\eta^w}(z,\zeta,\lambda)\big)\\ &= \overline{\partial} f(\zeta)\wedge K^{\eta^w}(z,\zeta,\lambda)\\ &\quad + (-1)^{p+q} f(\zeta)\wedge(\overline{\partial}_\zeta+d_\lambda)K^{\eta^w}(z,\zeta,\lambda),\end{aligned}$$

but now by Lemma 3.2, i), $(\overline{\partial}_{z,\zeta}+d_\lambda)K^{\eta^w}(z,\zeta,\lambda)=0$ and hence

$$\begin{aligned}&(d_\zeta+d_\lambda)(f(\zeta)\wedge K^{\eta^w}(z,\zeta,\lambda))\\ &\quad= \overline{\partial} f(\zeta)\wedge K^{\eta^w}(z,\zeta,\lambda) + (-1)^{p+q+1} f(\zeta)\wedge \overline{\partial}_z K^{\eta^w}(z,\zeta,\lambda)\\ &\quad= \overline{\partial} f(\zeta)\wedge K^{\eta^w}(z,\zeta,\lambda) - \overline{\partial}_z\big(f(\zeta)\wedge K^{\eta^w}(z,\zeta,\lambda)\big).\end{aligned}$$

We therefore get

$$(3.5)\quad \int_{(\zeta,\lambda)\in\partial D\times[0,1]} (d_\zeta+d_\lambda)\big(f(\zeta)\wedge K^{\eta^w}(z,\zeta,\lambda)\big) = \int_{(\zeta,\lambda)\in\partial D\times[0,1]} \overline{\partial} f(\zeta)\wedge K^{\eta^w}(z,\zeta,\lambda) - \overline{\partial}_z \int_{(\zeta,\lambda)\in\partial D\times[0,1]} f(\zeta)\wedge K^{\eta^w}(z,\zeta,\lambda).$$

Formula (3.1) then follows from (3.3), (3.4) and (3.5). □

Corollary 3.4. *Let D be a bounded domain in $\mathbb{C}^n$ with $\mathcal{C}^1$ boundary and let $w(z,\zeta)$ be a Leray section for D depending holomorphically on z such that all its partial derivatives with respect to z are continuous on $D \times U_{\partial D}$. For any continuous differential (p,q)-form f on $\overline{D}$ such that $\overline{\partial} f$ is also continuous on $\overline{D}$, where $0 \leqslant p \leqslant n$ and $1 \leqslant q \leqslant n$, we then have*

$$f = \overline{\partial} T_q^p f + T_{q+1}^p \overline{\partial} f \quad \text{on } D, \tag{3.6}$$

where $T_q^p = (-1)^{p+q} T$ and T is the operator defined in Theorem 3.3.

If $\overline{\partial} f = 0$ then $u = T_q f$ is a continuous solution of the equation $\overline{\partial} u = f$ on D. Moreover, this solution u is $\mathcal{C}^\alpha$ on D for any $\alpha \in]0,1[$ and if f is $\mathcal{C}^k$ on D for some $1 \leqslant k \leqslant +\infty$ then u is $\mathcal{C}^{k+\alpha}$ on D for any $\alpha \in]0,1[$.

Proof. By (3.1) it is enough to prove that under the hypotheses of the corollary, $\int_{\zeta \in \partial D} f(\zeta) \wedge K^w(z,\zeta) = 0$ for any $z \in D$ to establish (3.6). But for any $z \in D$,

$$\int_{\zeta \in \partial D} f(\zeta) \wedge K^w(z,\zeta) = \int_{\zeta \in \partial D} f(\zeta) \wedge (K^w)_q^p(z,\zeta),$$

where $(K^w)_q^p$ is the bidegree (p,q) part of K^w with respect to z, since $\dim \partial D = n-2$. As $q \geqslant 1$, $(K^w)_q^p = 0$ by definition of K^w since w is holomorphic in z, i.e. $\overline{\partial}_z w = 0$, and the result follows. If $\overline{\partial} f = 0$ then (3.6) implies that $f = \overline{\partial} T_q^p f$ on D and hence $u = T_q^p f$ is a solution of the equation $\overline{\partial} u = f$ on D. The regularity of u follows from Proposition 2.1 of Chapter III and the fact that since w is $\mathcal{C}^\infty$ with respect to z the function

$$\int_{(\zeta,\lambda) \in \partial D \times [0,1]} f(\zeta) \wedge K^{\eta^w}(z,\zeta,\lambda)$$

is $\mathcal{C}^\infty$ on D. □

When the domain D is convex with $\mathcal{C}^2$ boundary it is easy to construct a Leray section for D which is holomorphic in the variable z. Let r be a $\mathcal{C}^2$ defining function for D. We set $w(z,\zeta) = \left(\frac{\partial r}{\partial \zeta_1}(\zeta), \ldots, \frac{\partial r}{\partial \zeta_n}(\zeta)\right)$ for any ζ in a neighbourhood of ∂D. The map w is $\mathcal{C}^1$ on $\mathbb{C}^n \times U_{\partial D}$ and independent of z: in particular, it is holomorphic in z. If $\zeta \in \partial D$ then the condition $\langle w(z,\zeta), \zeta - z \rangle = 0$ characterises the complex tangent space $T_\zeta^{\mathbb{C}} \partial D$ to ∂D at ζ which does not meet D because D is convex. It follows that $\langle w(z,\zeta), \zeta - z \rangle \neq 0$ for any $z \in D$ and $\zeta \in \partial D$ and w is therefore a Leray section for D.

It is immediate from this construction and Corollary 3.4 that for any $\overline{\partial}$-closed differential (p,q)-form f, where $0 \leqslant p \leqslant n$, $1 \leqslant q \leqslant n$, which is $\mathcal{C}^1$ on $\overline{D}$, we can solve the equation $\overline{\partial} u = f$ on any bounded convex open set D with $\mathcal{C}^2$ boundary. More precisely:

Theorem 3.5. *Let D be a bounded open convex set in $\mathbb{C}^n$ with $\mathcal{C}^2$ boundary and let r be a $\mathcal{C}^2$ defining function for D. For any continuous differential*

(p,q)-form f on $\overline{D}$ such that $\overline{\partial} f$ is also continuous on $\overline{D}$, where $0 \leqslant p \leqslant n$, $1 \leqslant q \leqslant n$, we have

$$f = \overline{\partial} T_q^p f + T_{q+1}^p \overline{\partial} f \quad \text{on } D,$$

where

$$T_q^p f(z) = (-1)^{p+q}\Big(\int_{(\zeta,\lambda)\in\partial D\times[0,1]} f(\zeta)\wedge K^{\eta^w}(z,\zeta,\lambda) + \int_{\zeta\in D} f(\zeta)\wedge B(z,\zeta)\Big)$$

and
$$w(z,\zeta) = \big(\frac{\partial r}{\partial \zeta_1}(\zeta),\ldots,\frac{\partial r}{\partial \zeta_n}(\zeta)\big).$$

In particular, if $\overline{\partial} f = 0$ then $f = \overline{\partial} T_q^p f$. Moreover, if f is $\mathcal{C}^k$ on D then $u = T_q^p f$ is $\mathcal{C}^{k+\alpha}$ on D for any $\alpha \in]0,1[$.

In particular, this theorem proves that we can always solve $\overline{\partial}$ with improving regularity locally (by taking small balls, for example). This result is called the Poincaré lemma for $\overline{\partial}$ or *Dolbeault's Lemma.*

We end this section by proving Dolbeault's Lemma for currents.

Theorem 3.6. *Let D be an open convex bounded set in $\mathbb{C}^n$ and let T be a $\overline{\partial}$-closed current of bidegree (p,q) on D, where $0 \leqslant p \leqslant n$ and $1 \leqslant q \leqslant n$. For any open set $U \Subset D$ there is a current S on U such that $\overline{\partial} S = T$ on U.*

Proof. Let χ be a $\mathcal{C}^\infty$ function with compact support on $\mathbb{C}^n$ such that $\chi \equiv 1$ in a neighbourhood of $\overline{U}$ and $\operatorname{supp}\chi \Subset D$. Theorem 3.4 of Chapter III then gives us a representation

$$\chi T = -\big(\widetilde{B}'\overline{\partial}(\chi T) + \overline{\partial}\widetilde{B}'(\chi T)\big) = -\big(\widetilde{B}'(\overline{\partial}\chi \wedge T) + \overline{\partial}\widetilde{B}'(\chi T)\big)$$

since $\overline{\partial} T = 0$. To complete the proof it is then enough to solve the equation $\overline{\partial} u = \widetilde{B}'(\overline{\partial}\chi \wedge T)$ on U. By Proposition 3.3 of Chapter III, $\widetilde{B}'(\overline{\partial}\chi\wedge T)$ is $\mathcal{C}^\infty$ in a neighbourhood of $\overline{U}$. Without loss of generality we can assume that U is convex with $\mathcal{C}^2$ boundary and we can then solve the equation $\overline{\partial} u = \widetilde{B}'(\overline{\partial}\chi\wedge T)$ on U by applying Theorem 3.5. □

4 The Dolbeault isomorphism

Let X be a complex analytic manifold of dimension n. We denote by Ω^p the sheaf of germs of holomorphic p-forms on X and by $H^q(X,\Omega^p)$ the Čech cohomology groups of Ω^p on X. (The sheaf theory used in this chapter is summarised in Appendix B.)

We denote by $Z^\alpha_{p,q}(X)$, $0 \leqslant \alpha \leqslant \infty$, the subspace of differential forms $f \in \mathcal{C}^\alpha_{p,q}(X)$ such that $\overline{\partial} f = 0$ and by $E^\alpha_{p,q}(X)$ the subspace of differential forms $f \in Z^\alpha_{p,q}(X)$ such that $f = \overline{\partial} g$ for some $g \in \mathcal{C}^\alpha_{p,q-1}(X)$. We set $H^{p,q}_\alpha(X) = Z^\alpha_{p,q}(X)/E^\alpha_{p,q}(X)$. With this notation $H^{p,q}_\infty(X)$ is simply the Dolbeault cohomology group $H^{p,q}(X)$ defined in Chapter II. Letting $Z^{-\infty}_{p,q}(X)$ denote the subspace of $\mathcal{D}'_{p,q}(X)$ consisting of $\overline{\partial}$-closed currents and $E^{-\infty}_{p,q}(X)$ denote the subspace of $Z^{-\infty}_{p,q}(X)$ consisting of $\overline{\partial}$-exact currents, we set $H^{p,q}_{-\infty}(X) = Z^{-\infty}_{p,q}(X)/E^{-\infty}_{p,q}(X)$. We denote by Φ the map $Z^{\alpha}_{p,q}(X) \to Z^{-\infty}_{p,q}(X)$ which sends a differential form f to the current T_f defined by f.

Theorem 4.1. *Let X be a complex analytic manifold of dimension n. Then,*

$$H^q(X,\Omega^p) = 0 \quad \text{for any } p,q \geqslant n+1 \tag{4.1}$$

and for any pair (p,q) such that $0 \leqslant p,q \leqslant n$ there are isomorphisms

$$\delta^{p,q}_{\alpha} : H^{p,q}_{\alpha}(X) \longrightarrow H^q(X,\Omega^p), \quad 0 \leqslant \alpha \leqslant \infty \tag{4.2}$$

$$\delta^{p,q}_{-\infty} : H^{p,q}_{-\infty}(X) \longrightarrow H^q(X,\Omega^p) \tag{4.3}$$

and the diagram

$$\begin{array}{ccc} H^{p,q}_{\alpha}(X) & & \\ \Big\downarrow{\scriptstyle \Phi^*} & \overset{\delta^{p,q}_{\alpha}}{\searrow} & H^q(X,\Omega^p) \\ & \underset{\delta^{p,q}_{-\infty}}{\nearrow} & \\ H^{p,q}_{-\infty}(X) & & \end{array} \tag{4.4}$$

is commutative for any $\alpha \in [0,+\infty]$. In particular, for any $0 \leqslant \alpha \leqslant \infty$ and $0 \leqslant p,q \leqslant n$, the homomorphism

$$\Phi^* : H^{p,q}_{\alpha}(X) \longrightarrow H^{p,q}_{-\infty}(X) \tag{4.5}$$

induced by Φ is an isomorphism.

Proof. We consider the sheaf $Z^{\alpha}_{p,q}$ of $\overline{\partial}$-closed $\mathcal{C}^{\alpha}$ differential forms on X, the sheaf $\mathcal{D}'_{p,q}$ of currents on X, the sheaf $Z^{-\infty}_{p,q}$ of $\overline{\partial}$-closed currents on X and the sheaf $\widetilde{\mathcal{C}}^{\alpha}_{p,q}$ of $\mathcal{C}^{\alpha}$ differential forms whose $\overline{\partial}$ is also $\mathcal{C}^{\alpha}$ (i.e. if $U \subset X$ is an open set then $\widetilde{\mathcal{C}}^{\alpha}_{p,q}(U) = \{f \in \mathcal{C}^{\alpha}_{p,q}(U) \mid \overline{\partial} f \in \mathcal{C}^{\alpha}_{p,q+1}(U)\}$.).

We note that these sheaves are $\mathcal{E}$-module sheaves, where $\mathcal{E}$ is the sheaf of germs of $\mathcal{C}^{\infty}$ functions on X. By Proposition 3 of Appendix B it follows that

$$H^r(X,\widetilde{\mathcal{C}}^{\alpha}_{p,q}) = 0 \quad \text{for any } r \geqslant 1 \tag{4.6}$$

and

$$H^r(X,\mathcal{D}'_{p,q}) = 0 \quad \text{for any } r \geqslant 1. \tag{4.7}$$

It follows by the regularity of $\overline{\partial}$ (Chap. III, Th. 3.5 and Remark 3.6) that the map

$$\Phi : \Omega^p(X) \longrightarrow Z^{-\infty}_{p,0}(X)$$

is an isomorphism and if $0 \leqslant \alpha \leqslant \infty$ then $\Omega^p(X) = Z^\alpha_{p,0}(X)$. The theorem therefore holds for $q = 0$.

Suppose that $q \geqslant 1$. By Dolbeault's Lemma for differential forms the sequence

$$0 \longrightarrow Z^\alpha_{p,s-1} \longrightarrow \widetilde{\mathcal{C}}^\alpha_{p,s-1} \xrightarrow{\bar\partial} Z^\alpha_{p,s} \longrightarrow 0$$

is exact for any $s = 1, \dots, n$. Consider the long exact cohomology sequence associated to this exact sequence (cf. Appendix B, Theorem 5)

$$0 \longrightarrow Z^\alpha_{p,s-1}(X) \longrightarrow \widetilde{\mathcal{C}}^\alpha_{p,s-1}(X) \xrightarrow{\bar\partial} Z^\alpha_{p,s}(X) \xrightarrow{\delta^0_{\alpha,(p,s)}}$$

$$H^1(X, Z^\alpha_{p,s-1}) \longrightarrow H^1(X, \widetilde{\mathcal{C}}^\alpha_{p,s-1}) \xrightarrow{\bar\partial} H^1(X, Z^\alpha_{p,s}) \xrightarrow{\delta^1_{\alpha,(p,s)}} \cdots\cdots$$

$$\cdots\cdots\cdots\cdots\cdots\cdots\cdots\cdots\cdots\cdots\cdots\cdots$$

$$H^r(X, Z^\alpha_{p,s-1}) \longrightarrow H^r(X, \widetilde{\mathcal{C}}^\alpha_{p,s-1}) \xrightarrow{\bar\partial} H^r(X, Z^\alpha_{p,s}) \xrightarrow{\delta^{r+1}_{\alpha,(p,s)}} \cdots\cdots$$

It follows from (4.6) that the connecting maps

$$\delta^r_{\alpha,(p,s)} : H^r(X, Z^\alpha_{p,s}) \longrightarrow H^{r+1}(X, Z^\alpha_{p,s-1}) \tag{4.8}$$

are isomorphisms for any $r \geqslant 1$ and the connecting map $\delta^0_{\alpha,(p,s)}$ induces an isomorphism

$$\widehat{\delta}^0_{\alpha,(p,s)} : H^{p,s}_\alpha(X) \longrightarrow H^1(X, Z^\alpha_{p,s-1}), \quad 1 \leqslant s \leqslant n. \tag{4.9}$$

If $q \geqslant n+1$ then the isomorphisms (4.8) for any $r \geqslant 1$ imply that

$$H^q(X, Z^\alpha_{p,0}) \simeq H^{q-n}(X, Z^\alpha_{p,n}).$$

Since $Z^\alpha_{p,n} = \widetilde{\mathcal{C}}^\alpha_{p,n}$ and $Z^\alpha_{p,0} = \Omega^p$ we therefore have

$$H^q(X, \Omega^p) \simeq H^{q-n}(X, \widetilde{\mathcal{C}}^\alpha_{p,n}) \quad \text{for any } q \geqslant n+1$$

and by (4.6) this establishes (4.1).

Assume that $1 \leqslant q \leqslant n$ and set

$$\delta^{p,q}_\alpha = \delta^{q-1}_{\alpha,(p,1)} \circ \cdots \circ \delta^1_{\alpha,(p,q-1)} \circ \widehat{\delta}^0_{\alpha,(p,q)}.$$

Since the maps (4.9) and (4.8) are isomorphisms for any $r \geqslant 1$ and $Z^\alpha_{p,0} = \Omega^p$ we get an isomorphism

$$\delta^{p,q}_\alpha : H^{p,q}_\alpha(X) \longrightarrow H^q(X, \Omega^p).$$

Dolbeault's Lemma for currents (Theorem 3.6) then gives us exact sequences

$$0 \longrightarrow Z^{-\infty}_{p,s-1} \longrightarrow \mathcal{D}'_{p,s-1} \xrightarrow{\bar\partial} Z^{-\infty}_{p,s} \longrightarrow 0 \quad 1 \leqslant s \leqslant n.$$

By the regularity of $\overline{\partial}$, i.e. the isomorphism between $\Omega^p(X)$ and $Z^{-\infty}_{p,0}(X)$, and (4.7) we can repeat the above proof and we get an isomorphism

$$\delta^{p,q}_{-\infty} : H^{p,q}_{-\infty}(X) \longrightarrow H^q(X, \Omega^p).$$

Moreover, we have the following commutative diagram of short exact sequences

$$\begin{array}{ccccccccc} 0 & \longrightarrow & Z^{\alpha}_{p,s-1} & \longrightarrow & \widetilde{\mathcal{C}}^{\alpha}_{p,s-1} & \xrightarrow{\overline{\partial}} & Z^{\alpha}_{p,s} & \longrightarrow & 0 \\ & & \Phi \downarrow & & \Phi \downarrow & & \downarrow \Phi & & \\ 0 & \longrightarrow & Z^{-\infty}_{p,s-1} & \longrightarrow & \mathcal{D}'_{p,s-1} & \xrightarrow{\overline{\partial}} & Z^{-\infty}_{p,s} & \longrightarrow & 0. \end{array}$$

By the second part of the Snake Lemma (cf. Appendix B, Lemma 7) each of the diagrams

$$\begin{array}{ccc} H^r(X, Z^{\alpha}_{p,s}) & \xrightarrow{\delta^r_{\alpha,(p,s)}} & H^{r+1}(X, Z^{\alpha}_{p,s-1}) \\ \Phi^* \downarrow & & \downarrow \Phi^* \\ H^r(X, Z^{-\infty}_{p,s}) & \xrightarrow{\delta^r_{-\infty,(p,s)}} & H^{r+1}(X, Z^{-\infty}_{p,s-1}) \end{array} \qquad r \geqslant 0,\ 1 \leqslant s \leqslant n$$

is commutative and by definition of $\delta^{p,q}_{\alpha}$ and $\delta^{p,q}_{-\infty}$ it follows that diagram (4.4) is commutative. □

The isomorphisms in diagram (4.4) are called the *Dolbeault isomorphisms.*

Remarks. We can deduce the following results from Theorem 4.1.

i) If f is a $\mathcal{C}^{\alpha}$ differential form for some $0 \leqslant \alpha \leqslant \infty$ and there is a current T such that $\overline{\partial}T = T_f$ then there is a $\mathcal{C}^{\alpha}$ differential form g such that $\overline{\partial}g = f$. (This is the injectivity of (4.5).)
ii) For any $\overline{\partial}$-closed current T there is a current S such that $T - \overline{\partial}S$ is a $\mathcal{C}^{\infty}$ differential form. (This is the surjectivity of (4.5) for $\alpha = \infty$.)

More precisely:

Corollary 4.2. *Let X be a complex analytic manifold.*

i) *For any $\overline{\partial}$-closed current T on X and any neighbourhood U of the support of T there is a current S on X such that $T - \overline{\partial}S$ is a $\overline{\partial}$-closed $\mathcal{C}^{\infty}$ differential form with* $\operatorname{supp} S \subset U$.
ii) *If f is a $\mathcal{C}^k$ differential form on X, $k = 0, 1, \ldots, \infty$, and T is a current on X such that $\overline{\partial}T = T_f$ then for any neighbourhood U of the support of T there is a differential form $g \in \bigcap_{0<\alpha<1} \mathcal{C}^{k+\alpha}(X, E)$ such that $\overline{\partial}g = f$ on X and* $\operatorname{supp} g \subset U$.

Proof. Let us prove i). Choose a neighbourhood V of the support of T such that $\overline{V} \subset U$ and let χ_0 and χ_1 be $\mathcal{C}^\infty$ functions on X such that $\chi_0 = 1$ on a neighbourhood of $X \setminus V$, $\chi_0 = 0$ on a neighbourhood of the support of T, $\chi_1 = 1$ in a neighbourhood of $X \setminus U$ and $\chi_1 = 0$ on a neighbourhood of $\overline{V}$.

Since T is $\overline{\partial}$-closed and the morphism (4.5) is surjective, there is a current S_0 on X and a $\overline{\partial}$-closed $\mathcal{C}^\infty$ differential form f_0 on X such that $T - \overline{\partial} S_0 = T_{f_0}$ on X. Then $T_{f_0} = -\overline{\partial} S_0$ on $X \setminus \operatorname{supp} T$ and as the morphism (4.5) is injective there is a $\mathcal{C}^\infty$ form g_0 on $X \setminus \operatorname{supp} T$ such that $f_0 = \overline{\partial} g_0$ on $X \setminus \operatorname{supp} T$. On $X \setminus \overline{V}$ we then have $\overline{\partial}(S_0 + T_{\chi_0 g_0}) = T_{-f_0 + \overline{\partial} g_0} = 0$. The surjectivity of (4.5) then implies that there is a current R on $X \setminus \overline{V}$ and a $\overline{\partial}$-closed $\mathcal{C}^\infty$ differential form g_1 such that $S_0 + T_{\chi_0 g_0} - \overline{\partial} R = T_{g_1}$ on $X \setminus V$. The current $S = S_0 + T_{\chi_0 g_0 - \chi_1 g_1} - \overline{\partial}(\chi_1 R)$ then satisfies i).

We now prove ii). Denote by $\mathcal{F}$ the sheaf on X defined in the following way:

If V is an open set in X then $\mathcal{F}(V)$ is the vector space of differential forms $g \in \bigcap_{0<\alpha<1} \mathcal{C}^{k+\alpha}_\bullet(V)$ such that $\overline{\partial} g \in \bigcap_{0<\alpha<1} \mathcal{C}^{k+\alpha}_\bullet(V)$. This is a sheaf of $\mathcal{E}$-modules and it follows by Proposition 3 of Appendix B that

$$H^1(X, \mathcal{F}) = 0. \tag{4.10}$$

Let $V = (V_i)_{i \in I}$ and $W = (W_i)_{i \in I}$ be open covers of X such that for any i, $V_i \Subset W_i$ and V_i and W_i are biholomorphic to the unit ball in $\mathbb{C}^n$. Let $f \in \mathcal{C}^k_\bullet(X)$ be $\overline{\partial}$-closed. By Dolbeault's Lemma for forms (Theorem 3.5), we can find a family of forms $g_i \in \bigcap_{0<\alpha<1} \mathcal{C}^{k+\alpha}_\bullet(W_i)$ such that $\overline{\partial} g_i = f$ on W_i. If T is a current on X such that $\overline{\partial} T = T_f$, the injectivity of (4.5) implies there is a $\mathcal{C}^k$ differential form u on X such that $\overline{\partial} u = f$ on X. We set $h_i = g_i - u$ on V. Then $h_i \in Z^k(W_i)$ and applying Dolbeault's Lemma a second time we get a family of forms $k_i \in \bigcap_{0<\alpha<1} \mathcal{C}^{k+\alpha}_\bullet(V_i)$ such that $\overline{\partial} k_i = h_i$ on V_i. It follows that $k_i - k_j \in \mathcal{F}(V_i \cap V_j)$ and by (4.10) there is a family of forms $u_i \in \mathcal{F}(V_i)$ such that $k_i - k_j = u_i - u_j$ on $V_i \cap V_j$. Setting $v = g_i - \overline{\partial} u_i$ on V_i, we get a form $v \in \bigcap_{0<\alpha<1} \mathcal{C}^{k+\alpha}(X)$ such that $\overline{\partial} v = f$ on X.

In the above we have constructed a solution with the required regularity: it remains to show that the support condition is satisfied. Since $\overline{\partial} v = f$ on X, the current $T - T_v$ is $\overline{\partial}$-closed on X and the surjectivity of (4.5) for $\alpha = \infty$ implies that there is a current S on X and a $\overline{\partial}$-closed $\mathcal{C}^\infty$ differential form g_∞ on X such that $T - T_v + \overline{\partial} S = T_{g_\infty}$. Then $\overline{\partial} S = T_{v + g_\infty}$ on $X \setminus \operatorname{supp} T$. By the first part of the proof there is therefore a differential form $h \in \bigcap_{0<\alpha<1} \mathcal{C}^{k+\alpha}_\bullet$ $(X \setminus \operatorname{supp} T)$ such that $\overline{\partial} h = v + g_\infty$ on $X \setminus \operatorname{supp} T$. Choose a $\mathcal{C}^\infty$ function χ on X such that $\chi = 0$ in a neighbourhood of the support of T and $\chi = 1$ in a neighbourhood of $X \setminus U$. The differential form $g = v + g_\infty - \overline{\partial}(\chi h)$ then satisfies ii). □

The following result which will be useful in Section 5 of this chapter follows immediately from i) of Corollary 4.2.

Corollary 4.3. *Let X be a complex analytic manifold of dimension n and let Θ be one of the support families defined in Chapter II, § 7. We assume* *that* $H^{p,q}_{\Theta}(X) = 0$ *for some pair of integers* (p,q), $0 \leqslant p \leqslant n$, $1 \leqslant q \leqslant n$. *It follows that if* T *is a* $\overline{\partial}$*-closed* (p,q)*-current on* X *whose support is a member of* Θ *then there is a current* S *whose support is a member of* Θ *such that* $\overline{\partial}S = T$ *on* X.

5 Bochner's theorem and extensions of CR functions on manifolds

We will extend the results obtained in Section 2 to CR functions of class $\mathcal{C}^k$.

Theorem 5.1. *Let* X *be a non-compact complex analytic manifold and let* K *be a compact set in* X. *Assume that* $H^{0,1}_{\Phi}(X \smallsetminus K) = 0$. *For any relatively compact domain* D *in* X *such that*

1) $\partial D \smallsetminus K$ *is a* $\mathcal{C}^k$ *submanifold of* $X \smallsetminus K$ *for some* $k \geqslant 1$.
2) $D \smallsetminus K = \mathrm{Int}(\overline{D} \smallsetminus K)$.
3) $X \smallsetminus (\overline{D} \cup K)$ *is connected*

and any CR *function of class* $\mathcal{C}^k$ f *on* $\partial D \smallsetminus K$, *where* $0 \leqslant s \leqslant k$, *there is a* $\mathcal{C}^s$ *function* F *on* $\overline{D} \smallsetminus K$, *holomorphic on* $D \smallsetminus K$, *such that* $F|_{\partial D \smallsetminus K} = f$.

Proof. Consider the current $T = f[\partial D \smallsetminus K]^{0,1}$. This current has bidegree $(0,1)$ on $X \smallsetminus K$ and its support is contained in $\partial D \smallsetminus K$ which is an element of Φ. As the function f is CR, the current T is $\overline{\partial}$-closed (indeed, if φ is a $\mathcal{C}^\infty$ differential $(n, n-2)$-form with compact support in $X \smallsetminus K$ then $\langle \overline{\partial}T, \varphi\rangle = \langle T, \overline{\partial}\varphi\rangle = \int_{\partial D \smallsetminus K} f \wedge \overline{\partial}\varphi = 0$). The hypothesis $H^{0,1}_{\Phi}(X \smallsetminus K) = 0$ and Dolbeault's isomorphism then imply the existence of a current S whose support is a member of Φ such that $\overline{\partial}S = T$ (cf. Corollary 4.3). Since the support of T is contained in $\partial D \smallsetminus K$, $\overline{\partial}S$ vanishes on $X \smallsetminus (\partial D \cup K)$; S is therefore a holomorphic function on $(D \smallsetminus K) \cup (X \smallsetminus (\overline{D} \cup K))$ (cf. Chap. III, Th. 4.5). As the support of S is an element of Φ, S vanishes on an open set in $X \smallsetminus (\overline{D} \cup K)$ since X is not compact and $\overline{D} \cup K$ is compact. The connectedness of $X \smallsetminus (\overline{D} \cup K)$ and the principle of analytic continuation then imply that S vanishes on $X \smallsetminus (\overline{D} \cup K)$.

It remains to determine the behaviour of S close to $\partial D \smallsetminus K$. This follows from a local calculation. Indeed, given a point ξ in $\partial D \smallsetminus K$, a neighbourhood V_ξ of ξ in $X \smallsetminus K$ on which the equation $\overline{\partial}R = f[\partial D \smallsetminus K]^{0,1}$ has a solution and a current S_ξ of degree 0 such that $\overline{\partial}S_\xi = f[\partial D \smallsetminus K]^{0,1}$ on V_ξ, $\overline{\partial}(S - S_\xi) = 0$ on V_ξ and $S - S_\xi$ will then be a holomorphic (and hence $\mathcal{C}^\infty$) function on V_ξ. The currents S and S_ξ are then equal on V_ξ up to addition of a $\mathcal{C}^\infty$ function.

Consider $\xi \in \partial D \smallsetminus K$ and choose a neighbourhood U_ξ of ξ in $X \smallsetminus K$ such that

a) U_ξ is contained in a chart domain U of X.

b) There are local coordinates on U such that the image of U_ξ under the chart map is a bounded open convex set in $\mathbb{C}^n$.
c) $U_\xi \setminus (\partial D \cup K)$ has exactly two connected components.

Without loss of generality, we can identify U_ξ with its image in $\mathbb{C}^n$. By Theorem 3.6, for any convex open set $V_\xi \Subset U_\xi$ there is a current S_ξ such that $\overline{\partial} S_\xi = T$ on V_ξ. Moreover, we can write $S_\xi = u_\xi + \widetilde{B}'(\chi T)$ where χ is a $\mathcal{C}^\infty$ function with compact support in U_ξ which is equal to 1 in a neighbourhood of $\overline{V}_\xi$ and u_ξ is a $\mathcal{C}^\infty$ function such that $\overline{\partial} u_\xi = \widetilde{B}'(\overline{\partial}\chi \wedge T)$ on V_ξ. On V_ξ the current S_ξ is therefore equal to

$$\widetilde{B}'(\chi T) = \int_{\partial D \setminus K} \chi(\zeta) f(\zeta) B(z,\zeta)$$

up to addition of a $\mathcal{C}^\infty$ function. In other words, up to addition of a $\mathcal{C}^\infty$ function, S_ξ is equal to the Bochner–Martinelli transform of χf, which we studied in Chapter IV, § 1. As the current S vanishes on $V_\xi \cap (X \setminus (\overline{D} \cup K))$, $S_\xi|_{V_\xi \cap (X \setminus (\overline{D} \cup K)}$ can be extended continuously to S_ξ^- on $V_\xi \cap (X \setminus (D \cup K))$, as can the Bochner–Martinelli transform of χf. As f is continuous, $S_\xi|_{V_\xi \cap (D \setminus K)}$ can be continuously extended to S_ξ^+ on $V_\xi \cap (\overline{D} \setminus K)$ and $S_\xi^+ - S_\xi^- = f$ on $(\partial D \setminus K) \cap V_\xi$. It follows that $S|_{D \setminus K}$ can be extended to a continuous function F on $\overline{D} \setminus K$ such that $F|_{\partial D \setminus K} = f$, since $S|_{X \setminus (\overline{D} \cup K)} = 0$.

To prove the $\mathcal{C}^k$ regularity of S up to the boundary, it is enough to consider $F_\xi(z) = \int_{V_\xi \cap \partial D} f(\zeta) B(z,\zeta)$ since

$$\widetilde{B}'(\chi T) - \int_{V_\xi \cap \partial D} f(\zeta) B(z,\zeta) = \int_{\partial D \setminus (K \cup V_\xi)} \chi(\zeta) f(\zeta) B(z,\zeta)$$

is $\mathcal{C}^\infty$ on V_ξ. Assume that $V_\xi \cap \partial D$ is an open set with $\mathcal{C}^1$ boundary. Repeating the proof of Theorem 3.3 of Chapter IV, we get, for any $z \in V_\xi \setminus \partial D$,

$$\frac{\partial F_\xi}{\partial z_j}(z) = -\int_{\zeta \in \partial D \cap V_\xi} \widetilde{f}(\zeta) d_\zeta B_j(z,\zeta) \quad \text{for any } j = 1, \dots, n,$$

where $\widetilde{f}$ is an extension of f satisfying the conclusions of Lemma 2.3 of Chapter IV. Applying Stokes' formula, we get

$$\begin{aligned}
\frac{\partial F_\xi}{\partial z_j}(z) &= \int_{\zeta \in \partial D \cap V_\xi} d_\zeta \widetilde{f}(\zeta) \wedge B_j(z,\zeta) - \int_{\zeta \in \partial D \cap \partial V_\xi} \widetilde{f}(\zeta) \wedge B_j(z,\zeta) \\
&= \int_{\zeta \in \partial D \cap V_\xi} \frac{\partial \widetilde{f}}{\partial \zeta_j}(\zeta) \wedge B(z,\zeta) - \int_{\zeta \in \partial D \cap \partial V_\xi} \widetilde{f}(\zeta) \wedge B_j(z,\zeta),
\end{aligned}$$

where the last integral on the right-hand side is a $\mathcal{C}^\infty$ function on V_ξ because B_j is $\mathcal{C}^\infty$ on $\mathbb{C}^n \times \mathbb{C}^n \setminus \Delta$. As $F_\xi|_{V_\xi \setminus D}$ differs from $S|_{X \setminus (\overline{D} \cup K)} = 0$ by a $\mathcal{C}^\infty$ function on V_ξ, $\frac{\partial F_\xi}{\partial z_j}|_{V_\xi \setminus \overline{D}}$ can be continuously extended to $V_\xi \cap \overline{D}$. Induction on k completes the proof as in Theorem 3.3 of Chapter IV. □

Corollary 5.2 (Bochner's theorem). *Let X be a non-compact complex analytic manifold such that $H_c^{0,1}(X) = 0$. For any relatively compact domain D with $\mathcal{C}^k$ boundary in X such that $X \setminus \overline{D}$ is connected and any* CR *function f of class $\mathcal{C}^s$ on ∂D, where $0 \leqslant s \leqslant k$, there is a $\mathcal{C}^s$ function F on $\overline{D}$, holomorphic on D, such that $F\big|_{\partial D} = f$.*

Proof. Simply apply Theorem 5.1 with $K = \varnothing$, since the families c and Φ are equal in this case. □

Remark. Theorems 2.5 and 2.7 give cohomological conditions under which the conclusion of Theorem 5.1 is valid.

Comments

The study of extensions of CR functions defined on part of the boundary of a domain is mainly due to G. Lupacciolu [Lu1, Lu2]. There is an excellent presentation of this material in [Ci/St]. Poincaré's Lemma for $\overline{\partial}$ is due to P. Dolbeault and A. Grothendieck: it is the key point in the representation of cohomology groups of analytic sheaves in terms of the $\overline{\partial}$ complex (cf. [Do1, Do2]). In this chapter, we prove Dolbeault's Lemma using a resolution of $\overline{\partial}$ in convex domains via the Cauchy–Fantappié formula. Another proof of Dolbeault's Lemma using a resolution of $\overline{\partial}$ in polydiscs is given in [Ho2]. The proof of the Dolbeault isomorphism given here can be found in [He/Le2].

VI

Domains of holomorphy and pseudoconvexity

At the end of Chapter I and in Chapter III we met open sets in $\mathbb{C}^n$ on which any holomorphic function can be extended to a larger open set. The open sets which do not have this property are called domains of holomorphy: in this chapter we study such open sets. We start by giving a characterisation of domains of holomorphy in terms of holomorphic convexity (the Cartan–Thullen theorem). We then introduce the notion of pseudoconvexity in order to get a more analytic characterisation of domains of holomorphy. This requires us to define plurisubharmonic functions. We then prove that every domain of holomorphy is pseudoconvex: the converse, which is known as the Levi problem, is studied in Chapter VII.

1 Domains of holomorphy and holomorphic convexity

Whilst studying Hartogs' phenomenon in Chapter III we saw that there are open sets Ω in $\mathbb{C}^n$ such that any holomorphic function on Ω can be extended holomorphically to a larger domain. This leads us to define domains of holomorphy as follows.

Definition 1.1. An open set Ω in $\mathbb{C}^n$ is called a *domain of holomorphy* if there is no pair of open sets Ω_1 and Ω_2 in $\mathbb{C}^n$ with the following properties:

a) $\varnothing \neq \Omega_1 \subset \Omega_2 \cap \Omega$;
b) Ω_2 is connected and is not contained in Ω;
c) for any $f \in \mathcal{O}(\Omega)$ there is a function $f_2 \in \mathcal{O}(\Omega_2)$ such that $f = f_2$ on Ω_1.

Domains of holomorphy in $\mathbb{C}^n$ are only really interesting if $n \geqslant 2$ since any open set Ω in $\mathbb{C}$ is a domain of holomorphy (consider the function $f(z) = \frac{1}{z-z_0}$ for any $z_0 \in \partial\Omega$).

Examples.

1) It follows from Hartogs' theorem (Chapter III, Corollary 4.3) that if Ω is a domain in $\mathbb{C}^n$, $n \geqslant 2$, and K is a compact set in Ω then $\Omega \setminus K$ is not a domain of holomorphy.

C. Laurent-Thiébaut, *Holomorphic Function Theory in Several Variables: An Introduction*, Universitext, DOI 10.1007/978-0-85729-030-4_6, © Springer-Verlag London Limited 2011

2) The Euclidean ball $B(0,R)$ with centre 0 and radius R in $\mathbb{C}^n$ is a domain of holomorphy (for any $z_0 \in \partial B$, consider the function $f(z) = 1/R^2 - \sum_{j=1}^n z_j \overline{z}_j^0$ where $z_0 = (z_1^0, \ldots, z_n^0)$).
3) Hartogs' figure in $\mathbb{C}^n$, $n \geqslant 2$

$$Q = \Big\{ (z_1, \ldots, z_n) \in \mathbb{C}^n \mid |z_i| < r/2,\ 1 \leqslant i \leqslant n-1 \text{ or } r/2 \leqslant |z_n| \leqslant r \Big\}$$

is not a domain of holomorphy (cf. Chap. I, Th. 6.9).

Lemma 1.2. *Any open convex set in $\mathbb{C}^n$ is a domain of holomorphy.*

Proof. Given a point z_0 in $\partial\Omega$ the convexity of Ω implies there is an $\mathbb{R}$-linear map ℓ from $\mathbb{C}^n$ to $\mathbb{R}$ such that $\ell(z) < \ell(z_0)$ for any $z \in \Omega$. We can write $\ell(z) = \sum_{j=1}^n \alpha_j z_j + \beta_j \overline{z}_j$ and since ℓ is real valued we have $\beta_j = \overline{\alpha}_j$, $j = 1, \ldots, n$. It follows that $\ell(z) = \mathrm{Re}(h(z))$, where $h(z) = 2\sum_{j=1}^n \alpha_j z_j$ for any $z \in \mathbb{C}^n$. The holomorphic function on Ω given by $f(z) = 1/\big(h(z) - h(z_0)\big)$ then has no holomorphic extension to a neighbourhood of z_0, which proves the lemma. □

We will now try to characterise domains of holomorphy in terms of convexity with respect to holomorphic functions.

Definition 1.3. Let Ω be an open set in $\mathbb{C}^n$ and let K be a compact set in Ω. We define $\widehat{K}_\Omega$, the *holomorphically convex hull* of K by

$$\widehat{K}_\Omega = \big\{ z \in \Omega \mid |f(z)| \leqslant \sup_K |f|,\ \forall f \in \mathcal{O}(\Omega) \big\}.$$

Lemma 1.4. *Let Ω be an open set in $\mathbb{C}^n$ and let K be a compact set in Ω. The holomorphically convex hull $\widehat{K}_\Omega$ is then contained in the convex hull of K.*

Proof. We recall that the convex hull of K is the intersection of all the real half-spaces containing K. Moreover, a real hyperplane in $\mathbb{C}^n$ is defined by an equation $\mathrm{Re}\langle z, \xi \rangle = \alpha$, where $\xi \in \mathbb{C}^n$ is some complex vector and $\alpha \in \mathbb{R}$. If a point $z_0 \in \mathbb{C}^n$ is not in the convex hull of K then there is a $\xi \in \mathbb{C}^n$ such that $\mathrm{Re}\langle w, \xi \rangle < \mathrm{Re}\langle z_0, \xi \rangle$ for every $w \in K$. Consider the holomorphic function on $\mathbb{C}^n$ given by $f(z) = \exp(\langle z - z_0, \xi \rangle)$. This function then has the property that $|f(z_0)| = 1$ and $|f(w)| < 1$ for every $w \in K$, so $z_0 \notin \widehat{K}_\Omega$. □

Corollary 1.5. *If Ω is an open set in $\mathbb{C}^n$ and K is a compact set in Ω then $\widehat{K}_\Omega$ is bounded in $\mathbb{C}^n$.*

Proof. This follows immediately from Lemma 1.4 because the convex hull of a bounded set is bounded. □

Proposition 1.6. *Let Ω be an open set in $\mathbb{C}^n$ and let K be a compact set in Ω. Then:*

i) *$K \subset \widehat{K}_\Omega$ and $\widehat{\widehat{K}}_\Omega = \widehat{K}_\Omega$,*
ii) *if $K' \subset K$ then $\widehat{K}'_\Omega \subset \widehat{K}_\Omega$,*
iii) *if D is an open set in $\mathbb{C}^n$ such that $\Omega \subset D$ then $\widehat{K}_\Omega \subset \widehat{K}_D$,*
iv) *$\widehat{K}_\Omega$ is closed in Ω,*
v) *for any $M > 0$, $\varepsilon > 0$ and $z \in \Omega \setminus \widehat{K}_\Omega$ there is an $f \in \mathcal{O}(\Omega)$ such that $\sup_K |f| < \varepsilon$ and $f(z) > M$.*

Proof. Claims i), ii) and iii) are immediate consequences of Definition 1.3.

Let us prove iv). We note that $\widehat{K}_\Omega = \bigcap_{f\in\mathcal{O}(\Omega)} A_f$, where $A_f = \{z \in \Omega \mid |f(z)| \leqslant \sup_K |f|\}$. As the functions in $\mathcal{O}(\Omega)$ are continuous on Ω, the sets A_f are closed in Ω and $\widehat{K}_\Omega$ is therefore closed in Ω.

Let us prove v). For any $z \in \Omega \setminus \widehat{K}_\Omega$ there is a function $h \in \mathcal{O}(\Omega)$ such that $\sup_K |h| < |h(z)|$. After multiplying by a constant, we can assume that $\sup_K |h| < 1 < |h(z)|$. We then set $f = h^\ell$ for large enough ℓ. □

Remark. We have seen that $\widehat{K}_\Omega$ is always bounded and closed in Ω, but it is not in general a compact set in Ω.

Theorem 1.7. *Let Ω be a domain of holomorphy in $\mathbb{C}^n$. For any K in Ω we have*

$$\operatorname{dist}(\widehat{K}_\Omega, \partial\Omega) = \operatorname{dist}(K, \partial\Omega),$$

where $\operatorname{dist}(K, \partial\Omega) = \inf\{|w - z| \mid w \in K,\ z \in \partial\Omega\}$ *is the Euclidean distance between K and $\partial\Omega$.*

If $P(0, r)$ is the polydisc of centre O and multiradius $(r_1, \dots, r_n)$ and a is a point of Ω then we set

$$\delta^r_\Omega(a) = \sup\{\lambda > 0 \mid a + \lambda P(0, r) \subset \Omega\}.$$

Then $\operatorname{dist}(a, \partial\Omega) = \inf\{\delta^r_\Omega(a) \mid r > 0,\ \Sigma r_j^2 = 1\}$.

Lemma 1.8. *Let Ω be an open set in $\mathbb{C}^n$, let K be a compact set in Ω and let r be a multiradius. If $\eta > 0$ has the property that $\delta^r_\Omega(z) \geqslant \eta$ for any $z \in K$ then for any $a \in \widehat{K}_\Omega$ and $f \in \mathcal{O}(\Omega)$ the Taylor series of f at a converges on $P(a, \eta r)$.*

Proof (of the lemma). Fix a function $f \in \mathcal{O}(\Omega)$ and a real number η' such that $0 < \eta' < \eta$. The set $Q = \overline{\bigcup_{a\in K} P(a, \eta' r)}$ is then a compact set in Ω and there is therefore an $M \in \mathbb{R}$ such that $\sup_Q |f| \leqslant M$. By the Cauchy inequalities, $\sup_K |D^\alpha f| \leqslant \alpha! M(\eta' r)^{-\alpha}$ for any $\alpha \in \mathbb{N}^n$. As $D^\alpha f \in \mathcal{O}(\Omega)$,

$$D^\alpha f(a) \leqslant \alpha! M(\eta' r)^{-\alpha} \quad \text{for any } a \in \widehat{K}_\Omega$$

by definition of $\widehat{K}_\Omega$ and the Taylor series of f at any point $a \in \widehat{K}_\Omega$ converges to $P(a, \eta' r)$ by Abel's lemma. As this holds for any $\eta' < \eta$, this Taylor series converges on $P(a, \eta r)$. □

Proof (of the theorem). We note first that the inclusion $K \subset \widehat{K}_\Omega$ implies that

$$\mathrm{dist}(\widehat{K}_\Omega, \partial\Omega) \leqslant \mathrm{dist}(K, \partial\Omega).$$

Assume that $\eta = \mathrm{dist}(K, \partial\Omega) > \mathrm{dist}(\widehat{K}_\Omega, \partial\Omega)$: we will prove that Ω cannot then be a domain of holomorphy. Consider a point $a \in \widehat{K}_\Omega$ such that $\mathrm{dist}(a, \partial\Omega) < \eta$. There is a multiradius r such that $\sum_{j=1}^n r_j^2 = 1$ and $\delta_\Omega^r(a) < \eta \leqslant \delta_\Omega^r(z)$ for any $z \in K$. It then follows from Lemma 1.8 that the restriction of any holomorphic function on Ω to the connected component of $\Omega \cap P(a, \eta r)$ containing a has a holomorphic extension to $P(a, \eta r)$. As $\delta_\Omega^r(a) < \eta$, $P(a, \eta r)$ is not contained in Ω so the open set Ω cannot be a domain of holomorphy. It follows that if Ω is a domain of holomorphy then $\mathrm{dist}(\widehat{K}_\Omega, \partial\Omega) \geqslant \mathrm{dist}(K, \partial\Omega)$. □

It is therefore natural to introduce the following definition.

Definition 1.9. Let Ω be an open set in $\mathbb{C}^n$. We say that Ω is *holomorphically convex* if $\widehat{K}_\Omega$ is relatively compact in Ω for any compact set K in Ω.

If K is a compact set in Ω then K is said to be $\mathcal{O}(\Omega)$*-convex* if and only if $K = \widehat{K}_\Omega$.

Examples. Consider the open set $\Omega = \{z \in \mathbb{C}^n \mid 1/2 < |z| < 2\}$ and the compact set $K = \{z \in \mathbb{C}^n \mid |z| = 1\}$ in Ω.

1) If $n = 1$, then we can prove that $\widehat{K}_\Omega = K$ by using the holomorphic functions $f(z) = 1/z$ and $f(z) = z$ on Ω, which implies that K is $\mathcal{O}(\Omega)$-convex.
2) If $n \geqslant 2$ then Hartogs' phenomenon implies that any holomorphic function f on Ω can be extended to a holomorphic function $\widetilde{f}$ on the ball $B(0, 2)$. Applying the maximum principle to $\widetilde{f}$ we get $|\widetilde{f}(z)| = |f(z)| \leqslant \sup_K |f|$ for any z such that $\frac{1}{2} < |z| < 1$ and it follows that $B(0,1) \cap \Omega \subset \widehat{K}_\Omega$ which proves that $\widehat{K}_\Omega$ is not relatively compact in Ω. The open set Ω is therefore not holomorphically convex.

With this terminology, Theorem 1.7 says that any domain of holomorphy is holomorphically convex. In the rest of this section, we will prove that domains of holomorphy and holomorphically convex open sets are the same thing.

Lemma 1.10. *Let Ω be a holomorphically convex open set in $\mathbb{C}^n$. The set Ω then has an exhaustion $(K_j)_{j \in \mathbb{N}}$ by $\mathcal{O}(\Omega)$-convex compact sets K_j.*

Proof. We construct the sequence $(K_j)_{j \in \mathbb{N}}$ recursively starting with any exhaustion $(Q_\ell)_{\ell \in \mathbb{N}}$ of Ω by arbitrary compact sets. We set $K_1 = \widehat{Q}_1$: this is a compact set in Ω because Ω is holomorphically convex. Assume we have constructed $K_1, \ldots, K_j$ and consider some $\ell_j \geqslant j$ such that $K_j \subset \mathring{Q}_{\ell_j}$. We then simply set $K_{j+1} = \widehat{Q}_{\ell_j}$. □

Lemma 1.11. *Let Ω be a holomorphically convex open set in $\mathbb{C}^n$, let $(K_j)_{j\in\mathbb{N}}$ be an exhaustion of Ω by $\mathcal{O}(\Omega)$-convex compact sets and let $(p_j)_{j\in\mathbb{N}}$ be a sequence of points such that $p_j \in K_{j+1} \setminus K_j$. There is then a function $f \in \mathcal{O}(\Omega)$ such that $\lim_{j\to\infty} |f(p_j)| = +\infty$.*

Proof. The function f will be given as a series $\sum_{\nu=1}^{\infty} f_\nu$ such that $f_\nu \in \mathcal{O}(\Omega)$, $\sup_{K_\nu} |f_\nu| < 2^{-\nu}$ and $|f_j(p_j)| > j + 1 + \sum_{\nu=1}^{j-1} |f_\nu(p_j)|$ for all $j \geqslant 2$. Note that if such a sequence $(f_\nu)_{\nu\in\mathbb{N}^*}$ exists then the series $\sum_{\nu=1}^{\infty} f_\nu$ converges uniformly on any compact set K in Ω and hence its sum f is holomorphic on Ω. Moreover, if $j \geqslant 2$ then

$$|f(p_j)| \geqslant |f_j(p_j)| - \sum_{\nu\neq j} |f_\nu(p_j)| > j + 1 - \sum_{\nu>j} |f_\nu(p_j)|.$$

This implies that $|f(p_j)| > j$ since $\sum_{\nu>j} |f_\nu(p_j)| < \sum_{\nu=1}^{\infty} 2^{-\nu} \leqslant 1$. We construct the series $(f_\nu)_{\nu\in\mathbb{N}}$ recursively. We set $f_1 \equiv 0$ and given $f_1, \dots, f_{\ell-1}$ it follows from v) of Proposition 1.6 that there is a $f_\ell \in \mathcal{O}(\Omega)$ such that $\sup_{K_\ell} |f_\ell| < 2^{-\ell}$ and $|f_\ell(p_\ell)| > \ell + 1 \sum_{\nu=1}^{\ell-1} |f_\nu(p_\ell)|$, since $p_\ell \notin \widehat{K}_\ell$. □

Proposition 1.12. *An open set Ω in $\mathbb{C}^n$ is holomorphically convex if and only if for any sequence $(p_\nu)_{\nu\in\mathbb{N}} \subset \Omega$ without accumulation points in Ω there is a function $f \in \mathcal{O}(\Omega)$ such that $\sup_{\nu\in\mathbb{N}} |f(p_\nu)| = +\infty$.*

Proof. Suppose that Ω is holomorphically convex and consider an exhaustion $(K_j)_{j\in\mathbb{N}}$ of Ω by $\mathcal{O}(\Omega)$-convex compact sets. Such an exhaustion exists by Lemma 1.10. If $(p_\nu)_{\nu\in\mathbb{N}}$ is a sequence of points in Ω without accumulation points in Ω then there are sequences $(\nu_k)_{k\in\mathbb{N}}$ and $(j_k)_{k\in\mathbb{N}}$ such that $p_{\nu_k} \in K_{j_{k+1}} \setminus K_{j_k}$. The existence of f then follows from Lemma 1.11.

Conversely, let K be a compact set in Ω. By definition of holomorphic convexity, it will be enough to prove that any sequence $(p_\nu)_{\nu\in\mathbb{N}}$ of points in $\widehat{K}_\Omega$ has an accumulation point in $\widehat{K}_\Omega$. Let $(p_\nu)_{\nu\in\mathbb{N}}$ be a sequence of points in $\widehat{K}_\Omega$; for any $f \in \mathcal{O}(\Omega)$ we then have $\sup_{\nu\in\mathbb{N}} |f(p_\nu)| \leqslant \sup_K |f| < +\infty$. By hypothesis for any a sequence (p_ν) without accumulation points in Ω we can find a holomorphic function on Ω which is not bounded on (p_ν), so $(p_\nu)_{\nu\in\mathbb{N}}$ has an accumulation point in Ω. As $\widehat{K}_\Omega$ is closed in Ω this accumulation point lies in $\widehat{K}_\Omega$. □

Application. Let Ω be an open set in $\mathbb{C}^n$. The set Ω is holomorphically convex if for any point $p \in \partial\Omega$ there is a function $f_p \in \mathcal{O}(\Omega)$ such that $\lim_{\substack{z\to p \\ z\in\Omega}} |f_p(z)| = +\infty$. Indeed, if $(p_\nu)_{\nu\in\mathbb{N}}$ is a sequence of points in Ω without accumulation points in Ω, then either (p_ν) is an unbounded sequence and $f(z) = z$ is then unbounded on (p_ν) or (p_ν) has an accumulation point $p \in \partial\Omega$ and f_p is then unbounded on $(p_\nu)_{\nu\in\mathbb{N}}$. By Proposition 1.12, Ω is then holomorphically convex. This proves that any open set in $\mathbb{C}$ is holomorphically convex, as is any convex open set in $\mathbb{C}^n$.

We can now state and prove a theorem due to H. Cartan and P. Thullen characterising domains of holomorphy in terms of holomorphic convexity.

Theorem 1.13. *Let Ω be an open set in $\mathbb{C}^n$. The following are equivalent:*

i) *Ω is a domain of holomorphy,*
ii) $\operatorname{dist}(\widehat{K}_\Omega, \partial\Omega) = \operatorname{dist}(K, \partial\Omega)$ *for any compact set K in Ω,*
iii) *Ω is holomorphically convex,*
iv) *There is a function $f \in \mathcal{O}(\Omega)$ such that for any pair of open sets Ω_1 and Ω_2 in $\mathbb{C}^n$ satisfying a) and b) of Definition 1.1 there is no function $\widetilde{f} \in \mathcal{O}(\Omega_2)$ such that $\widetilde{f} = f$ on Ω_1.*

Proof. The facts that ii) $\Longrightarrow$ iii) and iv) $\Longrightarrow$ i) are immediate consequences of Definitions 1.9 and 1.1 respectively. The fact that i) $\Longrightarrow$ ii) follows from Theorem 1.7. It remains only to prove that iii) $\Longrightarrow$ iv).

Lemma 1.14. *Let Ω be an open set in $\mathbb{C}^n$, let p be a point in $\partial\Omega$, let U be a connected neighbourhood of p and let D be a connected component of $U \cap \Omega$. Then $\partial D \cap (U \cap \partial\Omega) \neq \varnothing$.*

Proof. As D is a connected component of $U \cap \Omega$, D is open in $\mathbb{C}^n$ and closed in $U \cap \Omega$. As the open set U is connected and D is strictly contained in U, D is not closed in U. Consider a point q on the boundary of D in U. As D is closed in $U \cap \Omega$, the point q must be contained in $\partial\Omega$. □

Lemma 1.15. *Let Ω be an open set in $\mathbb{C}^n$ and let $(K_\nu)_{\nu\in\mathbb{N}}$ be an exhaustion of Ω by compact sets. There is then a sequence of natural numbers $(\nu_j)_{j\in\mathbb{N}}$ and a sequence $(p_j)_{j\in\mathbb{N}}$ of points in Ω such that the following hold:*

1) $p_j \in K_{\nu_{j+1}} \setminus K_{\nu_j}$ *for any $j \in \mathbb{N}$*
2) *for any $p \in \partial\Omega$ and any connected neighbourhood U of p every connected component D of $U \cap \Omega$ contains an infinite number of points in the sequence $(p_j)_{j\in\mathbb{N}}$.*

Proof. Let $(a_\nu)_{\nu\in\mathbb{N}}$ be a sequence running through the set of all points of Ω with rational coordinates, let r_ν be the distance from a_ν to $\partial\Omega$ and let B_ν be the ball $B(a_\nu, r_\nu) \subset \Omega$. Let $(Q_j)_{j\in\mathbb{N}}$ be a sequence whose elements are the balls B_ν and which contains each B_ν infinitely often: for example, the sequence $B_1, B_1, B_2, B_1, B_2, B_3, B_1, B_2, B_3, B_4, \dots$ will do.

Set $K_{\nu_0} = K_0$ and suppose we have already constructed points $p_1, \dots, p_{\ell-1}$ and compact sets $K_{\nu_1}, \dots, K_{\nu_\ell}$ satisfying 1) for $j = 1, \dots, \ell-1$. As Q_j is not contained in any compact subset of Ω because it is one of the balls B_ν there is a $p_\ell \in Q_\ell \setminus K_{\nu_\ell}$. We then choose $\nu_{\ell+1}$ such that $p_\ell \in K_{\nu_{\ell+1}}$ and 1) holds for $j = \ell$. Let us prove that 2) also holds for the sequences thus constructed. Consider a point $p \in \partial\Omega$, let U be a connected neighbourhood of p and let D be a connected component of $U \cap \Omega$. By Lemma 1.14, there is a $q \in \partial D \cap U \cap \partial\Omega$. As the sequence $(a_\nu)_{\nu\in\mathbb{N}}$ is dense in $\mathbb{C}^n$, there is a point $a_\mu \in D$ close enough to q that $B_\mu \subset D$. But as B_μ appears infinitely often in the sequence $(Q_j)_{j\in\mathbb{N}}$ and $p_j \in Q_j$ the set $U \cap D$ contains infinitely many points p_j. □

End of the proof of Theorem 1.13. Suppose that Ω is holomorphically convex. By Lemma 1.10 it has a exhaustion $(K_\nu)_{\nu\in\mathbb{N}}$ by $\mathcal{O}(\Omega)$-convex compact sets. We can then apply Lemma 1.11 to the sequences $(p_j)_{j\in\mathbb{N}}$ and $(K_{\nu_j})_{j\in\mathbb{N}}$ associated to the sequence $(K_\nu)_{\nu\in\mathbb{N}}$ by Lemma 1.15 and this proves that there is a function $f \in \mathcal{O}(\Omega)$ such that $\lim_{j\to\infty} |f(p_j)| = \infty$. Consider a point $p \in \partial\Omega$, let U be a connected neighbourhood of p, let D be a connected component of $U \cap \Omega$ and suppose there is a $h \in \mathcal{O}(U)$ such that $h = f$ on D. Let $U' \Subset U$ be a neighbourhood of p and let D' be the connected component of $U' \cap \Omega$ meeting D. Then $\sup_{D'} |h| \leqslant \sup_{U'} |h| < +\infty$ and f must therefore be bounded on D'. But this is impossible since by construction the open set D' contains an infinity of the points p_j and $\lim_{j\to\infty} |f(p_j)| = +\infty$. Condition iv) therefore holds for the function f. □

We end this section by giving some new examples of domains of holomorphy and some stability properties of domains of holomorphy under set operations.

Proposition 1.16. *Let Ω be a domain of holomorphy in $\mathbb{C}^n$ and let $f_1,\dots,f_N \in \mathcal{O}(\Omega)$ be holomorphic functions on Ω. The set*

$$\Omega_f = \{z \in \Omega \mid |f_j(z)| < 1,\ j = 1,\dots,N\}$$

is then a domain of holomorphy.

Proof. Let K be a compact set in Ω_f and choose a real number $r < 1$ such that $|f_j| \leqslant r$ on K for any $j = 1,\dots,N$. Of course, this inequality is still valid on $\widehat{K}_\Omega$ and $\widehat{K}_\Omega$ is therefore contained in Ω_f. But $\widehat{K}_\Omega$ is compact because Ω is a domain of holomorphy and contains $\widehat{K}_{\Omega_f}$. This implies that $\widehat{K}_{\Omega_f}$ is compact and proves the proposition. □

Remark. If $\Omega_f \Subset \Omega$ then Ω_f is a domain of holomorphy even if Ω is not because $\widehat{K}_{\Omega_f}$ is closed in Ω since $\widehat{K}_{\Omega_f} \subset \{z \in \Omega \mid |f_j(z)| \leqslant r\} \subset \Omega_f$.

Examples.

1) If $\varphi \in \mathcal{O}(\mathbb{C}^n)$ then $\Omega = \{z \in \mathbb{C}^n \mid -A < \operatorname{Re}\varphi(z) < A\}$ is a domain of holomorphy.
2) An open set $\Omega \Subset \mathbb{C}^n$ is said to be an analytic polyhedron if there is a neighbourhood U of $\overline{\Omega}$ and functions $f_1,\dots,f_N \in \mathcal{O}(U)$ such that $\Omega = \{z \in U \mid |f_j(z)| < 1,\ j = 1,\dots,N\}$. Any analytic polyhedron is a domain of holomorphy.

Proposition 1.17. *The interior of an arbitrary intersection of domains of holomorphy is a domain of holomorphy.*

Proof. Let $(\Omega_i)_{i\in I}$ be a family of domains of holomorphy and let Ω be the interior of $\bigcap_{i\in I} \Omega_i$. Assume that $\Omega \neq \varnothing$. If K is a compact set in Ω then

$0 < d = \mathrm{dist}(K, \partial\Omega) \leqslant \mathrm{dist}(K, \partial\Omega_i)$ for any $i \in I$. As the open sets Ω_i are domains of holomorphy it follows from Theorem 1.7 that $d \leqslant \mathrm{dist}(\widehat{K}_{\Omega_i}, \partial\Omega_i)$ for any $i \in I$. As $\widehat{K}_\Omega$ is contained in every set of the form $\widehat{K}_{\Omega_i}$ the distance from $\widehat{K}_\Omega$ to the boundary of the domains Ω_i is at least d, as is the distance from $\widehat{K}_\Omega$ to the boundary of Ω. The open set Ω is therefore holomorphically convex and by Theorem 1.13 it is a domain of holomorphy. □

Proposition 1.18. *Let $\Omega_1 \subset \mathbb{C}^m$ and $\Omega_2 \subset \mathbb{C}^n$ be two domains of holomorphy. The cartesian product $\Omega_1 \times \Omega_2$ is then a domain of holomorphy.*

Proof. It will be enough to prove that, for any compact set K in $\Omega_1 \times \Omega_2$ of the form $K_1 \times K_2$, where K_i is a compact set in Ω_i for $i = 1, 2$, the set $\widehat{K}_{\Omega_1\times\Omega_2}$ is relatively compact in $\Omega_1 \times \Omega_2$. If $f \in \mathcal{O}(\Omega_i)$ for $i = 1$ or 2 then f defines a holomorphic function on $\Omega_1 \times \Omega_2$ and it follows that

$$\widehat{K}_{\Omega_1\times\Omega_2} \subset (\widehat{K}_1)_{\Omega_1} \times \Omega_2 \cap \Omega_1 \times (\widehat{K}_2)_{\Omega_2}.$$

It follows that $\widehat{K}_{\Omega_1\times\Omega_2}$ is contained in $(\widehat{K}_1)_{\Omega_1} \times (\widehat{K}_2)_{\Omega_2}$, which is relatively compact in $\Omega_1 \times \Omega_2$ because the Ω_i are holomorphically convex. □

If an open set Ω in $\mathbb{C}^n$ is not a domain of holomorphy then it is natural to ask whether there is a largest element $E(\Omega)$ in the set of open sets D containing Ω such that every holomorphic function on Ω can be extended to a holomorphic function on D.

Examples.

1) Let Θ be a domain of holomorphy in $\mathbb{C}^n$ for some $n \geqslant 2$ and let K be a compact set in Θ such that $\Theta \setminus K$ is connected. Set $\Omega = \Theta \setminus K$. It follows from Hartogs' phenomenon and the definition of a domain of holomorphy that $E(\Omega)$ exists and $E(\Omega) = \Theta$.
2) Let φ be a holomorphic function on $\mathbb{C}^n$ for some $n \geqslant 2$. Let Θ be the domain of holomorphy $\Theta = \{z \in \mathbb{C}^n \mid \mathrm{Re}\,\varphi(z) < 0\}$. If K is a closed subset of Θ such that $K \Subset \mathbb{C}^n$ and $\Theta \setminus K$ is connected then we set $\Omega : \Theta \setminus K$. The extension theorem given in Section 6 of Chapter V and Hartogs' phenomenon then imply that $E(\Omega)$ exists and $E(\Omega) = \Theta$.

As the following example shows, $E(\Omega)$ does not exist in general if we restrict ourselves to the class of open sets in $\mathbb{C}^n$. When it exists the set $E(\Omega)$ is called the *envelope of holomorphy of Ω*.

Consider the subset γ in $\mathbb{C}$ defined by $\gamma = \{x \in \mathbb{R} \mid 1 \leqslant x \leqslant 2\} \cup \{z = 2e^{i\theta} \mid 0 \leqslant \theta \leqslant \pi\} \cup \{x \in \mathbb{R} \mid -2 \leqslant x \leqslant 0\}$. We denote by Γ the subset of $\mathbb{C}^2$ defined by $\Gamma = \gamma \times \{0\}$. Set $W_1 = \{z \in \mathbb{C}^2 \mid 0 \leqslant |z_1| < 1/2,\ 0 \leqslant |z_2| < 1\} \cup \{z \in \mathbb{C}^2 \mid 0 \leqslant |z_1| < 1,\ 1/2 < |z_2| < 1\}$. W_1 is a Hartogs figure and by Theorem 6.9 of Chapter I, any holomorphic function on W_1 can be extended to a holomorphic function on the bidisc $D(0,1) \times D(0,1)$. We set $W_2 = D(1, 1/4) \times D(0, 1/4)$,

$W_3 = \{z \in \mathbb{C}^2 \mid \mathrm{dist}(z, \Gamma) < 1/4\}$ and $\Omega = W_1 \cup W_2 \cup W_3$. The open set Ω in $\mathbb{C}^2$ is connected because each of the sets W_i, $i = 1, 2, 3$ is connected and $\mathbb{C} \times \{0\} \cap \Omega$ is connected. Consider a function $f \in \mathcal{O}(\Omega)$ which is a square root of $(z_1 - i)$ on W_1 (f exists and is unique because Ω is connected; it can be constructed by analytic continuation). If the envelope of holomorphy $E(\Omega)$ of Ω existed as an open set in $\mathbb{C}^2$ then $E(\Omega)$ would contain the open set $\widetilde{\Omega} = D(0,1) \times D(0,1) \cup W_2 \cup W_3$, which is impossible. Indeed, let F be an extension of f to $E(\Omega)$: the function $z_1 \mapsto F|_{\widetilde{\Omega}}(z_1, 0)$ would then be a holomorphic function on $(\mathbb{C} \times \{0\}) \cap \widetilde{\Omega}$, which is a ring enclosing $(i, 0)$, and would be equal to f on $(\mathbb{C} \times \{0\}) \cap \Omega$. This would mean that $\sqrt{z_1 - i}$ can be defined on a ring enclosing i, which is impossible.

2 Plurisubharmonic functions

In this section we will generalise subharmonic functions to several complex variables. These new functions will be used to define pseudoconvexity in Section 3.

A. Harmonic and subharmonic functions

We start by recalling the definition and most important properties of harmonic functions. For more details the reader may wish to consult Section 1.4 of [He/Le1] or Chapter I of [Kr].

The Laplace operator Δ is defined in $\mathbb{C}$ by

$$\Delta = \frac{\partial^2}{\partial x^2} + \frac{\partial^2}{\partial y^2} = 4\frac{\partial^2}{\partial z \partial \overline{z}}.$$

Definition 2.1. A $\mathcal{C}^2$ function u on a domain D in $\mathbb{C}$ is said to be *harmonic* if and only if $\Delta u = 0$ on D.

Properties.

1) A real-valued function u defined on an open set D in $\mathbb{C}$ is harmonic if and only if u is locally the real part of a holomorphic function. In particular, any harmonic function is $\mathcal{C}^\infty$ and even real analytic.
2) *Mean property.* If u is a harmonic function on $D \subset \mathbb{C}$ then

$$u(a) = \frac{1}{2\pi} \int_0^{2\pi} u(a + re^{i\theta}) d\theta$$

whenever $\{z \mid |z - a| \leqslant r\} \subset D$.
3) *Maximum principle.* Let u be a real-valued harmonic function on an open set D in $\mathbb{C}$. It is then the case that:

iii) If u has a local maximum at $a \in D$ then u is constant on some neighbourhood of a and u is therefore constant on the connected component of D containing a,

iii) If D is relatively compact in $\mathbb{C}$ and $u \in \mathcal{C}(\overline{D})$ then $u(z) \leqslant \max_{\partial D} u$ for any $z \in D$,

4) *Dirichlet problem.* Let Ω be a domain with $\mathcal{C}^1$ boundary in $\mathbb{C}$. If f is a continuous function on the boundary of Ω then there is a unique function $F \in \mathcal{C}(\overline{\Omega})$, harmonic on Ω, such that $F|_{\partial\Omega} = f$.

Definition 2.2. A function u defined on an open set D in $\mathbb{C}$ with values in $[-\infty, +\infty[$ is said to be *subharmonic* if and only if

i) u is upper semicontinuous (u.s.c.), i.e. $\{z \in D \mid u(z) < s\}$ is open for any $s \in \mathbb{R}$ or alternatively $\overline{\lim}_{z\to a} u(z) \leqslant u(a)$ for any $a \in D$.

ii) For any compact set $K \subset D$ and continuous function h on K, harmonic on $\overset{\circ}{K}$, such that $h \geqslant u$ on ∂K, we have $h \geqslant u$ on K.

Remarks.

- If $u \equiv -\infty$ then u is subharmonic.
- In $\mathbb{R}$ we have $\Delta = \partial^2/\partial x^2$, the harmonic functions are the linear functions and the subharmonic functions are therefore the convex functions.

Proposition 2.3. *Let D be an open set in $\mathbb{C}$.*

i) *If u is subharmonic on D then λu is subharmonic on D for any $\lambda > 0$.*

ii) *If $(u_\alpha)_{\alpha\in A}$ is a family of subharmonic functions on D such that $u = \sup_{\alpha\in A} u_\alpha$ is finite and upper semicontinuous then u is subharmonic on D.*

iii) *If $(u_n)_{n\in\mathbb{N}}$ is a decreasing family of subharmonic functions on D then $u = \lim_{n\to\infty} u_n$ is subharmonic.*

Proof. Claims i) and ii) follow immediately from Definition 2.2.

Let us prove iii). We note that if $u = \lim_{n\to\infty} u_n$ then $\{z \in D \mid u(z) < s\} = \bigcup_{n\in\mathbb{N}} \{z \in D \mid u_n(z) < s\}$ for any $s \in \mathbb{R}$. This set is therefore open and u is therefore upper semicontinuous. Let K be a compact set in D and let h be a continuous function on K, harmonic on $\overset{\circ}{K}$, such that $h \geqslant u$ on ∂K. For any $\varepsilon > 0$, we set $E_j = \{z \in \partial K \mid u_j(z) \geqslant h(z) + \varepsilon\}$ for any $j \in \mathbb{N}$. The sets E_j are closed in ∂K since the functions u_j are u.s.c., so they are compact sets. Moreover, the sequence $(E_j)_{j\in\mathbb{N}}$ is decreasing and their intersection is empty. There is therefore an integer $\ell \in \mathbb{N}$ such that $E_\ell = \varnothing$, which implies that $u_\ell \leqslant h + \varepsilon$ on ∂K. But $u_\ell \leqslant h + \varepsilon$ on K as the function u_ℓ is subharmonic, and it follows that $u \leqslant h + \varepsilon$ on K since the sequence $(u_n)_{n\in\mathbb{N}}$ is decreasing. As this holds for any $\varepsilon > 0$ we have $u \leqslant h$ on K. □

Corollary 2.4. *The function*

$$u(z) = -\log(\mathrm{dist}(z, \partial D))$$

is subharmonic on D for any open set $D \subset \mathbb{C}$.

Proof. If $D = \mathbb{C}$, $u \equiv -\infty$ and hence u is subharmonic. If $D \neq \mathbb{C}$ then u is continuous and $u(z) = \sup_{\zeta \in \partial D}(-\log|z-\zeta|)$. But $-\log|z-\zeta|$ is a harmonic function, since it is locally the real part of a choice of $-\log(z-\zeta)$, and it follows from ii) of Proposition 2.3 that u is subharmonic. □

We will give several different characterisations of subharmonicity. We recall that if μ is a positive Borel measure on a compact set K and $u : K \to \mathbb{R} \cup \{-\infty\}$ is an upper semicontinuous function then the integral $\int_K u d\mu$ is meaningful and is defined by

$$\int_K u d\mu = \inf\Big\{ \int_K \varphi d\mu \mid \varphi \in \mathcal{C}(K),\ \varphi \geqslant u \Big\}.$$

Moreover, $u \in L^1(K, \mu)$ if and only if $\int_K u d\mu > -\infty$.

Theorem 2.5. *Let D be an open set in $\mathbb{C}$ and let u be an upper semicontinuous function from D to $\mathbb{R} \cup \{-\infty\}$. The following are then equivalent:*

i) *u is subharmonic.*
ii) *For any disc $\Delta \Subset D$ and any holomorphic polynomial f such that $u \leqslant \operatorname{Re} f$ on $\partial\Delta$ we have $u \leqslant \operatorname{Re} f$ on Δ.*
iii) *For any $a \in D$ there is an r_a such that $0 < r_a < \operatorname{dist}(a, \partial D)$ and*

$$u(a) \leqslant \frac{1}{2\pi}\int_0^{2\pi} u(a + re^{i\theta})d\theta \quad \text{for any } r \in \,]0, r_a[. \tag{2.1}$$

Remark. Property iii) is both local and additive, so it follows that:

1) If u_1 and u_2 are subharmonic functions on D then $u_1 + u_2$ is subharmonic on D.
2) A function u defined on an open set D in $\mathbb{C}$ is subharmonic if and only if every point of D has a neighbourhood on which u is subharmonic.

Lemma 2.6. *If u is an upper semicontinuous function on an open set D in $\mathbb{C}$ such that iii) of Theorem 2.5 (the submean property) holds then u satisfies the maximum principle.*

Proof. We argue by contradiction. Assume that the submean property holds for u and u has a local maximum at $a \in D$, i.e. there is a $\rho > 0$ such that $u(z) \leqslant u(a)$ for all $z \in D$ such that $|z - a| < \rho$. If u is not constant in a neighbourhood of a then there is a $z_0 \in D$ such that $|z_0 - a| = r < \min(\rho, r_a)$ and $u(z_0) < u(a)$. Consider the set $\{\theta \in [0, 2\pi] \mid u(a+re^{i\theta}) < u(a)\}$: this is an open set in $[0, 2\pi]$ because u is u.s.c. and it is non-empty because $|z_0 - a| = r$. We deduce that

$$\int_0^{2\pi} u(a + re^{i\theta})d\theta < \int_0^{2\pi} u(a)d\theta = 2\pi u(a)$$

which contradicts the fact the u satisfies iii) of Theorem 2.5. □

Proof (of Theorem 2.5).

The fact that i) $\Longrightarrow$ ii) follows immediately from the definition of subharmonic functions.

Let us prove that ii) $\Longrightarrow$ iii). Consider a point $a \in D$ and let $r > 0$ be such that $\Delta = \{z \in \mathbb{C} \mid |z-a| < r\} \Subset D$. Consider a function $\varphi \in \mathcal{C}(\partial\Delta)$ such that $\varphi \geqslant u$ on $\partial\Delta$. We can assume that φ is in fact continuous on $\overline{\Delta}$ and harmonic on Δ by the solution to the Dirichlet problem. For any $\tau < 1$, we set $\varphi_\tau(z) = \varphi(a + \tau(z-a))$. The functions thus defined are harmonic on some neighbourhood of $\overline{\Delta}$ and the family $(\varphi_\tau)_{\tau<1}$ converges uniformly to φ on $\overline{\Delta}$ as τ tends to 1. Moreover, there are holomorphic functions f_τ defined on some neighbourhood of $\overline{\Delta}$ such that $\varphi_\tau = \operatorname{Re} f_\tau$ since $\overline{\Delta}$ is simply connected. For any $\varepsilon > 0$ there is therefore a holomorphic polynomial f such that

$$u \leqslant \varphi \leqslant \operatorname{Re} f \leqslant \varphi + \varepsilon$$

(we simply take the initial terms of the Taylor series of f_τ for some τ close enough to 1). It follows from ii) and the mean property for $\operatorname{Re} f$ that

$$\begin{aligned} u(a) \leqslant \operatorname{Re} f(a) &= \frac{1}{2\pi}\int_0^{2\pi} \operatorname{Re} f(a + re^{i\theta})d\theta \\ &\leqslant \frac{1}{2\pi}\int_0^{2\pi} \varphi(a + re^{i\theta})d\theta + \varepsilon. \end{aligned}$$

As the real number $\varepsilon > 0$ is arbitrary we get

$$u(a) \leqslant \frac{1}{2\pi}\int_0^{2\pi} \varphi(a + re^{i\theta})d\theta$$

for any $\varphi \in \mathcal{C}(\partial\Delta)$ such that $u \leqslant \varphi$ on $\partial\Delta$, and the result follows by definition of the integral of an u.s.c. function.

We complete the proof of the theorem by proving that iii) implies i). Let $K \subset D$ be a compact set and let $h \in \mathcal{C}(K)$ be a harmonic function on $\overset{\circ}{K}$ such that $u \leqslant h$ on ∂K. We will prove that if iii) holds then $u \leqslant h$ on K. Consider the function $u - h$: it follows from iii) and the mean property for h that $(u-h)$ satisfies iii) on $\overset{\circ}{K}$ and by Lemma 2.6, $u - h$ therefore satisfies the maximum principle. It follows that $(u-h)(z) \leqslant \sup_{\partial K}(u-h) \leqslant 0$ for any $z \in K$, or in other words, $u \leqslant h$ on K. □

Proposition 2.7. *Let D be an open set in $\mathbb{C}$. If $f \in \mathcal{O}(D)$ then $\log|f|$ is subharmonic on D.*

Proof. Consider a point $a \in D$. If $f(a) = 0$ then $\log|f|(a) = -\infty$ and iii) of Theorem 2.5 holds; if $f(a) \neq 0$ then there is a simply connected neighbourhood of a on which f does not vanish and $\log|f|$ is harmonic on this neighbourhood so iii) again holds. Since $\log|f|$ is clearly u.s.c. the proposition holds. □

Proposition 2.8. *Let φ be an increasing convex function on $\mathbb{R}$. We set $\varphi(-\infty) = \lim_{x\to-\infty}\varphi(x)$. If u is a subharmonic function on an open set D in $\mathbb{C}$ then $\varphi\circ u$ is subharmonic on D.*

Proof. Since the function φ is convex and increasing, for any $x_0 \in \mathbb{R}$ there is a real number k such that $\varphi(x) \geqslant \varphi(x_0) + k(x - x_0)$. This implies that, for any $z \in D$ and $r > 0$ such that $D(z,r) \Subset D$,

$$\frac{1}{2\pi}\int_0^{2\pi}\varphi(u(z+re^{i\theta}))d\theta \geqslant \varphi(x_0) + k\Big(\frac{1}{2\pi}\int_0^{2\pi}u(z+re^{i\theta})d\theta - x_0\Big)$$

Let us take $x_0 = \frac{1}{2\pi}\int_0^{2\pi}u(z+re^{i\theta})d\theta$. By the subharmonicity of u and the fact that φ is increasing we have

$$\varphi(u(z)) \leqslant \varphi\Big(\frac{1}{2\pi}\int_0^{2\pi}u(z+re^{i\theta})d\theta\Big) \leqslant \frac{1}{2\pi}\int_0^{2\pi}\varphi(u(z+re^{i\theta})d\theta$$

which proves the proposition since $\varphi\circ u$ is clearly u.s.c.. □

Examples. If $f \in \mathcal{O}(\Delta)$ then both $|f|$ and more generally $|f|^\alpha$ for any $\alpha > 0$ are subharmonic functions.

Proposition 2.9. *Let u be a subharmonic function on an open set D in $\mathbb{C}$. Assume that u is not equal to $-\infty$ on any connected component of D. The function u is then integrable on any compact subset of D and in particular $u > -\infty$ almost everywhere with respect to the Lebesgue measure.*

Proof. Let $z \in D$ be such that $u(z) > -\infty$ and let Δ be a closed disc of centre z contained in D. As the function u is u.s.c., it is bounded above on $\overline{\Delta}$ so

$$\int_\Delta u d\lambda < +\infty.$$

If δ denotes the radius of the disc Δ then the submean property implies that

$$\int_\Delta u d\lambda = \int_0^\delta\Big(\int_0^{2\pi}u(z+re^{i\theta})d\theta\Big)r\ dr \geqslant \pi\delta^2 u(z) > -\infty.$$

It follows that u is integrable on Δ. We denote by E the set of points z such that u is integrable in a neighbourhood of z. This is an open set and it follows from the above that $u = -\infty$ in a neighbourhood of any point in $D \setminus E$. Since u is u.s.c., $D \setminus E$ is also open. It follows that $D \setminus E$ is a union of connected components of D, which must be empty by hypothesis since $u = -\infty$ on $D \setminus E$. □

We end this section with a new characterisation of subharmonic functions.

Theorem 2.10. *Let u be a subharmonic function on an open set D in $\mathbb{C}$ which is not identically $-\infty$ on any connected component of D. For any positive $v \in \mathcal{D}(D)$ we then have*

$$\int u\Delta v \, d\lambda \geqslant 0, \tag{2.2}$$

where λ is the Lebesgue measure on $\mathbb{C}$.

Proof. Consider a real number $r \in \mathbb{R}$ such that $0 < r < \operatorname{dist}(\operatorname{supp} v, \complement D)$. For any $z \in \operatorname{supp} v$,

$$2\pi u(z) \leqslant \int_0^{2\pi} u(z + re^{i\theta})d\theta$$

since u is subharmonic. Multiply this equation by v and integrate with respect to λ, which is possible by Proposition 2.9; we then get

$$\int u(z)\Big(\int_0^{2\pi} v(z - re^{i\theta})d\theta - 2\pi v(z)\Big)d\lambda(z) \geqslant 0. \tag{2.3}$$

Taking the Taylor series to second order of v we get

$$\begin{aligned} v(z + h + ik) &= v(z) + h\frac{\partial v}{\partial x}(z) + k\frac{\partial v}{\partial y}(z) \\ &\quad + \frac{1}{2}\Big(h^2\frac{\partial^2 v}{\partial x^2}(z) + 2hk\frac{\partial^2 v}{\partial x \partial y}(z) + k^2\frac{\partial^2 v}{\partial y^2}(z)\Big) + O(|h+ik|^3). \end{aligned}$$

Replacing $h + ik$ by $-(r\cos\theta + ir\sin\theta)$ and integrating with respect to θ we get

$$\int_0^{2\pi} v(z - re^{i\theta})d\theta - 2\pi v(z) = \frac{\pi r^2}{2}\Delta v(z) + O(r^3).$$

Multiplying (2.3) by $2/\pi r^2$ and letting r tend to 0 we get (2.2). □

Note that if u is a $\mathcal{C}^2$ subharmonic function then Theorem 2.10 says that $\Delta u \geqslant 0$. We now prove the converse for $\mathcal{C}^2$ functions. This converse still holds for locally integrable functions in the following form (cf. [Ho2], Th. 1.6.11):

> Let $u \in \mathcal{L}^1_{\text{loc}}(D)$ be a function such that $\int u\Delta v d\lambda \geqslant 0$ for any positive $v \in \mathcal{D}(D)$. There is then a unique subharmonic function $\widetilde{u}$ on D which is equal to u almost everywhere.

We shall not prove this version of the theorem.

Theorem 2.11. *Let u be a $\mathcal{C}^2$ function on an open set D in $\mathbb{C}$ such that $\Delta u \geqslant 0$ on D. The function u is then subharmonic on D.*

Proof. Let K be a compact subset of D and let h be a continuous function on K, harmonic on $\mathring{K}$, such that $v = u - h \leqslant 0$ on ∂K.

Suppose first that $\Delta u > 0$ and argue by contradiction. If there is a $z \in K$ such that $v(z) > 0$ then v will have a maximum at some point $z_0 \in \overset{\circ}{K}$ and it follows that $\Delta v(z_0)$ must be negative or zero. Indeed, if $g(t) = v(x_0 + it)$, where $x_0 = \operatorname{Re} z_0$, then g has a maximum at $t = y_0 = \operatorname{Im} z_0$ and it follows that $\partial^2 v/\partial y^2(z_0) = \partial^2 g/\partial t^2(y_0) \leqslant 0$. We can show in a similar way that $\partial^2 v/\partial x^2(z_0) \leqslant 0$ and hence $\Delta v(z_0) \leqslant 0$. But this contradicts the fact that $\Delta v = \Delta u - \Delta h > 0$ since $\Delta h = 0$.

We return to the case where $\Delta u \geqslant 0$ and set $u_j(z) = u(z) + |z|^2/j$. Then $\Delta u_j = \Delta u + 4/j > 0$ for any $j \in \mathbb{N}^*$ and u_j is therefore subharmonic by the above. The sequence $(u_j)_{j\in\mathbb{N}}$ is a decreasing sequence of subharmonic functions which converges to u, so it follows from Proposition 2.3 iii) that u is subharmonic. □

We end this section with a lemma on the mean values of subharmonic functions which will be very useful in the rest of this section.

Lemma 2.12. *Let u be a subharmonic function on the disc $\{z \in \mathbb{C} \mid |z-a| < \rho\}$. The function*

$$A(u,r) = \frac{1}{2\pi}\int_0^{2\pi} u(a + re^{i\theta})d\theta$$

is then an increasing function of r on $]0,\rho[$.

Proof. We set $\Delta(r) = \{z \in \mathbb{C} \mid |z-a| < r\}$ and we consider two real numbers r_1 and r_2 such that $0 < r_1 < r_2 < \rho$. Consider a function $\varphi \in \mathcal{C}(\partial\Delta(r_2))$ such that $\varphi \geqslant u$ on $\partial\Delta(r_2)$. By the solution to the Dirichlet problem we can assume that φ is continuous on $\overline{\Delta(r_2)}$ and harmonic on $\Delta(r_2)$. The mean property for harmonic functions then says that $A(\varphi, r) = \varphi(a)$ for any $r \leqslant r_2$. Since u is subharmonic on $\Delta(\rho)$ we have $u \leqslant \varphi$ on $\Delta(r_2)$ and hence

$$A(u,r_1) \leqslant A(\varphi,r_1) = A(\varphi,r_2),$$

so finally we get

$$A(u,r_1) \leqslant \inf\{A(\varphi,r_2) \mid \varphi \in \mathcal{C}(\partial\Delta(r_2)),\ \varphi \geqslant u\} \leqslant A(u,r_2).$$ □

B. Plurisubharmonic functions

Let D be an open set in $\mathbb{C}^n$.

Definition 2.13. A function u from D to $\mathbb{R} \cup \{-\infty\}$ is said to be *plurisubharmonic* (psh) on D if u is upper semicontinuous and for any $a \in D$ and $w \in \mathbb{C}^n$ the function $\lambda \mapsto u(a + \lambda w)$ is subharmonic on the open set $\{\lambda \in \mathbb{C} \mid a + \lambda w \in D\}$.

We denote the set of plurisubharmonic functions on D by $\mathrm{PSH}(D)$.

A certain number of the properties of plurisubharmonic functions follow directly from the corresponding properties of subharmonic functions:

i) Let $(u_\alpha)_{\alpha\in A}$ be a family of plurisubharmonic functions on D. If $u = \sup_{\alpha\in A} u_\alpha$ is finite and upper semicontinuous on D then u is plurisubharmonic on D.
ii) If $(u_n)_{n\in\mathbb{N}}$ is a decreasing sequence of plurisubharmonic functions on D then $u = \lim_{n\to\infty} u_n$ is plurisubharmonic.
iii) The set $\mathrm{PSH}(D)$ is stable under addition and multiplication by positive constants.
iv) A function $u \in \mathrm{PSH}(D)$ if and only if u is plurisubharmonic in a neighbourhood of every point $a \in D$.
v) If $f \in \mathcal{O}(D)$ then $\log|f|$ is plurisubharmonic on D as is $|f|^\alpha$ for any $\alpha > 0$.

Remark. If D is an open set in $\mathbb{C}^n$ then the function $u(z) = -\log(\operatorname{dist}(z, \partial D))$ is not necessarily plurisubharmonic. Indeed, consider the open set $D = \mathbb{C}^2 \setminus \{0\}$ and points $a = (1,0) \in D$ and $w = (0,1)$. Then

$$u(a+\lambda w) = -\log(\operatorname{dist}(a+\lambda w, \partial D) = -\log\sqrt{1+|\lambda|^2}.$$

This function has an absolute maximum at the point $\lambda = 0$ and therefore cannot be subharmonic, which proves that u is not plurisubharmonic.

Theorem 2.14. *Let D be an open set in $\mathbb{C}^n$ and let u be a real-valued $\mathcal{C}^2$ function on D. The function $u \in \mathrm{PSH}(D)$ if and only if the complex Hessian of u at the point z*

$$L_z u(\zeta) = \sum_{j,k=1}^{n} \frac{\partial^2 u}{\partial z_j \partial \overline{z}_k}(z)\zeta_j\overline{\zeta}_k$$

is a positive semi-definite hermitian form on $\mathbb{C}^n$ for every $z \in D$.

Proof. A straightforward calculation proves that

$$\frac{\partial^2}{\partial\lambda\partial\overline{\lambda}} u(a+\lambda w) = L_{a+\lambda w}u(w).$$

The theorem then follows from Theorems 2.10 and 2.11 which characterise $\mathcal{C}^2$ subharmonic functions by the positivity of their Laplacian. □

Corollary 2.15. *Let D be an open set in $\mathbb{C}^n$, let D' be an open set in $\mathbb{C}^m$ and let F be a holomorphic map from D to D'. If $u \in \mathrm{PSH}(D') \cap \mathcal{C}^2(D')$ then $u \circ F \in \mathrm{PSH}(D)$.*

Proof. For any $a \in D$ and $w \in \mathbb{C}^n$ we have $L_a(u\circ F)(w) = L_{F(a)}u(F'(a)w)$. We then simply apply Theorem 2.14. □

Definition 2.16. Let D be an open set in $\mathbb{C}^n$ and let u be a $\mathcal{C}^2$ function on D. The *Levi form of* u at $z \in D$ is the complex Hessian $L_z u$ of u at z, i.e. the Hermitian form

$$\zeta \longmapsto L_z u(\zeta) = \sum_{j,k=1}^{n} \frac{\partial^2 u}{\partial z_j \partial \overline{z}_k}(z)\zeta_j\overline{\zeta}_k.$$

Definition 2.17. A function $u \in \mathrm{PSH}(D) \cap \mathcal{C}^2(D)$ is said to be *strictly plurisubharmonic* on D if and only if for any $z \in D$ the Levi form $L_z u$ of u at z is a positive definite Hermitian form.

We want to extend Corollary 2.15 to arbitrary plurisubharmonic functions. To do this we start by proving a regularisation theorem.

Lemma 2.18. *Let u be a plurisubharmonic function on an open set D in $\mathbb{C}^n$. Assume there is no connected component of D on which u is identically $-\infty$. The function u is then integrable on any compact set in D and in particular $u > -\infty$ almost everywhere.*

Proof. Repeat the proof of Proposition 2.9 replacing the disc Δ by a polydisc. □

Theorem 2.19. *Let D be an open set in $\mathbb{C}^n$: set $D_j = \{z \in D \mid |z| < j$ and $\operatorname{dist}(z, \partial D) > 1/j\}$. Let u be a plurisubharmonic function on D which is not identically $-\infty$ on any connected component of D. There is then a sequence $(u_j)_{j\in\mathbb{N}}$ of functions in $\mathcal{C}^\infty(D)$ such that*

i) *u_j is strictly plurisubharmonic on D_j.*
ii) *$u_j(z) \geqslant u_{j+1}(z)$ for any $z \in D_j$.*
iii) *$\lim_{j\to\infty} u_j(z) = u(z)$ for any $z \in D$.*
iv) *if u is continuous then the sequence $(u_j)_{j\in\mathbb{N}}$ converges to u uniformly on all compact sets in D.*

Proof. Let $\theta \in \mathcal{D}(\mathbb{R})$ be a positive $\mathcal{C}^\infty$ function supported on $[-1,1]$ such that $\int_{\mathbb{C}^n} \theta(|z|)d\lambda(z) = 1$. Since D_j is relatively compact on D the function u is integrable on D_j and we can consider the function v_j defined by

$$v_j(z) = \int_{D_j} u(\zeta)\theta\big(j|z-\zeta|\big)j^{2n}d\lambda(\zeta)$$

which is a $\mathcal{C}^\infty$ function on $\mathbb{C}^n$. We now prove that v_j has the submean property for every complex line segment contained in D_j, which implies that v_j is plurisubharmonic on D_j. Note that $v_j(z) = \int_{\mathbb{C}^n} u(z-\zeta/j)\theta(|\zeta|)d\lambda(\zeta)$ for any $z \in D_j$. For any $a \in D_j$ and $\omega \in \mathbb{C}^n$,

$$\begin{aligned}\frac{1}{2\pi}\int_0^{2\pi} v_j(a+re^{i\theta}\omega)d\theta &= \int_{\mathbb{C}^n}\Big[\frac{1}{2\pi}\int_0^{2\pi} u(a+re^{i\theta}\omega-\zeta/j)d\theta\Big]\theta(|\zeta|)d\lambda(\zeta)\\ &\geqslant \int_{\mathbb{C}^n} u\big(a-\zeta/j\big)\theta(|\zeta|)d\lambda(\zeta) = v_j(a),\end{aligned}$$

since u is plurisubharmonic. By Lemma 2.12 applied to the subharmonic function $\lambda \mapsto u(z-\lambda\zeta)$, the integral

$$\frac{1}{2\pi}\int_0^{2\pi} u\big(z-e^{it}\zeta/j\big)dt$$

is a decreasing function of j. Moreover,

$$v_j(z) = \int_{\mathbb{C}^n} u(z - \zeta/j)\theta(|\zeta|)d\lambda(\zeta) = \int_{\mathbb{C}^n} u(z - \zeta e^{it}/j)\theta(|\zeta|)d\lambda(\zeta)$$

for any $t \in \mathbb{R}$ and hence

$$v_j(z) = \int_{\mathbb{C}^n} \Big[\frac{1}{2\pi}\int_0^{2\pi} u(z - e^{it}\zeta/j)dt\Big]\theta(|\zeta|)d\lambda(\zeta).$$

It follows that the sequence $(v_j)_{j\in\mathbb{N}}$ is decreasing. Applying the submean property to the subharmonic function $\lambda \mapsto u(z - \lambda\zeta)$, we get

$$v_j(z) \geqslant \int_{\mathbb{C}^n} u(z)\theta(|\zeta|)d\lambda(\zeta) = u(z).$$

Since u is upper semicontinuous for any $\varepsilon > 0$ there is a $\delta > 0$ such that $B(z,\delta) \subset \{\zeta \in D \mid u(\zeta) < u(z) + \varepsilon\}$. It follows that, for any $j > 1/\delta$,

$$u(z - \zeta/j) < u(z) + \varepsilon \quad \text{for } |\zeta| \leqslant 1$$

whence

$$v_j(z) = \int_{\mathbb{C}^n} u(z - \zeta/j)\theta(|\zeta|)d\lambda(\zeta) < u(z) + \varepsilon \quad \text{because} \int_{\mathbb{C}^n} \theta(|\zeta|)d\lambda(\zeta) = 1.$$

For any $j > 1/\delta$ we therefore have $u(z) < v_j(z) < u(z) + \varepsilon$.

If moreover u is continuous the properties of regularising operators imply that $(v_j)_{j\in\mathbb{N}}$ converges uniformly to u on all compact sets.

To complete the proof we simply set $u_j(z) = v_j(z) + |z|^2/j$. The sequence $(u_j)_{j\in\mathbb{N}}$ then satisfies conditions i)-iv). □

Theorem 2.20. *Let D be an open set in $\mathbb{C}^n$, let D' be an open set in $\mathbb{C}^m$ and let F be a holomorphic map from D to D'. If $u \in \mathrm{PSH}(D')$ then $u \circ F \in \mathrm{PSH}(D)$.*

Proof. We can assume without loss of generality that Ω is connected and that u is not the constant map $-\infty$. Let $(u_j)_{j\in\mathbb{N}}$ be a sequence of functions converging to u which satisfies the conclusions of Theorem 2.19. If Ω is relatively compact in D then for large enough j the functions $u_j \circ F$ are plurisubharmonic on Ω by Corollary 2.15. The sequence $(u_j \circ F)_{j\in\mathbb{N}}$ is then a decreasing sequence of functions which converges to $u \circ F$, so $u \circ F$ is plurisubharmonic on D. □

Remark. It follows from the above theorem that if F is a biholomorphic map from D to D' and u is a function on D' then $u \in \mathrm{PSH}(D')$ if and only if $u \circ F \in \mathrm{PSH}(D)$. This implies that it is possible to define plurisubharmonic functions on complex analytic manifolds.

It is clear that any strictly convex $\mathcal{C}^2$ function u on D is strictly plurisubharmonic. We now prove that if $du \neq 0$ then the converse is true up to holomorphic change of local coordinates.

Definition 2.21. Let D be an open set in $\mathbb{C}^n$ and let u be a real-valued $\mathcal{C}^2$ function on D. For any $z, \zeta \in D$ we set

$$\widehat{F}_u(z,\zeta) = -\left[2\sum_{j=1}^{n} \frac{\partial u}{\partial \zeta_j}(\zeta)(z_j - \zeta_j) + \sum_{j,k=1}^{n} \frac{\partial^2 u}{\partial \zeta_j \partial \zeta_k}(z_j - \zeta_j)(z_k - \zeta_k)\right].$$

The function $\widehat{F}_u(z,\zeta)$ is called the *Levi polynomial* of u.

The following lemma links the Levi form and Levi polynomial of u to the second-order Taylor series of u.

Lemma 2.22. *Let D be an open set in $\mathbb{C}^n$ and let u be a real-valued $\mathcal{C}^2$ function on D. For any $\zeta \in D$ and any z close enough to ζ,*

$$u(z) = u(\zeta) - \operatorname{Re}\widehat{F}_u(z,\zeta) + L_\zeta u(z-\zeta) + o(|\zeta - z|^2).$$

Proof. Let $(x_j(\zeta))_{1\leqslant j\leqslant 2n}$ be the real coordinates of $\zeta \in \mathbb{C}^n$ such that $\zeta_j = x_j(\zeta) + ix_{j+n}(\zeta)$. A straightforward calculation proves that

$$\sum_{j=1}^{2n} \frac{\partial u}{\partial x_j}(\zeta)(x_j(z) - x_j(\zeta)) = 2\operatorname{Re}\left[\sum_{j=1}^{n} \frac{\partial u}{\partial \zeta_j}(z_j - \zeta_j)\right]$$

and

$$\begin{aligned}
&\frac{1}{2}\sum_{j,k=1}^{2n} \frac{\partial^2 u}{\partial x_j \partial x_k}\big(x_j(z) - x_j(\zeta)\big)\big(x_k(z) - x_k(\zeta)\big)\\
&= \sum_{j,k=1}^{n} \frac{\partial^2 u}{\partial \zeta_j \partial \overline{\zeta}_k}(\zeta)(z_j - \zeta_j)(\overline{z}_k - \overline{\zeta}_k) + \operatorname{Re}\left[\sum_{j,k=1}^{n} \frac{\partial^2 u}{\partial \zeta_j \partial \zeta_k}(\zeta)(z_j - \zeta_j)(z_k - \zeta_k)\right].
\end{aligned}$$

The lemma then follows from the definitions of $\widehat{F}_u$ and $L_\zeta u$ on taking the Taylor series of u at ζ to order 2. □

Theorem 2.23. *Let u be a strictly plurisubharmonic $\mathcal{C}^2$ function on a neighbourhood of $0 \in \mathbb{C}^n$. If $du(0) \neq 0$ then there is a biholomorphic map h from a neighbourhood U of 0 in $\mathbb{C}^n$ to a neighbourhood W of 0 in $\mathbb{C}^n$ such that $u \circ h^{-1}$ is strictly convex on W, i.e.*

$$\sum_{j,k=1}^{2n} \frac{\partial^2 u \circ h^{-1}}{\partial x_j \partial x_k}(\zeta) t_j t_k > 0 \quad \text{for any } \zeta \in W \text{ and } t \in \mathbb{R}^{2n} \setminus \{0\},$$

where $x_j = x_j(\zeta)$ are the real coordinates of ζ given by $\zeta_j = x_j(\zeta) + ix_{j+n}(\zeta)$.

Proof. Since $du(0) \neq 0$ and u is real-valued, $\partial u(0) \neq 0$. Assume that $\partial u/\partial \zeta_1(0) \neq 0$. The map $h(z) := (\widehat{F}_u(z,0), z_2, \ldots, z_n)$ is then a biholomorphism from U to a neighbourhood V of $0 = h(0)$. We set

$$f(\zeta) = (f_1(\zeta), \ldots, f_n(\zeta)) = h^{-1}(\zeta)$$

for any $\zeta \in V$. For any ζ close enough to 0 we then have

$$u \circ f(\zeta) = u(0) - \operatorname{Re} \widehat{F}_u(f(\zeta), 0) + L_0 u(f(\zeta)) + o(|f(\zeta)|^2)$$

by Lemma 2.22. But $\widehat{F}_u(f(\zeta), 0) = \zeta_1$ by definition of $f = h^{-1}$, which implies that

$$u \circ f(\zeta) = u(0) - \operatorname{Re} \zeta_1 + L_0 u(f(\zeta)) + o(|f(\zeta)|^2)$$

as ζ tends of 0. Let $F = (F_1, \ldots, F_n)$ be the differential of f at the origin, i.e. the linear map $F : \mathbb{C}^n \to \mathbb{C}^n$ such that $f(\zeta) = F(\zeta) + O(|\zeta|^2)$ as ζ tends to 0. Then

$$u \circ f(\zeta) = u(0) - \operatorname{Re} \zeta_1 + L_0 u(F(\zeta)) + o(|\zeta|^2)$$

as ζ tends to 0, which implies that

$$\sum_{j,k=1}^{2n} \frac{\partial^2 u \circ f}{\partial x_j \partial x_k}(0) x_j(\zeta) x_k(\zeta) = L_0 u(F(\zeta)) \quad \text{for any } \zeta \in \mathbb{C}^n$$

by the uniqueness of the Taylor series of u at 0. As $F(\zeta) \neq 0$ for any $\zeta \neq 0$ since f is biholomorphic at 0 and u is strictly plurisubharmonic in a neighbourhood of 0 we get

$$\sum_{j,k=1}^{2n} \frac{\partial^2 u \circ f}{\partial x_j \partial x_k}(0) t_j t_k > 0 \quad \text{for any } t \in \mathbb{R}^{2n} \setminus \{0\}.$$

As the function u is $\mathcal{C}^2$ this remains true if we replace 0 by a point ζ in some neighbourhood W of 0. □

We end this section with a lemma proving that strictly plurisubharmonic functions are stable under small perturbations.

Lemma 2.24. *Let ρ be a strictly plurisubharmonic $\mathcal{C}^2$ function defined in a neighbourhood of a compact set K in $\mathbb{C}^n$. There is then a real number $\varepsilon > 0$ such that, for any $\mathcal{C}^2$ function φ defined in a neighbourhood of K such that $\left|\frac{\partial^2 \varphi(z)}{\partial z_j \partial \overline{z}_k}\right| < \varepsilon$ for any $z \in K$ and $1 \leqslant j, k \leqslant n$, the function $\rho + \varphi$ is again strictly plurisubharmonic in a neighbourhood of K.*

Proof. By definition of strictly plurisubharmonic functions we simply take

$$\varepsilon = \frac{1}{n^2} \min_{\substack{z \in K, \omega \in \mathbb{C}^n \\ |\omega| = 1}} \sum_{j,k=1}^{n} \frac{\partial^2 \rho}{\partial z_j \partial \overline{z}_k}(z) \omega_j \overline{\omega}_k. \qquad \square$$

3 Pseudoconvexity

We saw in the previous section that if Ω is an arbitrary open set in $\mathbb{C}^n$ then the function $-\log(\mathrm{dist}(z, \partial\Omega))$ is not necessarily plurisubharmonic. In this section we will study a new class of open sets in $\mathbb{C}^n$, pseudoconvex open sets, which are characterised by the fact that the function $-\log$(the distance to the boundary) is plurisubharmonic. We will prove that this class contains the class of domains of holomorphy. (Indeed, we will see in Chapter VII that these two classes are the same.)

Theorem 3.1. *If Ω is a domain of holomorphy in $\mathbb{C}^n$ then*

$$-\log(\mathrm{dist}(z, \partial\Omega))$$

is a continuous plurisubharmonic function.

Proof. The function $u(z) = -\log(\mathrm{dist}(z, \partial\Omega)$ is continuous on Ω. To prove it is also plurisubharmonic we need the following lemma

Lemma 3.2. *Let Ω be a domain of holomorphy in $\mathbb{C}^n$ and let K be a compact set in Ω. If $f \in \mathcal{O}(\Omega)$ is a function which has the property that $|f(z)| \leqslant \mathrm{dist}(z, \partial\Omega)$ for any $z \in K$ then $|f(z)| \leqslant \mathrm{dist}(z, \partial\Omega)$, for any $z \in \widehat{K}_\Omega$.*

Proof. If $|f(z)| \leqslant \mathrm{dist}(z, \partial\Omega)$ then $|f(z)| \leqslant \delta^r_\Omega(z)$ for any multiradius $r > 0$ such that $\Sigma r_j^2 = 1$ (cf. Section 1 for the definition of $\delta^r_\Omega(z)$). If $t \in]0, 1[$ then the set

$$D = \{w \in \mathbb{C}^n \mid |w_j - z_j| \leqslant t r_j |f(z)|, \, j = 1, \dots, n, \, z \in K\}$$

is a compact set in Ω. For any $u \in \mathcal{O}(\Omega)$ there is then a constant M such that $|u(w)| \leqslant M$ for any $w \in D$, and by the Cauchy inequalities it follows that

$$|D^\alpha u(z)| t^{|\alpha|} r^\alpha |f(z)|^{|\alpha|} \frac{1}{\alpha!} \leqslant M, \quad \text{for any } z \in K. \tag{3.1}$$

Since the function $f^{|\alpha|} D^\alpha u$ is holomorphic on Ω, (3.1) still holds if $z \in \widehat{K}_\Omega$. We have therefore proved that the Taylor series of u at any point $\zeta \in \widehat{K}_\Omega$ converges on the polydisc $\zeta + |f(\zeta)| P(0, r)$. As Ω is a domain of holomorphy this polydisc must be contained in Ω, which implies that $|f(\zeta)| \leqslant \mathrm{dist}(\zeta, \partial\Omega)$ for any $\zeta \in \widehat{K}_\Omega$. □

End of the proof of Theorem 3.1. Fix points $z_0 \in \Omega$ and $w \in \mathbb{C}^n \setminus \{0\}$ and choose $r > 0$ small enough that $D = \{z_0 + \tau w \mid \tau \in \mathbb{C}, |\tau| \leqslant r\}$ is contained in Ω. We set $bD = \{z_0 + \tau w \mid \tau \in \mathbb{C}, |\tau| = r\}$. Let f be a holomorphic polynomial such that

$$-\log\big(\mathrm{dist}(z_0 + \tau w, \partial\Omega)\big) \leqslant \mathrm{Re}\, f(\tau) \quad \text{whenever } |\tau| = r. \tag{3.2}$$

Consider a holomorphic polynomial F on $\mathbb{C}^n$ such that $F(z_0+\tau w)=f(\tau)$; condition (3.2) can then be written as

$$|e^{-F(z)}| \leqslant \operatorname{dist}(z,\partial\Omega) \quad \text{for any } z\in bD.$$

By the maximum principle the holomorphically convex hull of bD relative to Ω contains D. It then follows from Lemma 3.2 that

$$-\log\big(\operatorname{dist}(z_0+\tau w,\partial\Omega)\big) \leqslant \operatorname{Re} f(\tau) \quad \text{whenever } |\tau|\leqslant r.$$

Theorem 2.5 then implies that the function

$$\tau \longmapsto -\log(\operatorname{dist}(z_0+\tau w,\partial\Omega))$$

is subharmonic on the open set of $\mathbb{C}$ on which it is defined. It follows that the function $-\log(\operatorname{dist}(z,\partial\Omega))$ is plurisubharmonic on Ω. □

We now give some other conditions which are equivalent to the hypothesis of Theorem 3.1.

Definition 3.3. Let Ω be an open set in $\mathbb{C}^n$. For any compact subset K of Ω we define the *psh-convex hull* of K relative to Ω by

$$\widehat{K}^p_\Omega=\{z\in\Omega \mid u(z)\leqslant \sup_K u, \forall\, u\in \mathrm{PSH}(\Omega)\}.$$

Remark. As $f\in\mathcal{O}(\Omega)$ implies that $|f|\in\mathrm{PSH}(\Omega)$ it is clear that $\widehat{K}^p_\Omega\subset\widehat{K}_\Omega$.

An *analytic disc* in $\mathbb{C}^n$ is a non-constant holomorphic map $\varphi:\Delta\to\mathbb{C}^n$ where Δ is the unit disc in $\mathbb{C}$. If φ can be extended continuously to $\overline{\Delta}$ then we will say that $\varphi(\overline{\Delta})$ is a *closed analytic disc* and that $\varphi(\partial\Delta)$ is the *boundary* of the disc.

Theorem 3.4. *Let Ω be an open set in $\mathbb{C}^n$. The following conditions are then equivalent:*

i) $-\log(\operatorname{dist}(z,\partial\Omega))$ *is continuous and plurisubharmonic on Ω.*
ii) *There is a continuous plurisubharmonic function u on Ω such that, for any $c\in\mathbb{R}$,*
$$\Omega_c=\{z\in\Omega \mid u(z)<c\}\Subset\Omega.$$
iii) *For any compact set K in Ω we have $\widehat{K}^p_\Omega\Subset\Omega$.*
iv) *Let $(\delta_\alpha)_{\alpha\in A}$ be a family of analytic discs contained in Ω. If $\bigcup_{\alpha\in A} b\delta_\alpha\Subset\Omega$ then $\bigcup_{\alpha\in A}\delta_\alpha\Subset\Omega$. (This result is called the "Kontinuitätssatz".)*

Proof. We prove first that i) implies ii). We set $u(z)=|z|^2-\log(\operatorname{dist}(z,\partial\Omega))$: the function u is then continuous and plurisubharmonic on Ω by i) and it is clear that it satisfies ii).

Suppose that ii) holds. Let K be a compact set in D and let u be the function whose existence is guaranteed by ii). Set $c = \max_K u$. Then $u \leqslant c$ on $\widehat{K}^p_\Omega$ by definition of the psh-convex hull and hence

$$\widehat{K}^p_\Omega \subset \{z \in \Omega \mid u(z) \leqslant c\} \Subset \Omega.$$

Let us now prove that iii) implies iv). Let δ be a closed analytic disc in Ω and let $u \in \mathrm{PSH}(\Omega)$ be a plurisubharmonic function on Ω. Consider a parameterisation $\varphi : \overline{\Delta} \to \Omega$ of δ. The function $u \circ \varphi$ is then subharmonic and it follows that, for any $z \in \overline{\Delta}$,

$$u \circ \varphi(z) \leqslant \sup_{\zeta \in \partial\Delta} u \circ \varphi(\zeta).$$

It follows that, for any $p \in \delta$,

$$u(p) \leqslant \sup_{\xi \in \partial\delta} u(\xi),$$

which implies that $\delta \subset \widehat{b\delta}^p_\Omega$. It follows that if $(\delta_\alpha)_{\alpha\in A}$ is a family of closed analytic discs in Ω then $\bigcup_{\alpha\in A} \delta_\alpha \subset \widehat{\left(\bigcup_{\alpha\in A} b\delta_\alpha\right)}^p_\Omega$, which proves that iii) implies iv).

We end by proving that iv) implies that the function $-\log(\mathrm{dist}(z, \partial\Omega))$ is plurisubharmonic. As in the proof of Theorem 3.1, fix points $z_0 \in \Omega$ and $w \in \mathbb{C}^n \setminus \{0\}$ and choose $r > 0$ small enough that $D = \{z_0 + \tau w \mid \tau \in \mathbb{C}, |\tau| \leqslant r\}$ is contained in Ω. Let f be a holomorphic polynomial such that

$$-\log\big(\mathrm{dist}(z_0 + \tau w, \partial\Omega)\big) \leqslant \operatorname{Re} f(\tau) \quad \text{whenever } |\tau| = r,$$

or in other words

$$|e^{-f(\tau)}| \leqslant \mathrm{dist}(z_0 + \tau w, \partial\Omega) \quad \text{whenever } |\tau| = r. \tag{3.3}$$

We want to prove that this inequality still holds if $|\tau| \leqslant r$, which will prove the subharmonicity of $\tau \mapsto -\log(\mathrm{dist}(z_0 + \tau w, \partial\Omega))$, which in turn will establish i). Consider a point $a \in \mathbb{C}^n$ such that $|a| < 1$ and consider the map $\tau \mapsto z_0 + \tau w + a e^{-f(\tau)}$ defined whenever $|\tau| \leqslant r$. We denote the image of this map by D_a. The sets D_a are closed analytic discs in $\mathbb{C}^n$; moreover, $bD_a \Subset \Omega$ so $D_a \Subset \Omega$ for any a such that $|a| < 1$ by iv) applied to the family consisting of the discs D_a. Therefore,

$$z_0 + \tau w + a e^{-f(\tau)} \in \Omega \quad \text{whenever } |a| < 1 \text{ and } |\tau| \leqslant r.$$

Letting $|a|$ tend to 1 we get

$$|e^{-f(\tau)}| \leqslant \mathrm{dist}(z_0 + \tau w, \partial\Omega) \quad \text{whenever } |\tau| \leqslant r,$$

or in other words $-\log(\mathrm{dist}(z_0 + \tau w, \partial\Omega)) \leqslant \operatorname{Re} f(\tau)$ whenever $|\tau| \leqslant r$. □

Definition 3.5. An open set Ω in $\mathbb{C}^n$ is said to be *pseudoconvex* if one of the equivalent conditions of Theorem 3.4 holds.

Definition 3.6. A continuous real-valued function φ defined on an open set D in $\mathbb{C}^n$ is an *exhaustion function for* D if for any $c \in \mathbb{R}$ the set $D_c = \{z \in D \mid \varphi(z) < c\}$ is relatively compact in D.

Remarks.

1) An exhaustion function φ has the property that $\varphi(z) \to \infty$ as z approaches the boundary of D.
2) A domain D is therefore pseudoconvex if and only if it has a continuous plurisubharmonic exhaustion function.

Corollary 3.7. *If Ω is a domain of holomorphy in $\mathbb{C}^n$ then Ω is pseudoconvex.*

Proof. This follows immediately from Theorem 3.1 and the definition of pseudoconvexity. □

Remarks.

1) The converse of Corollary 3.7 holds and will be proved in Chapter VII.
2) If Ω is a holomorphically convex set in $\mathbb{C}^n$ then it is clearly pseudoconvex since for any compact set K in Ω we have $\widehat{K}^{P}_{\Omega} \subset \widehat{K}_{\Omega}$. By this method we can reprove Corollary 3.7 using Theorem 1.13 instead of Theorem 3.1.

Theorem 3.8. *Let $(\Omega_\alpha)_{\alpha \in A}$ be a family of open pseudoconvex sets in $\mathbb{C}^n$. The interior Ω of the intersection of the sets Ω_α is then also pseudoconvex.*

Proof. This follows from i) of Theorem 3.4 and the fact that the upper bound of a family of plurisubharmonic functions is plurisubharmonic if it is continuous. In our case,

$$-\log(\operatorname{dist}(z, \partial\Omega)) = \sup_{\alpha \in A} -\log(\operatorname{dist}(z, \partial\Omega_\alpha))$$

is indeed continuous. □

We will now prove that pseudoconvexity is in fact a local property of the boundary.

Theorem 3.9. *An open set Ω in $\mathbb{C}^n$ is pseudoconvex if and only if every point $\xi \in \overline{\Omega}$ has a neighbourhood U_ξ such that $U_\xi \cap \Omega$ is pseudoconvex.*

Proof. Necessity. Simply take U_ξ to be any convex neighbourhood of ξ since $U_\xi \cap \Omega$ is then pseudoconvex because it is the intersection of two pseudoconvex domains.

Sufficiency. We start by considering the case where Ω is bounded. Consider a point $\xi \in \partial\Omega$ and let U_ξ be a neighbourhood of ξ such that $U_\xi \cap \Omega$

is pseudoconvex. The function $-\log(\operatorname{dist}(z,\partial(U_\xi\cap\Omega)))$ is then plurisubharmonic on $U_\xi\cap\Omega$. As $\operatorname{dist}(z,\partial(U_\xi\cap\Omega))=\operatorname{dist}(z,\partial\Omega)$ for any z close enough to ξ, there is a neighbourhood U of $\partial\Omega$ such that the function $-\log(\operatorname{dist}(z,\partial\Omega))$ is plurisubharmonic on $U\cap\Omega$. As the open set Ω is bounded, $\Omega\setminus U$ is compact and the number $m=\sup_{\Omega\setminus U}(-\log(\operatorname{dist}(z,\partial\Omega))$ is finite. It follows that

$$\varphi(z)=\max(-\log(\operatorname{dist}(z,\partial\Omega)),|z|^2+m+1)$$

is a plurisubharmonic exhaustion function for Ω which implies that Ω is pseudoconvex.

If Ω is not bounded then we apply the above argument to the open set $\Omega_\nu=\Omega\cap B(0,\nu)$ which is pseudoconvex because it is a bounded set satisfying the conditions of the theorem. Since Ω is the increasing union of the sets Ω_ν, the sequence of functions $(-\log(\operatorname{dist}(z,\partial\Omega_\nu)))$ is a decreasing sequence of plurisubharmonic functions which converges to the function $-\log(\operatorname{dist}(z,\partial\Omega))$. This function is therefore plurisubharmonic. It follows that Ω is pseudoconvex. □

Theorem 3.10. *Let Ω be a pseudoconvex open set in $\mathbb{C}^n$ and let K be a compact set in Ω. For any neighbourhood U of $\widehat{K}^p_\Omega$ in Ω there is a function ρ such that*

i) *ρ is strictly plurisubharmonic and $\mathcal{C}^\infty$ on Ω.*
ii) *$\rho<0$ on K and $\rho>0$ on $\Omega\setminus U$.*
iii) *For any $c\in\mathbb{R}$, $\{z\in\Omega \mid \rho(z)<c\}\Subset\Omega$.*

Proof. The key to the proof of this theorem is the construction of a continuous plurisubharmonic function φ for which ii) and iii) hold. As the open set Ω is pseudoconvex, it has a continuous plurisubharmonic exhaustion function ψ. After adding a constant to ψ if necessary, we can suppose that $\psi<0$ on K. Consider the set $K'=\{z\in\Omega \mid \psi(z)\leqslant 0\}$; the set K' is then a compact set in Ω because ψ is an exhaustion function. Since U is a neighbourhood of $\widehat{K}^p_\Omega$ for any $z\in K'\cap(\Omega\setminus U)$ there is a function $\varphi_z\in\mathrm{PSH}(\Omega)$ such that $\varphi_z(z)>0$ and $\varphi_z<0$ on K. Moreover, the regularisation theorem 2.19 says that we can assume φ_z is continuous. As the function φ_z is continuous it is strictly positive on some neighbourhood of z and by the compactness of K' we can therefore find a finite number of strictly plurisubharmonic continuous functions $\varphi_1,\dots,\varphi_N$ on Ω such that

$$\max(\varphi_1,\dots,\varphi_N)>0 \quad \text{on } K'\cap(\Omega\setminus U)$$

and

$$\max(\varphi_1,\dots,\varphi_N)<0 \quad \text{on } K.$$

The function $\varphi=\max(\psi,\varphi_1,\dots,\varphi_N)$ is then plurisubharmonic and continuous on Ω and satisfies ii) and iii).

To get a $\mathcal{C}^\infty$ plurisubharmonic function ρ satisfying ii) and iii) we simply apply the following lemma.

Lemma 3.11. *Let u be a continuous plurisubharmonic exhaustion function on a domain $\Omega \subset \mathbb{C}^n$. For any compact set K in Ω and any real number $\varepsilon > 0$ there is a $\mathcal{C}^\infty$ strictly plurisubharmonic exhaustion function ρ on Ω such that*

$$u \leqslant \rho \text{ on } \Omega \text{ and } |\rho(z) - u(z)| < \varepsilon \text{ for any } z \in K.$$

Proof. For any $j \in \mathbb{N}$ we set $\Omega_j = \{z \in \Omega \mid u(z) < j\}$. Then $\Omega_j \Subset \Omega$ and, adding a constant to u if necessary, we can suppose that $K \subset \Omega_0$. Fix an $\varepsilon > 0$. By Theorem 2.19 there is a sequence $(u_j)_{j\in\mathbb{N}}$ of $\mathcal{C}^\infty$ functions on Ω such that the functions u_j are strictly plurisubharmonic on Ω_{j+2}, $u(z) < u_0(z) < u(z) + \varepsilon$ for any $z \in \overline{\Omega}_1$ and $u(z) < u_j(z) < u(z) + 1$ for any $z \in \Omega_j$, $j \geqslant 1$. It follows that $u_j - j + 1 < 0$ on Ω_{j-2} and $u_j - j + 1 > 0$ on $\overline{\Omega}_j \setminus \Omega_{j-1}$ for any $j \geqslant 2$. Let χ be a $\mathcal{C}^\infty$ function on $\mathbb{R}$ such that $\chi(t) = 0$ for any $t \leqslant 0$ and $\chi(t), \chi'(t)$ and $\chi''(t)$ are strictly positive for any $t > 0$. Then $\chi \circ (u_j - j + 1) \equiv 0$ on Ω_{j-2} and $\chi \circ (u_j - j + 1) \geqslant 0$ on $\Omega \setminus \Omega_{j-2}$. Calculating the Levi form we see that $\chi \circ (u_j - j + 1)$ is plurisubharmonic on Ω_{j+2} and strictly plurisubharmonic and positive on $\overline{\Omega}_j \setminus \Omega_{j-1}$. We then recursively choose integers m_j such that for any $\ell \geqslant 2$, $\rho_\ell = u_0 + \sum_{j=2}^{\ell} m_j \chi \circ (u_j - j + 1)$ is strictly plurisubharmonic on Ω_ℓ. We have therefore constructed a sequence of functions $(\rho_\ell)_{\ell \geqslant 2}$ such that $\rho_\ell = u_0$ on Ω_0, $\rho_\ell \geqslant u$ and $\rho_\ell = \rho_{\ell-1}$ on $\Omega_{\ell-2}$. The function $\rho = \lim_{\ell\to\infty} \rho_\ell$ then has the required properties. □

Remark. It follows from Lemma 3.11 and the definition of pseudoconvex open sets that an open set Ω in $\mathbb{C}^n$ is pseudoconvex if and only if it has a $\mathcal{C}^2$ strictly plurisubharmonic exhaustion function. By the following Morse lemma every open pseudoconvex set Ω in $\mathbb{C}^n$ has a $\mathcal{C}^2$ strictly pseudoconvex exhaustion function ρ such that the set of its critical points, i.e. $\{z \in \Omega \mid d\rho(z) = 0\}$, is discrete in Ω.

Lemma 3.12 (Morse lemma). *Let Ω be an open set in $\mathbb{C}^n$ and let ρ be a $\mathcal{C}^2$ strictly plurisubharmonic function on Ω. For any $\varepsilon > 0$ there is then an $\mathbb{R}$-linear form $L : \mathbb{C}^n \to \mathbb{R}$ such that $\max_{z\in\mathbb{C}^n, |z|=1} |L(z)| \leqslant \varepsilon$, the set* $\mathrm{Crit}(\rho + L) = \{z \in \Omega \mid d(\rho + L)(z) = 0\}$ *is discrete in Ω and $\rho + L$ is strictly plurisubharmonic on Ω.*

Proof. Since ρ is $\mathcal{C}^2$ the classical Morse lemma (cf. [Mi, § 2, Lemme A], [Ra, Appendix A] or [He/Le2], Appendix B) says that the critical points of $\rho + L$ are isolated for almost any $\mathbb{R}$-linear form $L : \mathbb{C}^n \to \mathbb{R}$. For any $\varepsilon > 0$ we can therefore find an $\mathbb{R}$-linear form $L : \mathbb{C}^n \to \mathbb{R}$ such that $\max_{z\in\mathbb{C}^n, |z|=1} |L(z)| \leqslant \varepsilon$ and $\mathrm{Crit}(\rho + L)$ is discrete. It follows from Definition 2.17 that $\rho + L$ is strictly plurisubharmonic because the second derivatives of L are zero. □

Corollary 3.13. *Under the hypothesis of Theorem 3.10 there is a function ρ such that i), ii) and iii) of Theorem 3.10 hold and* $\mathrm{Crit}(\rho) :=$ $\{z \in \Omega \mid d\rho(z) = 0\}$ *is discrete in Ω.*

Corollary 3.14. *If Ω is a pseudoconvex open set in $\mathbb{C}^n$ and K is a compact set in Ω then $\widehat{K}^p_\Omega = \widehat{K}^{p\cap\mathcal{C}^\infty}_\Omega$. In particular $\widehat{K}^p_\Omega$ is closed and is therefore compact in Ω.*

Proof. Since $\widehat{K}^{p\cap\mathcal{C}^\infty}_\Omega = \{z \in \Omega \mid u(z) < \sup_K u, \forall\, u \in \mathrm{PSH}(\Omega) \cap \mathcal{C}^\infty(\Omega)\}$ it is obvious that $\widehat{K}^p_\Omega \subset \widehat{K}^{p\cap\mathcal{C}^\infty}_\Omega$. Consider a point $z \in \Omega \setminus \widehat{K}^p_\Omega$; applying Theorem 3.10 to the compact set K and the neighbourhood $U = \Omega \setminus \{z\}$ in $\widehat{K}^p_\Omega$ we see that there is a function $\varphi \in \mathrm{PSH}(\Omega) \cap \mathcal{C}^\infty(\Omega)$ such that $\varphi(z) > 0$ and $\varphi < 0$ on $\widehat{K}^p_\Omega$. It follows that $\varphi(z) < 0$ on K and hence $z \notin \widehat{K}^{p\cap\mathcal{C}^\infty}_D$. □

Proposition 3.15. *Let Ω be an open set in $\mathbb{C}^n$. If there is a continuous plurisubharmonic function ρ defined on a neighbourhood $U_{\partial\Omega}$ of the boundary of Ω such that*

$$\Omega \cap U_{\partial\Omega} = \{z \in U_{\partial\Omega} \mid \rho(z) < 0\}$$

then Ω is pseudoconvex.

Proof. Consider a point $\xi \in \partial\Omega$ and let $\varepsilon > 0$ be small enough that $B(\xi,\varepsilon)$ is relatively compact in $U_{\partial\Omega}$. By Theorem 3.9, it is enough to prove that $\Omega \cap B(\xi,\varepsilon)$ is pseudoconvex. Consider the plurisubharmonic function defined in some neighbourhood of $\overline{\Omega\cap B(\xi,\varepsilon)}$ by

$$\varphi(z) = \max(|z-\xi| - \varepsilon, \rho(z)).$$

Then $\varphi = 0$ on $\partial(\Omega \cap B(\xi,\varepsilon))$ and $\varphi < 0$ on $\Omega \cap B(\xi,\varepsilon)$. Let K be a compact subset of $\Omega \cap B(\xi,\varepsilon)$. Then $\sup_K \varphi = \alpha < 0$ and

$$\widehat{K}^p_{\Omega\cap B(\xi,\varepsilon)} \subset \{z \in \Omega \cap B(\xi,\varepsilon) \mid \varphi(z) < \alpha\} \Subset \Omega \cap B(\xi,\varepsilon),$$

from which it follows that $\Omega \cap B(\xi,\varepsilon)$ is pseudoconvex. □

We now give a characterisation of pseudoconvex domains with $\mathcal{C}^2$ boundary.

Theorem 3.16. *Let Ω be an open set in $\mathbb{C}^n$ with $\mathcal{C}^2$ boundary and let ρ be a real-valued $\mathcal{C}^2$ function defined on a neighbourhood $U_{\partial\Omega}$ of the boundary of Ω such that $U_{\partial\Omega} \cap \Omega = \{z \in U_{\partial\Omega} \mid \rho(z) < 0\}$ and $d\rho(z) \neq 0$ for every $z \in \partial\Omega$. The open set Ω is then pseudoconvex if and only if*

$$L_z\rho(w) \geqslant 0 \quad \text{for every } z \in \partial\Omega \text{ and } w \in T^{\mathbb{C}}_z(\partial\Omega), \tag{3.4}$$

where $L_z\rho(w) = \sum_{j,k=1}^n \frac{\partial^2\rho}{\partial z_j \partial\overline{z}_k}(z) w_j \overline{w}_k$ is the Levi form of ρ at the point z and $T^{\mathbb{C}}_z(\partial\Omega)$ the complex tangent space $\{w \in \mathbb{C}^n \mid \sum_{j=1}^n \frac{\partial\rho}{\partial z_j}(z) w_j = 0\}$.

Proof. Let ρ_1 be another $\mathcal{C}^2$ defining function for Ω. We saw in Chapter II, Lemma 8.2 that there is a strictly positive $\mathcal{C}^1$ function h on $\partial\Omega$ such that

$\rho_1 = h\rho$. A straightforward calculation then proves that, for any $z \in \partial\Omega$ and $w \in T_z^{\mathbb{C}}(\partial\Omega)$,

$$L_z\rho_1(w) = h(z)L_z\rho(w)$$

and hence (3.4) does not depend on the choice of defining function.

Necessity. Set $\rho(z) = -\operatorname{dist}(z,\partial\Omega)$ if $z \in \Omega$ and $\rho(z) = \operatorname{dist}(z,\partial\Omega)$ if $z \in \mathbb{C}^n \setminus \Omega$. Since $\partial\Omega$ is $\mathcal{C}^2$ the function ρ is $\mathcal{C}^2$ in a neighbourhood $U_{\partial\Omega}$ of the boundary of Ω. As Ω is pseudoconvex the function $-\log(-\rho)$ is plurisubharmonic and $\mathcal{C}^2$ on Ω and hence

$$\sum_{j,k=1}^{n} \Big[\frac{1}{\rho^2}\frac{\partial\rho}{\partial z_j}(z)\frac{\partial\rho}{\partial\overline{z}_k}(z) - \frac{1}{\rho}\frac{\partial^2\rho}{\partial z_j\partial\overline{z}_k}(z)\Big]w_j\overline{w}_k \geqslant 0$$

for any $z \in U_{\partial\Omega}\cap\Omega$ and any $w \in \mathbb{C}^n$. If $z \in U_{\partial\Omega}\cap\Omega$ and $\sum_{j=1}^n \frac{\partial\rho}{\partial z_j}(z)w_j = 0$ then

$$\sum_{j,k=1}^{n} \frac{\partial^2\rho}{\partial z_j\partial\overline{z}_k}(z)w_j\overline{w}_k \geqslant 0.$$

This equation holds for $z \in \partial\Omega$ on passing to the limit, but the condition

$$\sum_{j=1}^{n} \frac{\partial\rho}{\partial z_j}(z)w_j = 0$$

then implies that $w \in T_z^{\mathbb{C}}(\partial\Omega)$.

Sufficiency. We argue by contradiction. Suppose that Ω is not pseudoconvex and let $U_{\partial\Omega}$ be a neighbourhood of the boundary of Ω on which the function ρ defined by

$$\rho(z) = \begin{cases} -\operatorname{dist}(z,\partial\Omega) & \text{if } z \in \Omega \\ \operatorname{dist}(z,\partial\Omega) & \text{if } z \in \mathbb{C}^n \setminus \Omega \end{cases}$$

is $\mathcal{C}^2$. As Ω is not pseudoconvex there is a point $\xi \in \Omega \cap U_{\partial\Omega}$ such that the function $-\log(-\rho)$ is not plurisubharmonic at ξ (cf. Theorem 3.9). This means that there is a $w \in \mathbb{C}^n \setminus \{0\}$ such that

$$\gamma := \frac{\partial^2}{\partial\lambda\partial\overline{\lambda}}\log(-\rho(\xi+\lambda w))\big|_{\lambda=0} > 0.$$

Letting λ tend to 0, Taylor's formula says that

$$\log(-\rho(\xi+\lambda w)) = \log|\rho(\xi)| + \operatorname{Re}(\alpha\lambda+\beta\lambda^2) + \gamma|\lambda|^2 + o(|\lambda|^2), \tag{3.5}$$

where α are β are constants (cf. Lemma 2.22). Choose $\eta \in \mathbb{C}^n$ such that $|\eta| = |\rho(\xi)|$ and $\xi+\eta \in \partial\Omega$. Set $\zeta_s(\lambda) = \xi + \lambda w + s\eta e^{\alpha\lambda+\beta\lambda^2}$ for any $0 < s \leqslant 1$. By (3.5) there is a $\varepsilon > 0$ such that if $|\lambda| < \varepsilon$ and $0 < s \leqslant 1$ then

$$\begin{aligned} \operatorname{dist}(\zeta_s(\lambda),\partial\Omega) &\geqslant -\rho(\xi+\lambda w) - s|\eta|\,|e^{\alpha\lambda+\beta\lambda^2}| \\ &\geqslant |\rho(\xi)|(e^{\gamma|\lambda|^2/2} - s)|e^{\alpha\lambda+\beta\lambda^2}| \end{aligned} \tag{3.6}$$

and it follows that

$$\operatorname{dist}(\zeta_s(\lambda), \partial\Omega) > 0, \quad \text{whenever } 0 < s < 1 \text{ and } |\lambda| \leqslant \varepsilon.$$

As $\zeta_s(0) = \xi + s\eta \in \Omega$ for any $0 < s < 1$ we have $\zeta_s(\lambda) \in \Omega$ for any $0 < s < 1$ and $|\lambda| \leqslant \varepsilon$. Passing to the limit as s tends to 1 we get $\zeta_1(\lambda) \in \overline{\Omega}$ whenever $|\lambda| \leqslant \varepsilon$. As $\xi \in U_{\partial\Omega}$, diminishing ε if necessary, we can assume that $\zeta_1(\lambda) \in \overline{\Omega} \cap U_{\partial\Omega}$ for $|\lambda| \leqslant \varepsilon$ and we then have $\rho(\zeta_1(\lambda)) = -\operatorname{dist}(\zeta_1(\lambda), \partial\Omega)$. It follows by (3.6) that

$$-\rho(\zeta_1(\lambda)) \geqslant \rho(\xi)(e^{\gamma|\lambda|^2/2} - 1)|e^{\alpha\lambda+\beta\lambda^2}|, \quad \text{for } |\lambda| \leqslant \varepsilon. \tag{3.7}$$

The right-hand side of (3.7) is a strictly convex positive function of λ on some neighbourhood of 0. Since $\rho(\zeta_1(0)) = 0$ the function $-\rho \circ \zeta_1$ is strictly convex at 0 and $d(\rho \circ \zeta_1)(0) = 0$. In particular

$$\frac{\partial^2 \rho(\zeta_1(\lambda))}{\partial\lambda\partial\overline{\lambda}}\Big|_{\lambda=0} < 0 \quad \text{and} \quad \frac{\partial \rho(\zeta_1(\lambda))}{\partial\lambda}\Big|_{\lambda=0} = 0.$$

Since $\zeta_1(\lambda)$ is a holomorphic function of λ it follows that

$$\sum_{j,k=1}^{n} \frac{\partial^2 \rho}{\partial z_j \partial \overline{z}_k}(\xi + \eta) w_j \overline{w}_k < 0 \quad \text{and} \quad \sum_{j=1}^{n} \frac{\partial \rho}{\partial z_j}(\xi + \eta) w_j = 0$$

which contradicts (3.4) because $\xi + \eta \in \partial\Omega$. □

Definition 3.17. Let Ω be a relatively compact open set in $\mathbb{C}^n$ with $\mathcal{C}^2$ boundary and let ρ be a real-valued $\mathcal{C}^2$ function defined on a neighbourhood $U_{\partial\Omega}$ of the boundary of Ω such that $U_{\partial\Omega} \cap \Omega = \{z \in U_{\partial\Omega} \mid \rho(z) < 0\}$ and $d\rho(z) \neq 0$ for every $z \in \partial\Omega$. We say that Ω is *strictly pseudoconvex* if

$$L_z\rho(w) > 0 \quad \text{for any } z \in \partial\Omega \text{ and } w \in T_z^{\mathbb{C}}(\partial\Omega) \setminus \{0\}. \tag{3.8}$$

Remark. Of course, condition (3.8) is independent of the choice of defining function ρ.

Theorem 3.18. *A relatively compact open set Ω in $\mathbb{C}^n$ with $\mathcal{C}^2$ boundary is strictly pseudoconvex if and only if it has a $\mathcal{C}^2$ strictly plurisubharmonic defining function. If Ω is strictly pseudoconvex and ρ is a $\mathcal{C}^2$ defining function for Ω, then the function $\widetilde{\rho} = e^{\lambda\rho} - 1$ is a $\mathcal{C}^2$ strictly plurisubharmonic defining function for large enough λ.*

Proof. Suppose that Ω has a $\mathcal{C}^2$ strictly plurisubharmonic defining function ρ. For any z contained in some neighbourhood of $\partial\Omega$ and any $w \in \mathbb{C}^n \setminus \{0\}$ we then have $L_z\rho(w) > 0$, so (3.8) holds.

Conversely suppose that Ω is strictly pseudoconvex. It will be enough to prove that for large enough λ the function $\widetilde{\rho} = e^{\lambda\rho} - 1$ is strictly plurisubharmonic on some neighbourhood $U_{\partial\Omega}$ of $\partial\Omega$ since $\Omega \cap U_{\partial\Omega} = \{z \in U_{\partial\Omega} \mid \widetilde{\rho}(z) < 0\}$ and $d\widetilde{\rho}(z) = \lambda e^{\rho} d\rho(z) \neq 0$ for any $z \in \partial\Omega$. Then

$$\frac{\partial^2 \widetilde{\rho}}{\partial z_j \partial \overline{z}_k}(z) = \lambda \frac{\partial^2 \rho}{\partial z_j \partial \overline{z}_k}(z) e^{\lambda\rho} + \lambda^2 \frac{\partial \rho}{\partial z_j}(z) \frac{\partial \rho}{\partial \overline{z}_k}(z) e^{\lambda\rho}.$$

It follows that, for any $z \in \partial\Omega$ and $w \in \mathbb{C}^n$,

$$L_z\widetilde{\rho}(w) = \lambda L_z\rho(w) + \lambda^2 \Big| \sum_{j=1}^{n} \frac{\partial \rho}{\partial z_j}(z) w_j \Big|^2. \tag{3.9}$$

Consider the set

$$K = \Big\{ (z, w) \in \partial\Omega \times \mathbb{C}^n \mid |w| = 1 \text{ and } \sum_{j,k=1}^{n} \frac{\partial^2 \rho}{\partial z_j \partial \overline{z}_k} w_j \overline{w}_k \leqslant 0 \Big\}.$$

Since K is compact and (3.8) says that $\big| \sum_{j=1}^{n} \frac{\partial \rho}{\partial z_j}(z) w_j \big| > 0$ for any $(z, w) \in K$ so if λ is large enough then

$$\max_{(z,w)\in K} |L_z\rho(w)| < \lambda \min_{(z,w)\in K} \Big| \sum_{j=1}^{n} \frac{\partial \rho}{\partial z_j}(z) w_j \Big|.$$

By (3.9) and Definition 2.17, $\widetilde{\rho}$ is strictly plurisubharmonic in a neighbourhood of $\partial\Omega$. □

Theorem 3.19. *Let $\Omega \Subset \mathbb{C}^n$ be a strictly pseudoconvex open set with $\mathcal{C}^2$ boundary. There is then a neighbourhood $U_{\overline{\Omega}}$ of $\overline{\Omega}$ and a $\mathcal{C}^2$ strictly plurisubharmonic function $\rho : U_{\overline{\Omega}} \to \mathbb{R}$ such that $d\rho(z) \neq 0$ for any $z \in \partial\Omega$ and*

$$\Omega = \{z \in U_{\overline{\Omega}} \mid \rho(z) < 0\}$$

and

$$\partial\Omega = \{z \in U_{\overline{\Omega}} \mid \rho(z) = 0\}.$$

Proof. By Theorem 3.18, the open set Ω has a $\mathcal{C}^2$ strictly plurisubharmonic defining function ρ_0. In other words, ρ_0 is defined on a neighbourhood $U_{\partial\Omega}$ of the boundary of Ω, $d\rho_0(z) \neq 0$ for any $z \in \partial\Omega$ and $\Omega \cap U_{\partial\Omega} = \{z \in U_{\partial\Omega} \mid \rho_0(z) < 0\}$. Let $\delta > 0$ be small enough that $K_\delta = \{z \in U_{\partial\Omega} \mid -\delta \leqslant \rho_0(z) \leqslant 0\}$ is a compact set in $U_{\partial\Omega}$, and choose a real-valued $\mathcal{C}^\infty$ function χ on $\mathbb{R}$ such that

$$\chi(t) = -\delta \quad \text{for any } t \leqslant -\delta,$$
$$\chi(0) = 0,$$
$$\frac{d^2\chi}{dt^2}(t) \geqslant 0 \quad \text{for any } t \in \mathbb{R}$$

and

$$\frac{d\chi}{dt}(t) > 0 \quad \text{whenever } -\delta < t < +\infty,$$

Define ρ_1 by $\rho_1 = -\delta$ on $\Omega \setminus K_\delta$ and $\rho_1 = \chi \circ \rho_0$ on $U_{\partial\Omega}$. It follows immediately from the properties of χ that ρ_1 is a $\mathcal{C}^2$ plurisubharmonic function on $\Omega \cup U_{\partial\Omega}$ which is strictly plurisubharmonic on $\{z \in U_{\partial\Omega} \mid \rho_0(z) > -\delta\}$, $d\rho_1(z) \neq 0$ for any $z \in \partial\Omega$ and

$$\Omega = \{z \in \Omega \cup U_{\partial\Omega} \mid \rho_1(z) < 0\}.$$

It follows from Theorem 3.10 that there is a $\mathcal{C}^\infty$ strictly plurisubharmonic function ρ_2 on Ω such that $\Omega_\alpha = \{z \in \Omega \mid \rho_2(z) < \alpha\} \Subset \Omega$ for any $\alpha \in \mathbb{R}$. Choose $\beta \in \mathbb{R}$ large enough that $\rho_1 > -\delta/2$ on $\Omega \setminus \Omega_\beta$ and choose a real-valued function $\psi \in \mathcal{C}^\infty(\mathbb{C}^n)$ such that $\psi = 1$ on some neighbourhood of $\overline{\Omega}_\beta$ and $\psi = 0$ on some neighbourhood of $\mathbb{C}^n \setminus \Omega$. Define $\widetilde{\rho}_2$ by $\widetilde{\rho}_2 = \psi\rho_2$ on Ω and $\widetilde{\rho}_2 = 0$ on $\mathbb{C}^n \setminus \Omega$. The function $\widetilde{\rho}_2$ is strictly plurisubharmonic on some neighbourhood of $\overline{\Omega}_\beta$ and hence $\rho_1 + c\rho_2$ is strictly plurisubharmonic on some neighbourhood of $\overline{\Omega}_\beta$ for any $c > 0$. As the function ρ_1 is also strictly plurisubharmonic on $U_{\partial\Omega} \setminus \Omega_\beta$ and $\widetilde{\rho}_2 = 0$ on $U_{\partial\Omega} \setminus \Omega$ the function $\widetilde{\rho} = \rho_1 + c\widetilde{\rho}_2$ is $\mathcal{C}^2$ and strictly plurisubharmonic on $U_{\overline{\Omega}} = U_{\partial\Omega} \cup \Omega$ for small enough c. For small enough c we also have $\Omega = \{z \in U_{\overline{\Omega}} \mid \widetilde{\rho}(z) < 0\}$. Choose a positive $\mathcal{C}^\infty$ function φ on $\mathbb{C}^n$ such that $\overline{\Omega} = \{z \in \mathbb{C}^n \mid \varphi(z) = 0\}$ (cf. Lemma 1.4.13 in [Na2], for example). Passing to a smaller $U_{\overline{\Omega}}$ if necessary, the function $\rho = \widetilde{\rho} + \varepsilon\varphi$ is then the function we seek provided ε is small enough. □

Corollary 3.20. *Let $\Omega \Subset \mathbb{C}^n$ be a strictly pseudoconvex open set with $\mathcal{C}^2$ boundary. For any compact set $K \subset \partial\Omega$ and any neighbourhood U_K of K in $\mathbb{C}^n$ there is a strictly pseudoconvex open set $\widetilde{\Omega}$ with $\mathcal{C}^2$ boundary such that*

$$\Omega \cup K \subset \widetilde{\Omega} \subset \Omega \cup U_K.$$

Proof. Consider an open set $U_{\overline{\Omega}}$ and a function ρ for which the conclusions of Theorem 3.19 hold. Choose a positive $\mathcal{C}^\infty$ function χ with compact support in $U_K \cap U_{\overline{\Omega}}$ which is strictly positive on K. The set $\widetilde{\Omega} = \{z \in U_{\overline{\Omega}} \mid \rho(z) - \varepsilon\chi(z) < 0\}$ is then the open set we seek provided ε is small enough. □

We will now study the link between strict convexity and strict pseudoconvexity.

Proposition 3.21. *Any strictly convex domain $\Omega \Subset \mathbb{C}^n$ with $\mathcal{C}^2$ boundary is strictly pseudoconvex.*

Proof. This follows from the fact that a strictly convex $\mathcal{C}^2$ function is strictly plurisubharmonic. □

We now prove that the converse is locally true up to holomorphic change of coordinates.

Lemma 3.22. *Let V be an open set in $\mathbb{C}^n$ and let ρ be a strictly convex $\mathcal{C}^2$ function on V. We set $D = \{z \in V \mid \rho(z) < 0\}$; for any compact convex set $K \subset \overline{D} \cap V$ and any neighbourhood U_K of K there is then a strictly convex open set Ω with $\mathcal{C}^2$ boundary such that $K \subset \overline{\Omega} \subset U_K \cap \overline{D}$.*

Proof. Since K is convex there is a strictly convex open set Ω_1^0 with $\mathcal{C}^\infty$ boundary such that $K \Subset \Omega_1^0 \Subset U_K \cap V$. Let ρ_1 be a strictly convex $\mathcal{C}^\infty$ function on a neighbourhood $U_1 \Subset U_K \cap V$ of Ω_1^0 such that $\Omega_1^0 = \{z \in U_1 \mid \rho_1(z) < 0\}$. Since no point of the boundary of Ω_1^0 is a local minimum of ρ_1 and ρ_1 is strictly convex $d\rho_1(\zeta) \neq 0$ for any $\zeta \in \partial\Omega_1^0$. It follows that if $\varepsilon > 0$ is small enough then $\Omega_1^\varepsilon = \{z \in U_1 \mid \rho_1(z) < \varepsilon\} \Subset U_1$. Choose $\mathcal{C}^\infty$ functions f and g from $\mathbb{R}$ to $\mathbb{R}$ such that:

$$\frac{\partial f}{\partial t}(t) > 0, \ \frac{\partial g}{\partial t}(t) \geqslant 0 \ \frac{\partial^2 f}{\partial t^2}(t) \geqslant 0 \text{ and } \frac{\partial^2 g}{\partial t^2}(t) \geqslant 0 \quad \text{for all } t \in \mathbb{R},$$
$$-1 < f(t) < 0 \text{ if } t < 0, \quad f(0) = 0, \quad \text{and } f(t) > 0 \text{ if } t > 0,$$
$$g(t) = 0 \text{ if } t \leqslant 0, \quad g(t) > 0 \text{ if } t > 0 \quad \text{and} \quad g(t) = 1 \text{ if } t > \varepsilon.$$

Set $\varphi(x,y) = f(x) + g(y)$ for any $(x,y) \in \mathbb{R}^2$. The function φ is then a convex $\mathcal{C}^\infty$ function on $\mathbb{R}^2$ such that

$$\frac{\partial \varphi}{\partial x}(x,y) > 0 \quad \text{and} \quad \frac{\partial \varphi}{\partial y}(x,y) \geqslant 0 \quad \text{on } \mathbb{R}^2,$$
$$\varphi(x,y) > 0 \text{ if } \max(x, y-\varepsilon) > 0 \quad \text{and} \quad \varphi(x,y) < 0 \text{ if } x < 0 \text{ and } y \leqslant 0.$$

The function $\psi(z) = \varphi(\rho(z), \rho_1(z))$ is then a strictly convex $\mathcal{C}^2$ function and if $\Omega = \{z \in U_1 \mid \psi(z) < 0\}$ then $\Omega_1^0 \cap D \subset \Omega \subset \Omega_1^\varepsilon \cap D$ and hence $K \subset \overline{\Omega} \subset U_K \cap \overline{D}$. Since ψ is strictly convex and no point of the boundary of Ω is a local minimum of ψ, $d\psi(z) \neq 0$ for any $z \in \partial\Omega$ and it follows that Ω is a strictly convex open set with $\mathcal{C}^2$ boundary. □

Theorem 3.23. *Let V be an open set in $\mathbb{C}^n$ and let ρ be a $\mathcal{C}^2$ strictly plurisubharmonic function on V such that $d\rho(z) \neq 0$ for any $z \in \Gamma = \{z \in V \mid \rho(z) = 0\}$. Set $D = \{z \in V \mid \rho(z) < 0\}$. For any point $\xi \in \Gamma$ there is then a neighbourhood U_ξ of ξ and a strictly pseudoconvex open set with $\mathcal{C}^2$ boundary, $\Omega \Subset \mathbb{C}^n$, such that*

i) *$U_\xi \cap D \subset \Omega \subset D$.*
ii) *There is a biholomorphic map h defined on a convex neighbourhood of $\overline{\Omega}$ such that $h(\Omega)$ is strictly convex.*

Proof. By Theorem 2.23 there is a biholomorphic map h from a convex neighbourhood V_ξ of ξ in $\mathbb{C}^n$ to an open set W in $\mathbb{C}^n$ such that $\rho \circ h^{-1}$ is strictly convex on the open set W. Let $U' \Subset W$ be a ball centred on $h(\xi)$. By Lemma 3.22 there is a strictly convex open set Ω' with $\mathcal{C}^2$ boundary such that $\overline{U}' \cap h(V_\xi \cap \overline{D}) \subset \overline{\Omega}' \subset h(V_\xi \cap \overline{D})$. Setting $U_\xi = h^{-1}(U')$ and $\Omega = h^{-1}(\Omega')$ proves the theorem. □

Corollary 3.24. *A relatively compact open set Ω in $\mathbb{C}^n$ with $\mathcal{C}^2$ boundary is strictly pseudoconvex if and only if for any point $\xi \in \partial\Omega$ there is a neighbourhood U_ξ of ξ in $\mathbb{C}^n$ and a biholomorphic map h_ξ defined on a convex neighbourhood of $\overline{U}_\xi$ such that $h_\xi(U_\xi \cap \Omega)$ is a strictly convex open set with $\mathcal{C}^2$ boundary.*

Proof. The condition is necessary by Theorem 3.23. It is sufficient by definition of strictly pseudoconvex open sets because condition (3.8) is invariant under holomorphic coordinate changes. □

Remark. The above proves that strict pseudoconvexity is the locally biholomorphically invariant version of strict convexity. A counter-example by J.J. Kohn and L. Nirenberg shows that there is no similar relationship between convexity and pseudoconvexity even in the case where the boundary is assumed $\mathcal{C}^2$ (cf. [Ko/Ni]).

Comments

Historically, the characterisation of existence domains of holomorphic functions started with the work of F. Hartogs and E.E. Levi at the beginning of the century. The characterisation of such domains in terms of holomorphic convexity is due to H. Cartan and P. Thullen [Ca/Th]. Plurisubharmonic functions were introduced by K. Oka and P. Lelong, who described their main properties [Lel1]. The plurisubharmonicity of the function $-\log \delta_\Omega$, suggested by Hartogs [Har], was proved by K. Oka [Ok], P. Lelong [Lel2] and H. Bremermann [Br2]. Levi's condition for domains of holomorphy was discovered in 1910 by E.E. Levi [Lev] in the two variable case.

All the concepts and theorems presented in this chapter can be found in most books on several complex variable function theory, such as [He/Le1], [Ho2], [Kr], [Na1] and [Ra]. We follow Range's presentation [Ra] closely for the Cartan–Thullen theorem and holomorphic convexity; for plurisubharmonic functions and pseudoconvexity we closely follow Hörmander [Ho2] and we closely follow Henkin and Leiterer [He/Le1] for strictly pseudoconvex domains with $\mathcal{C}^2$ boundary.

VII

The Levi problem and the resolution of $\overline{\partial}$ in strictly pseudoconvex domains

This chapter is devoted to solving the Levi problem – or in other words, to proving that any pseudoconvex open set in $\mathbb{C}^n$ is a domain of holomorphy. We proceed by studying $\overline{\partial}$ in pseudoconvex open sets using local integral representation formulas for strictly pseudoconvex domains and then applying H. Grauert's bumping technique.

We start by studying the Cauchy–Riemann equation in bounded strictly convex domains with $\mathcal{C}^2$ boundary in $\mathbb{C}^n$. Using the Cauchy–Fantappié formula introduced in Chapter V, we prove that the solution is Hölder continuous of order 1/2 if the data is continuous on the closure of the domain. This result together with the fact that locally strictly pseudoconvex domains with $\mathcal{C}^2$ boundary are the same thing as strictly convex domains enables us to prove a finiteness result for certain $\overline{\partial}$-cohomology groups using functional analysis techniques. In Section 4 we develop Grauert's bumping method and show that these $\overline{\partial}$-cohomology groups are isomorphic to the Dolbeault cohomology groups defined in Chapter II. And finally, a result of Laufer's presented in Section 5 proves the vanishing of Dolbeault cohomology groups on bounded strictly pseudoconvex open sets with $\mathcal{C}^2$ boundary in $\mathbb{C}^n$. Section 6 is devoted to the construction of a global integral formula for solving $\overline{\partial}$ in bounded strictly pseudoconvex open sets with $\mathcal{C}^2$ boundary in $\mathbb{C}^n$. We finally solve the Levi problem in Section 7. The last section then generalises the above results to complex analytic manifolds. We get a characterisation of Stein manifolds in terms of pseudoconvexity and prove vanishing theorems for Dolbeault cohomology with compact support which, together with the cohomological results in Chapter V, enable us to give geometric conditions which imply the existence of extensions of CR functions.

1 Solving $\overline{\partial}$ with Hölder estimates in strictly convex open sets

Let D be a bounded strictly convex open set with $\mathcal{C}^2$ boundary in $\mathbb{C}^n$. There is then a real-valued $\mathcal{C}^2$ function ρ on $\mathbb{C}^n$ such that

$$D = \{z \in \mathbb{C}^n \mid \rho(z) < 0\}, \qquad d\rho(z) \neq 0 \quad \text{for any } z \in \partial D,$$

and
$$\sum_{j,k=1}^{2n} \frac{\partial^2 \rho}{\partial x_j \partial x_k}(z) t_j t_k \geqslant \alpha |t|^2 \quad \text{for any } z \in \partial D \text{ and } t \in \mathbb{R}^{2n},$$

C. Laurent-Thiébaut, *Holomorphic Function Theory in Several Variables: An Introduction*, Universitext, DOI 10.1007/978-0-85729-030-4_7, © Springer-Verlag London Limited 2011

where α is some strictly positive real number and the coordinates $x_j = x_j(z)$ are the real coordinates of the point $z \in \mathbb{C}^n$ such that $z_j = x_j(z) + ix_{j+n}(z)$. As we saw in Section 3 of Chapter V, the function

$$w_\rho(z,\zeta) = \Big(\frac{\partial \rho}{\partial \zeta_1}(\zeta), \dots, \frac{\partial \rho}{\partial \zeta_n}(\zeta)\Big)$$

is a Leray section for the open set D. Since D is strictly convex we have the following, more precise, estimate:

Lemma 1.1. *There is a neighbourhood $U_{\partial D}$ of the boundary of D and there are strictly positive real numbers ε and β which have the property that, for every $\zeta \in U_{\partial D}$ and $z \in \mathbb{C}^n$ such that $|\zeta - z| \leqslant \varepsilon$,*

$$2\,\mathrm{Re}\langle w_\rho(\zeta), \zeta - z\rangle \geqslant \rho(\zeta) - \rho(z) + \beta|\zeta - z|^2. \tag{1.1}$$

Proof. Let $x_j = x_j(\zeta)$ be the real coordinates of $\zeta \in \mathbb{C}^n$ such that $\zeta_j = x_j(\zeta) + x_{j+n}(\zeta)$. Then

$$\begin{aligned} 2\,\mathrm{Re}\langle w_\rho(\zeta), \zeta - z\rangle &= \mathrm{Re}\sum_{j=1}^n \Big(\frac{\partial \rho}{\partial x_j}(\zeta) - i\frac{\partial \rho}{\partial x_{j+n}}(\zeta)\Big)\big(x_j(\zeta - z) + ix_{j+n}(\zeta - z)\big) \\ &= \sum_{j=1}^{2n} \frac{\partial \rho}{\partial x_j}(\zeta) x_j(\zeta - z). \end{aligned}$$

The Taylor–Young formula then says that

$$\begin{aligned} \rho(z) = \rho(\zeta) - 2\,\mathrm{Re}\langle w_\rho(\zeta), \zeta - z\rangle + \frac{1}{2}\sum_{j,k=1}^{2n} \frac{\partial^2 \rho}{\partial x_j \partial x_k}(\zeta) x_j(\zeta - z) x_k(\zeta - z) \\ + O\big(|\zeta - z|^2\big), \end{aligned}$$

which implies that for small enough ε and for ζ contained in some neighbourhood $U_{\partial D}$ of ∂D,

$$2\,\mathrm{Re}\langle w_\rho(\zeta), \zeta - z\rangle \geqslant \rho(\zeta) - \rho(z) + \frac{\alpha}{4}|\zeta - z|^2, \quad \text{if } |\zeta - z| \leqslant \varepsilon$$

since ρ is strictly convex. □

In Chapter V, we proved the following theorem.

Theorem 1.2. *Let D be a bounded convex open set in $\mathbb{C}^n$ with C^2 boundary and let ρ be a C^2 defining function for D. For any continuous differential (p,q)-form f on $\overline{D}$ such that $\overline{\partial} f$ is also continuous on $\overline{D}$, $0 \leqslant p \leqslant n$ and $0 \leqslant q \leqslant n$,*

$$\begin{aligned} f &= Lf + T_1^p \overline{\partial} f && \text{on } D \text{ if } q = 0 \\ f &= \overline{\partial} T_q^p f + T_{q+1}^p \overline{\partial} f && \text{on } D \text{ if } 1 \leqslant q \leqslant n, \end{aligned}$$

where $Lf = \int_{\zeta\in\partial D} f(\zeta)\wedge K^{w_\rho}(\cdot,\zeta)$ *and*

$$T_q^p f = (-1)^{p+q}\Big(\int_{(\zeta,\lambda)\in\partial D\times[0,1]} f(\zeta)\wedge K^{\eta^{w_\rho}}(\cdot,\zeta,\lambda) + \int_{\zeta\in D} f(\zeta)\wedge B(\cdot,\zeta)\Big).$$

In particular, if $\overline{\partial} f = 0$ *and* $q \geqslant 1$ *then* $u = T_q^p f$ *is a solution of the equation* $\overline{\partial} u = f$ *on* D. *Moreover, if* f *is* $\mathcal{C}^k$ *on* D *then* $u = T_q^p f$ *is* $\mathcal{C}^{k+\alpha}$ *on* D *for every* $\alpha \in \,]0,1[$.

The estimate given in Lemma 1.1 will provide us with an estimate for the order 1/2 Hölder norm of solutions of $\overline{\partial}$ on strictly convex domains with $\mathcal{C}^2$ boundary.

We proved in Chapter III, Proposition 2.1, that if D is a bounded open set in $\mathbb{C}^n$ and f is a bounded differential form on D then the function

$$\widetilde{B}_D f = \int_{\zeta\in D} f(\zeta)\wedge B(\cdot,\zeta)$$

has the property that

$$|\widetilde{B}_D f|_{\alpha,D} \leqslant C_\alpha |f|_{0,D},$$

for any $\alpha \in \,]0,1[$. We still have to deal with the quantity

$$R_{\partial D}^{w_\rho} f = \int_{(\zeta,\lambda)\in\partial D\times[0,1]} f(\zeta)\wedge K^{\eta^{w_\rho}}(\cdot,\zeta,\lambda).$$

Lemma 1.3. *Let* D *be a bounded open set in* $\mathbb{C}^n$ *with* $\mathcal{C}^1$ *boundary and let* $U_{\partial D}$ *and* $U_{\overline{D}}$ *be neighbourhoods of* ∂D *and* $\overline{D}$ *respectively. Let* $w(z,\zeta) = (w_1(z,\zeta),\dots,w_n(z,\zeta))$ *be a* $\mathcal{C}^1$ *map from* $U_{\overline{D}}\times U_{\partial D}$ *to* $\mathbb{C}^n$ *with the following properties:*

i) $w(z,\zeta)$ *depends holomorphically on* z *in* $U_{\overline{D}}$ *and* $d_\zeta w(z,\zeta)$ *therefore also depends holomorphically on* z *in* $U_{\overline{D}}$,
ii) $w(z,\zeta)$ *is a Leray section for* D,
iii) *for any point* $\xi\in\partial D$ *there is a neighbourhood* U_ξ *of* ξ *and real-valued* $\mathcal{C}^1$ *functions* $t_1(z,\zeta),\dots,t_{2n-1}(z,\zeta)$ *defined for any* z *and* ζ *in* U_ξ *such that*
   a) *for any* $z\in U_\xi$ *the functions* $t_1(z,\cdot),\dots,t_{2n-1}(z,\cdot)$ *are real coordinates on* $\partial D\cap U_\xi$,
   b) *there is a* $\delta>0$ *such that, for any* $z\in D\cap U_\xi$ *and any* $\zeta\in\partial D\cap U_\xi$,

$$(1.2)\qquad |\langle w(z,\zeta),\zeta-z\rangle| \geqslant \delta\big(|t_1(z,\zeta)| + |t(z,\zeta)|^2 + \operatorname{dist}(z,\partial D)\big),$$

with $|t(z,\zeta)|^2 = \sum_{j=1}^{2n-1}|t_j(z,\zeta)|^2$ *and* $\operatorname{dist}(z,\partial D) = \inf\{|z-x|, x\in\partial D\}$.

There is then a constant C *such that for any continuous differential form* f *on* $\overline{D}$

$$|R_{\partial D}^{w} f|_{1/2,D} \leqslant C|f|_{0,D}.$$

Proof. Note that if $n = 1$ then $R^w_{\partial D} = 0$. Assume from now on that $n \geqslant 2$ and set $\Phi(z,\zeta) = \langle w(z,\zeta), \zeta - z\rangle$. The determinantal expression then gives us that

$$(1.3)\quad (2i\pi)^n R^w_{\partial D} f(z)$$
$$= \int_{(\zeta,\lambda)\in\partial D\times[0,1]} f(\zeta) \wedge \det_{1,n-1}\left(\eta^w(z,\zeta,\lambda), (\overline{\partial}_{z,\zeta} + d_\lambda)\eta^w(z,\zeta,\lambda)\right) \wedge \omega(\zeta - z)$$

where

$$\eta^w(z,\zeta,\lambda) = (1-\lambda)\frac{w(z,\zeta)}{\Phi(z,\zeta)} + \lambda\frac{\overline{\zeta} - \overline{z}}{|\zeta - z|^2}$$

and

$$\det_{s_1\cdots s_m}(a_1,\ldots,a_m) = \det(\underbrace{a_1,\ldots,a_1}_{s_1},\ldots,\underbrace{a_m,\ldots,a_m}_{s_m}).$$

For simplicity, we assume that f has bidegree $(0,q)$; we can therefore replace $\omega(\zeta - z)$ by $\omega(\zeta)$ in (1.3). By hypothesis i), $\overline{\partial}_z w = 0$ and $\overline{\partial}_z \Phi = 0$ so that

$$(d_\lambda + \overline{\partial}_{z,\zeta})\eta^w(z,\zeta,\lambda) = \Big(\frac{\overline{\zeta} - \overline{z}}{|\zeta - z|^2} - \frac{w}{\Phi}\Big)d\lambda + (1-\lambda)\Big(\frac{\overline{\partial}_\zeta w}{\Phi} - \frac{w}{\Phi}\frac{\overline{\partial}_\zeta \Phi}{\Phi}\Big)$$
$$+ \lambda\Big(\frac{d\overline{\zeta} - d\overline{z}}{|\zeta - z|^2} - \frac{\overline{\zeta} - \overline{z}}{|\zeta - z|^2}\frac{\overline{\partial}_{z,\zeta}|\zeta - z|^2}{|\zeta - z|^2}\Big).$$

Developing the determinant in (1.3), we get

$$R^w_{\partial D} f(z)$$
$$= \int_{(\zeta,\lambda)\in\partial D\times[0,1]} f(\zeta) \wedge \sum_{s=0}^{n-2} p_s \det_{1,1,n-s-2,s}\Big(\frac{w}{\Phi}, \frac{\overline{\zeta} - \overline{z}}{|\zeta - z|^2}, \frac{\overline{\partial}_\zeta w}{\Phi}, \frac{d\overline{\zeta} - d\overline{z}}{|\zeta - z|^2}\Big) \wedge d\lambda \wedge \omega(\zeta)$$

where the functions p_s are polynomials in λ. Integrating with respect to λ gives us that

$$R^w_{\partial D} f(z) = \sum_{s=0}^{n-2} A_s \int_{\zeta\in\partial D} f(\zeta) \wedge \frac{\det_{1,1,n-s-2,s}(w, \overline{\zeta} - \overline{z}, \overline{\partial}_\zeta w, d\overline{\zeta} - d\overline{z})}{\Phi^{n-s-1}|\zeta - z|^{2s+2}} \wedge \omega(\zeta).$$

The coefficients of the differential form $R^w_{\partial D} f$ are therefore linear combinations of integrals of the following form:

$$(1.4)\qquad E(z) = \int_{\partial D} \frac{f_I \psi}{\Phi^{n-s-1}|\zeta - z|^{2s+2}} \bigwedge_{j\neq m} d\overline{\zeta}_j \wedge \omega(\zeta),$$

where $0 \leqslant s \leqslant n-2$ and $1 \leqslant m \leqslant n$, the function f_I is one of the coefficients of f and ψ is a product of functions of type w_j, $\overline{\zeta}_j - \overline{z}_j$ or $\partial w_j/\partial\zeta_k$, $j,k = 1,\ldots,n$. As ψ contains at least one factor of the form $\overline{\zeta}_j - \overline{z}_j$, there is a constant C_1 such that $|\psi| \leqslant C_1|\zeta - z|$.

To estimate the integrals appearing in (1.4) we use a lemma due to Hardy and Littlewood.

Lemma 1.4. *Let D be a bounded domain in $\mathbb{R}^n$ with $\mathcal{C}^1$ boundary and let $g \in \mathcal{C}^1(D)$ be a function such that*

$$|dg(x)| \leqslant C_g[\operatorname{dist}(x, \partial D)]^{\alpha-1},$$

where C_g is a positive constant and $\alpha \in\,]0,1[$. Then $g \in \Lambda^\alpha(D)$. Moreover, there is a compact set K in D and a constant $C > 0$ depending only on D and α such that

$$|g|_{\alpha,D} \leqslant C[C_g + |g|_K].$$

Proof. Let us start by proving the following result:

$$(1.5)\quad \begin{cases} \text{There is a real number } \varepsilon \in\,]0,1[\text{ with the property that if } g \\ \text{is any } \mathcal{C}^1 \text{ function on } D \text{ such that } |dg(x)| \leqslant C_g[\operatorname{dist}(x,\partial D)]^{\alpha-1} \\ \text{then } |g(x) - g(y)| \leqslant CC_g|x-y|^\alpha \\ \text{for any } x, y \in D \text{ such that } |x-y| < \varepsilon \end{cases}$$

Since ∂D is $\mathcal{C}^1$, there is a sufficiently small $\varepsilon > 0$ and a constant $C' > 0$ such that for any $x, y \in D$ which have the property that $\operatorname{dist}(x, \partial D) < \varepsilon$, $\operatorname{dist}(y, \partial D) < \varepsilon$ and $|x - y| < \varepsilon$ there is a $\mathcal{C}^1$ function $\gamma_{x,y} : [0, 3|x-y|] \to D$ such that

$$\begin{aligned}
&\gamma_{x,y}(0) = x, \quad \gamma_{x,y}(3|x-y|) = y, \\
&\Big|\frac{d}{d\lambda}\gamma_{x,y}(\lambda)\Big| \leqslant C' \quad \text{for any } \lambda \in [0, 3|x-y|], \\
&\operatorname{dist}(\gamma_{x,y}(\lambda), \partial D) \geqslant \begin{cases} \lambda & \text{if } \lambda \in [0, |x-y|], \\ |x-y| & \text{if } \lambda \in [|x-y|, 2|x-y|], \\ 3|x-y| - \lambda & \text{if } \lambda \in [2|x-y|, 3|x-y|]. \end{cases}
\end{aligned}$$

We therefore get

$$\begin{aligned}
|g(x) - g(y)| &= \Big|\int_0^{3|x-y|} \frac{d}{d\lambda} g\big(\gamma_{x,y}(\lambda)\big)d\lambda\Big| \\
&\leqslant \int_0^{3|x-y|} |dg\big(\gamma_{x,y}(\lambda)\big)|\,\Big|\frac{d}{d\lambda}\gamma_{x,y}(\lambda)\Big| d\lambda \\
&\leqslant C' \int_0^{3|x-y|} |dg\big(\gamma_{x,y}(\lambda)\big)| d\lambda \\
&\leqslant C'C_g\Big[\int_0^{|x-y|} \lambda^{\alpha-1} d\lambda + \int_{|x-y|}^{2|x-y|} |x-y|^{\alpha-1} d\lambda \\
&\qquad + \int_{2|x-y|}^{3|x-y|} (3|x-y| - \lambda)^{\alpha-1} d\lambda\Big] \\
&\leqslant \Big(\frac{2}{\alpha} + 1\Big)C'C_g|x-y|^\alpha.
\end{aligned}$$

We set $K = \{x \in D \mid \operatorname{dist}(x, \partial D) \geqslant \varepsilon/2\}$. The set K is compact because D is bounded. Moreover, as g is $\mathcal{C}^1$, for any $x, y \in K$ and $|x - y| < \varepsilon$,

$$|g(x) - g(y)| \leqslant C''|x - y| \leqslant C''|x - y|^\alpha \quad \text{for any } \varepsilon < 1$$

which proves (1.5). It then follows from (1.5) that $|g|_D \leqslant |g|_K + CC_g\varepsilon^\alpha < +\infty$ and hence if $|x - y| > \varepsilon$ then

$$\Delta_g(x, y) = \frac{|g(x) - g(y)|}{|x - y|^\alpha} \leqslant 2|g|_D\varepsilon^{-\alpha}. \tag{1.6}$$

It follows from (1.5) and (1.6) that

$$\sup_{x,y\in D} \Delta_g(x, y) \leqslant \max(CC_g, 2|g|_D\varepsilon^{-\alpha})$$

and the lemma follows. □

End of the proof of Lemma 1.3. By Lemma 1.4 it is enough to prove that for any $j = 1, \ldots, n$ and $z \in D$,

$$\left.\begin{array}{r} |\partial E/\partial z_j(z)| \\ \text{and } |\partial E/\partial \overline{z}_j(z)| \end{array}\right\} \leqslant C|f|_{0,D}[\operatorname{dist}(z, \partial D)]^{-1/2}, \tag{1.7}$$

where C is a constant independent of $z \in D$ and f. We have

$$\begin{aligned} \frac{\partial}{\partial z_j} \frac{\psi}{\Phi^{n-s-1}|\zeta - z|^{2s+2}} = \frac{\partial\psi/\partial z_j}{\Phi^{n-s-1}|\zeta - z|^{2s+2}} &- \frac{(n - s - 1)(\partial\Phi/\partial z_j)\psi}{\Phi^{n-s}|\zeta - z|^{2s+2}} \\ &- \frac{(s + 1)(\overline{\zeta}_j - \overline{z}_j)\psi}{\Phi^{n-s-1}|\zeta - z|^{2s+4}} \end{aligned}$$

and since $\partial\Phi/\partial\overline{z}_j = 0$,

$$\frac{\partial}{\partial \overline{z}_j} \frac{\psi}{\Phi^{n-s-1}|\zeta - z|^{2s+2}} = \frac{\partial\psi/\partial\overline{z}_j}{\Phi^{n-s-1}|\zeta - z|^{2s+2}} - \frac{(s + 1)(\zeta_j - z_j)\psi}{\Phi^{n-s-1}|\zeta - z|^{2s+4}}.$$

As $\partial\psi/\partial z_j, \partial\Phi/\partial z_j$ and $\partial\psi/\partial\overline{z}_j$ are bounded for $(z, \zeta) \in \overline{D} \times \partial D$ and $|\psi| \leqslant C_1|\zeta - z|$, there exists a constant C_2 such that

$$\begin{aligned} \left|\frac{\partial}{\partial z_j} \frac{\psi}{\Phi^{n-s-1}|\zeta - z|^{2s+2}}\right| &\leqslant C_2\Big[\frac{1}{|\Phi|^{n-s-1}|\zeta - z|^{2s+2}} + \frac{1}{|\Phi|^{n-s}|\zeta - z|^{2s+1}}\Big] \\ \left|\frac{\partial}{\partial \overline{z}_j} \frac{\psi}{\Phi^{n-s-1}|\zeta - z|^{2s+2}}\right| &\leqslant \frac{C_2}{|\Phi|^{n-s-1}|\zeta - z|^{2s+2}}. \end{aligned}$$

We can therefore find a constant C_3 such that $\partial E/\partial z_j(z)$ and $\partial E/\partial\overline{z}_j(z)$ are bounded by

$$C_3|f|_{0,D}\Big[\int_{\partial D} \frac{d\sigma_{2n-1}}{|\Phi|^{n-s-1}|\zeta - z|^{2s+2}} + \int_{\partial D} \frac{d\sigma_{2n-1}}{|\Phi|^{n-s}|\zeta - z|^{2s+1}}\Big], \tag{1.8}$$

where $d\sigma_{2n-1}$ is the volume form on ∂D.

Let us prove that for any $\xi \in \partial D$ there is a neighbourhood U_ξ of ξ and a constant C_ξ such that, for any $z \in D \cap U_\xi$,

$$\int_{\partial D \cap U_\xi} \frac{d\sigma_{2n-1}}{|\Phi|^{n-s}|\zeta - z|^{2s+1}} \leqslant C_\xi [\mathrm{dist}(z, \partial D)]^{-1/2} \tag{1.9}$$

$$\int_{\partial D \cap U_\xi} \frac{d\sigma_{2n-1}}{|\Phi|^{n-s-1}|\zeta - z|^{2s+2}} \leqslant C_\xi [\mathrm{dist}(z, \partial D)]^{-1/2}. \tag{1.10}$$

We start by proving (1.9). Fix $\xi \in \partial D$ and consider a neighbourhood U_ξ of ξ and coordinates $t_1(z,\zeta), \ldots, t_{2n-1}(z,\zeta)$ such that condition iii) of Lemma 1.3 holds. Let $\varphi(z,\zeta)$ be the function defined for any $z \in U_\xi$ and any $\zeta \in \partial D \cap U_\xi$ with the property that, for any $z \in U_\xi$,

$$d\sigma_{2n-1} = \varphi(z,\zeta) d_\zeta t_1(z,\zeta) \wedge \cdots \wedge d_\zeta t_{2n-1}(z,\zeta).$$

There is a $\gamma > 0$ such that $|\zeta - z| \geqslant \gamma |t(z,\zeta)|$ for any $z \in U_\xi$ and any $\zeta \in \partial D \cap U_\xi$, and hence for any $z \in D \cap U_\xi$

$$\int_{\partial D \cap U_\xi} \frac{d\sigma_{2n-1}}{|\Phi|^{n-s}|\zeta - z|^{2s+1}}$$
$$\leqslant \frac{1}{\delta^{n-s}\gamma^{2s+1}} \int_{\partial D \cap U_\xi} \frac{\varphi(z,\zeta) d_\zeta t_1(z,\zeta) \wedge \cdots \wedge d_\zeta t_{2n-1}(z,\zeta)}{\big(|t_1(z,\zeta)| + |t(z,\zeta)|^2 + \mathrm{dist}(z,\partial D)\big)^{n-s} |t(z,\zeta)|^{2s+1}}.$$

Restricting U_ξ if necessary, we can find constants Γ and R such that $|\varphi(z,\zeta)| \leqslant \Gamma$ for any $z \in U_\xi$ and $\zeta \in \partial D \cap U_\xi$ and for every $z \in U_\xi$ the surface $\partial D \cap U_\xi$ is mapped by the diffeomorphism $\zeta \mapsto (t_1(z,\zeta), \ldots, t_{2n-1}(z,\zeta))$ to an open set in the ball of centre 0 and radius R in $\mathbb{R}^{2n-1}$. Then

$$\int_{\partial D \cap U_\xi} \frac{d\sigma_{2n-1}}{|\Phi|^{n-s}|\zeta - z|^{2s+1}}$$
$$\leqslant \frac{\Gamma}{\delta^{n-s}\gamma^{2s+1}} \int_{\substack{x \in \mathbb{R}^{2n-1} \\ |x| < R}} \frac{dx_1 \wedge \cdots \wedge dx_{2n-1}}{\big(|x_1| + |x|^2 + \mathrm{dist}(z,\partial D)\big)^{n-s} |x|^{2s+1}}$$

but now

$$\big(|x_1| + |x|^2 + \mathrm{dist}(z,\partial D)\big)^{n-s} |x|^{2s+1} \geqslant \big(|x_1| + |x|^2 + \mathrm{dist}(z,\partial D)\big)^2 |x|^{2n-3}$$

and it follows that

$$\int_{\partial D \cap U_\xi} \frac{d\sigma_{2n-1}}{|\Phi|^{n-s}|\zeta - z|^{2s+1}} \tag{1.11}$$
$$\leqslant \frac{\Gamma}{\delta^{n-s}\gamma^{2s+1}} \int_{\substack{x \in \mathbb{R}^{2n-1} \\ |x| < R}} \frac{dx_1 \wedge \cdots \wedge dx_{2n-1}}{\big(|x_1| + |x|^2 + \mathrm{dist}(z,\partial D)\big)^2 |x|^{2n-3}}.$$

Similarly, to prove (1.10) we start by proving that

$$
\begin{aligned}
(1.12)\quad & \int_{\partial D \cap U_\xi} \frac{d\sigma_{2n-1}}{|\Phi|^{n-s-1}|\zeta - z|^{2s+2}} \\
& \leqslant \frac{\Gamma}{\delta^{n-s-1}\gamma^{2s+2}} \int_{\substack{x \in \mathbb{R}^{2n-1} \\ |x|<R}} \frac{dx_1 \wedge \cdots \wedge dx_{2n-1}}{(|x_1| + |x|^2 + \operatorname{dist}(z, \partial D))|x|^{2n-2}}.
\end{aligned}
$$

We now only have to estimate the right-hand sides of (1.11) and (1.12) to complete the proof of (1.9) and (1.10).

Lemma 1.5. *Consider numbers $n \geqslant 1$ and $R > 0$. There is a constant $C > 0$ such that, for any $\varepsilon > 0$,*

i) $\displaystyle \int_{\substack{x \in \mathbb{R}^n \\ |x|<R}} \frac{dx_1 \wedge \cdots \wedge dx_n}{(|x_1| + |x|^2 + \varepsilon)^2 |x|^{n-2}} \leqslant C\varepsilon^{-1/2},$

ii) $\displaystyle \int_{\substack{x \in \mathbb{R}^n \\ |x|<R}} \frac{dx_1 \wedge \cdots \wedge dx_n}{(|x_1| + |x|^2 + \varepsilon) |x|^{n-1}} \leqslant C\varepsilon^{-1/2}.$

Proof. Let us prove i). For $n = 1$,

$$
\int_{|x|<R} \frac{dx}{(|x| + |x|^2 + \varepsilon)^2 |x|^{-1}} \leqslant \int_0^R \frac{2x dx}{\varepsilon^2 + x^2} \leqslant C_D \log \varepsilon \leqslant C\varepsilon^{-1/2}.
$$

For $n \geqslant 2$ we set $x' = (x_2, \ldots, x_n)$ and integrating with respect to the variable x_1 then gives us

$$
\begin{aligned}
\int_{|x|<R} \frac{dx_1 \wedge \cdots \wedge dx_n}{(|x_1| + |x|^2 + \varepsilon)^2 |x|^{n-2}} &\leqslant C_1 \int_{|x'|\leqslant R} \frac{dx_2 \wedge \cdots \wedge dx_n}{(\varepsilon + |x'|^2)|x'|^{n-2}} \\
&\leqslant C_1 \int_0^R dr \int_{|x'|=r} \frac{d\sigma_{n-2}}{(\varepsilon + r^2) r^{n-2}} \\
&\leqslant C_2 \int_0^R \frac{dr}{\varepsilon + r^2} \leqslant C\varepsilon^{-1/2}.
\end{aligned}
$$

Let us now prove ii). We have

$$
\begin{aligned}
\int_{|x|<R} \frac{dx_1 \wedge \cdots \wedge dx_n}{(|x_1| + |x|^2 + \varepsilon)|x|^{n-1}} &\leqslant \int_{|x|\leqslant R} \frac{dx_1 \wedge \cdots \wedge dx_n}{(|x|^2 + \varepsilon)|x|^{n-1}} \\
&\leqslant \int_0^R dr \int_{|x|=r} \frac{d\sigma_{n-1}}{(\varepsilon + r^2) r^{n-1}} \\
&\leqslant C_0 \int_0^R \frac{dr}{\varepsilon + r^2} \leqslant C\varepsilon^{-1/2}.
\end{aligned}
$$

□

End of the proof of Lemma 1.3. For any $\xi \in \partial D$ let U_ξ be a neighbourhood of ξ such that (1.9) and (1.10) hold for any $z \in U_\xi \cap D$. If $V_\xi \Subset U_\xi$ is a neighbourhood of ξ then the integrals

$$\int_{\partial D \smallsetminus U_\xi} \frac{d\sigma_{2n-1}}{|\Phi|^{n-s}|\zeta - z|^{2s+1}} \quad \text{and} \quad \int_{\partial D \smallsetminus U_\xi} \frac{d\sigma_{2n-1}}{|\Phi|^{n-s-1}|\zeta - z|^{2s+2}}$$

are clearly bounded on $V_\xi \cap D$ and it follows that, for any $z \in V_\xi \cap D$,

$$\int_{\partial D} \frac{d\sigma_{2n-1}}{|\Phi|^{n-s}|\zeta - z|^{2s+1}} \leqslant C_\xi[\mathrm{dist}(z, \partial D)]^{-1/2} \tag{1.13}$$

$$\int_{\partial D} \frac{d\sigma_{2n-1}}{|\Phi|^{n-s-1}|\zeta - z|^{2s+2}} \leqslant C_\xi[\mathrm{dist}(z, \partial D)]^{-1/2}. \tag{1.14}$$

Since ∂D is compact there is a neighbourhood V of ∂D such that the left-hand sides of (1.13) and (1.14) are bounded by $C[\mathrm{dist}(z, \partial D)]^{-1/2}$ for any $z \in V \cap D$. As these two integrals are bounded on $D \smallsetminus V$ this completes the proof of the lemma. □

Theorem 1.6. *Let D be a bounded strictly convex domain with $\mathcal{C}^2$ boundary in $\mathbb{C}^n$. There is then a constant $C > 0$ such that, for any continuous differential form f on $\overline{D}$,*

$$\|R_{\partial D}^{w_\rho} f + \widetilde{B}_D f\|_{1/2,D} \leqslant C|f|_{0,D}. \tag{1.15}$$

In particular, if f is a continuous differential (p,q)-form on $\overline{D}$ where $0 \leqslant p \leqslant n$ and $1 \leqslant q \leqslant n$ such that $\overline{\partial} f = 0$ then the solution of the equation $\overline{\partial} u = f$ given on D by $u = (-1)^{p+q}(R_{\partial D}^{w_\rho} f + \widetilde{B}_D f)$ has the property that

$$|u|_{1/2,D} \leqslant C|f|_{0,D}. \tag{1.16}$$

Proof. By Theorem 1.2 and Lemma 1.3, (1.15) will be established if we can prove that the Leray section w_ρ associated to the domain D verifies the hypotheses of Lemma 1.3.

Let ρ be a strictly convex defining function for D on $\mathbb{C}^n$. The function

$$w_\rho(z,\zeta) = \Big(\frac{\partial \rho}{\partial \zeta_1}(\zeta), \dots, \frac{\partial \rho}{\partial \zeta_n}(\zeta)\Big)$$

is then a Leray section for D so ii) holds. The section w_ρ is independent of z, so i) also holds. It remains to prove iii). Let the coordinates $x_j = x_j(\zeta)$ be the real coordinates of $\zeta \in \mathbb{C}^n$ such that $\zeta_j = x_j(\zeta) + i x_{j+n}(\zeta)$. Set $t_1(z,\zeta) = \mathrm{Im}\langle w_\rho(\zeta), \zeta - z\rangle$. Then

$$t_1(z,\zeta) = \sum_{j=1}^n \frac{1}{2}\Big[\frac{\partial \rho}{\partial x_j}(\zeta) x_{j+n}(\zeta - z) - \frac{\partial \rho}{\partial x_{j+n}}(\zeta) x_j(\zeta - z)\Big]$$

and $$d_\zeta t_1(z,\zeta)\big|_{\zeta = z} = \sum_{j=1}^n \frac{1}{2}\Big[\frac{\partial \rho}{\partial x_j}(z) d_{x_{j+n}}(z) - \frac{\partial \rho}{\partial x_{j+n}}(z) dx_j(z)\Big].$$

But now $d_\zeta t_1(z,\zeta)\big|_{\zeta=z} \wedge d\rho(z) \neq 0$ for any $z \in \partial D$ because the coefficient of $dx_j(z) \wedge dx_{j+n}(z)$ in the differential form $d_\zeta t_1(z,\zeta)\big|_{\zeta=z} \wedge d\rho(z)$ is $-\frac{1}{2}\Big((\frac{\partial \rho}{\partial x_j}(z))^2 + (\frac{\partial \rho}{\partial x_{j+n}}(z))^2\Big)$, which cannot vanish for all $j = 1,\dots,n$ because $d\rho(z) \neq 0$. Choose a point $\xi \in \partial D$. We can then find a neighbourhood U_ξ of ξ and $\mathcal{C}^1$ functions $t_2,\dots,t_{2n-1}$ on U_ξ such that for any $z \in U_\xi$ the functions $(t_1(z,\cdot), t_2, \dots, t_{2n-1})$ are local coordinates on $\partial D \cap U_\xi$. Set $t_j(z,\zeta) = t_j(\zeta) - t_j(z)$ for any $z, \zeta \in U_\xi$ and $j = 2,\dots,2n-1$. For any $z \in U_\xi$ the functions $(t_1(z,\cdot), t_2(z,\cdot), \dots, t_{2n-1}(z,\cdot))$ are then local coordinates on $\partial D \cap U_\xi$. We now prove (1.2). Since $t_j(z,z) = 0$, on restricting U_ξ it follows from Taylor's formula that there is a $\delta_1 > 0$ such that $|\zeta - z| \geqslant \delta_1 |t(z,\zeta)|$ for any $z,\zeta \in U_\xi$. After again restricting U_ξ, Lemma 1.1 implies that there is a δ_2 such that, for any $z, \zeta \in U_\xi$,

$$2|\langle w_\rho(\zeta), \zeta - z\rangle| \geqslant \delta_2\big(|t_1(z,\zeta)| + \rho(\zeta) - \rho(z) + |t(\zeta - z)|^2\big),$$

which proves (1.2) since $\rho = 0$ on ∂D and on restricting U_ξ we can assume there is a $\delta_3 > 0$ such that $-\rho(z) \geqslant \delta_3 \operatorname{dist}(z, \partial D)$ for any $z \in D \cap U_\xi$. □

We end this discussion by proving – using a counter-example of E.M. Stein – that the order 1/2 Hölder bound proved above cannot be improved.

Let $D = \{(z_1,z_2) \in \mathbb{C}^n \mid |z_1|^2 + |z_2|^2 < 1\}$ be the unit ball in $\mathbb{C}^2$. Let log be the branch of the logarithm defined on $\mathbb{C} \setminus \mathbb{R}^+$ with argument between 0 and 2π. We set

$$f(z_1,z_2) = \begin{cases} \dfrac{d\overline{z}_2}{\log(z_1 - 1)} & \text{if } (z_1,z_2) \in \overline{D} \setminus \{(1,0)\} \\ 0 & \text{if } (z_1,z_2) = (1,0). \end{cases}$$

This is a $\mathcal{C}^\infty$ differential $(0,1)$-form on $\overline{D} \setminus \{(1,0)\}$ since $\log(z_1 - 1)$ does not vanish if $z_1 \notin (1,+\infty)$. Moreover, f is continuous on $\overline{D}$ because $|\log(z_1-1)|$ tends to ∞ as $z_1 \to 1$ and the fact that the function $z \mapsto 1/\log(z_1 - 1)$ is holomorphic on D implies that $\overline{\partial} f = 0$ on D. We will prove that if $\alpha > 1/2$ then there is no function u defined on D such that $\overline{\partial} u = f$ on D and $|u|_{\alpha,D} < +\infty$. Suppose that u is a solution of the equation $\overline{\partial} u = f$ such that $|u|_{\alpha,D} < +\infty$. Since $\overline{\partial}(\overline{z}_2/\log(z_1-1)) = f$, the function $u - \overline{z}_2/\log(z_1-1)$ is then holomorphic on D. Let ε be a real number such that $0 < 2\varepsilon < 1$; the circles

$$\{(z_1,z_2) \in \mathbb{C}^2 \mid z_1 = 1 - \varepsilon,\ |z_2| = \sqrt{\varepsilon}\}$$

and

$$\{(z_1,z_2) \in \mathbb{C}^2 \mid z_1 = 1 - 2\varepsilon,\ |z_2| = \sqrt{\varepsilon}\}$$

are then contained in D. Applying Cauchy's theorem to the function $z_2 \mapsto u(z_1, z_2) - \overline{z}_2/\log(z_1 - 1)$ on each of these circles, we get

$$\int_{|z_2|=\sqrt{\varepsilon}} u(1-\varepsilon, z_2)dz_2 = \int_{|z_2|=\sqrt{\varepsilon}} \frac{\overline{z}_2 dz_2}{\log(-\varepsilon)} = \frac{2i\pi\varepsilon}{\log(-\varepsilon)}$$

and
$$\int_{|z_2|=\sqrt{\varepsilon}} u(1-2\varepsilon, z_2)dz_2 = \int_{|z_2|=\sqrt{\varepsilon}} \frac{\overline{z}_2 dz_2}{\log(-2\varepsilon)} = \frac{2i\pi\varepsilon}{\log(-2\varepsilon)}.$$

Since $|u|_{\alpha,D} < +\infty$ there is a constant $C > 0$ such that, for any ε satisfying $0 < 2\varepsilon < 1$,

$$\left| \frac{1}{\log(-\varepsilon)} - \frac{1}{\log(-2\varepsilon)} \right| < C\varepsilon^{\alpha-1/2}.$$

But now $\log(-\varepsilon) = \operatorname{Log}|\varepsilon| + i\pi$ and $\log(-2\varepsilon) = \operatorname{Log} 2 + \operatorname{Log}|\varepsilon| + i\pi$ and it follows that $\operatorname{Log} 2 \leqslant C\varepsilon^{\alpha-1/2}|\log(-\varepsilon)\log(-2\varepsilon)|$, which is impossible.

Using Corollary 3.22 of Chapter VI, Theorem 1.6 gives us a local resolution result for bounded strictly pseudoconvex domains with $\mathcal{C}^2$ boundary.

Corollary 1.7. *Let Ω be a bounded strictly pseudoconvex open set with $\mathcal{C}^2$ boundary in $\mathbb{C}^n$ and let f be a differential (p,q)-form where $0 \leqslant p \leqslant n$ and $1 \leqslant q \leqslant n$ which is continuous on $\overline{\Omega}$ and $\overline{\partial}$-closed on Ω. For any $\xi \in \partial\Omega$ there is then a neighbourhood U_ξ of ξ in $\mathbb{C}^n$, a differential $(p,q-1)$-form $u_\xi \in \Lambda^{1/2}_{p,q-1}(U_\xi \cap \Omega)$ and a constant $C_\xi > 0$ such that $\overline{\partial} u_\xi = f$ on $U_\xi \cap \Omega$ and $|u_\xi|_{1/2, U_\xi\cap\Omega} \leqslant C_\xi |f|_{0,\Omega}$.*

Proof. If $\xi \in \partial\Omega$ then Corollary 3.24 of Chapter VI says that there is a neighbourhood U_ξ of ξ in $\mathbb{C}^n$ and a biholomorphic map h_ξ defined in a neighbourhood of $\overline{U}_\xi$ such that $h_\xi(U_\xi \cap \Omega)$ is a bounded strictly convex domain with $\mathcal{C}^2$ boundary. If we set $u_\xi = h^*_\xi T h_{\xi *} f$, where $T = R^{w_\rho}_{\partial h(U_\xi\cap\Omega)} + \widetilde{B}_{h(U_\xi\cap\Omega)}$, then u_ξ has the required properties by Theorem 1.6. □

2 Local uniform approximation of $\overline{\partial}$-closed forms in strictly pseudoconvex domains

We start by proving that any holomorphic function defined in a neighbourhood of some compact convex set in $\mathbb{C}^n$ can be uniformly approximated on this set by holomorphic functions on $\mathbb{C}^n$.

Proposition 2.1. *Let K be a convex compact set in $\mathbb{C}^n$ and let V be a neighbourhood of K. If $0 \leqslant p \leqslant n$ then any holomorphic $(p,0)$-form on V can be uniformly approximated on K by holomorphic $(p,0)$-forms defined on $\mathbb{C}^n$.*

Proof. Since K is a compact convex set in $\mathbb{C}^n$ and V is a neighbourhood of K there is a convex open set D with $\mathcal{C}^2$ boundary such that $K \subset D \Subset V$ and $D = \{z \in \mathbb{C}^n \mid \rho(z) < 0\}$. Here, ρ is a $\mathcal{C}^2$ convex function on $\mathbb{C}^n$ such that

$d\rho(z) \neq 0$ for any $z \in \mathbb{C}^n \setminus D$. Let h be a holomorphic $(p,0)$-form on V, set $D_\alpha = \{z \in \mathbb{C}^n \mid \rho(z) < \alpha\}$ and let α_0 be the supremum of the set of real numbers $\alpha \in \mathbb{R}$ such that h can be uniformly approximated on K by holomorphic $(p,0)$-forms defined on D_α. Since $K \subset D \Subset V$ we know that $\alpha_0 > 0$.

We start by proving that $\alpha_0 = +\infty$. Argue by contradiction and suppose that $\alpha_0 < +\infty$. For any $\alpha > 0$, consider the Leray section $w_\rho(\zeta) = \left(\frac{\partial \rho}{\partial \zeta_1}(\zeta), \ldots, \frac{\partial \rho}{\partial \zeta_n}(\zeta)\right)$ for D_α. The function $\Phi(z,\zeta) = \langle w_\rho(\zeta), \zeta - z\rangle$ is then holomorphic with respect to z on $\mathbb{C}^n$. Let $\delta > 0$ be a sufficiently small real number: by definition of α_0, the $(p,0)$-form h can be uniformly approximated on K by holomorphic $(p,0)$-forms defined on some neighbourhood of $\overline{D}_{\alpha_0-\delta}$. It will be enough to show that any holomorphic $(p,0)$-form defined on a neighbourhood of $\overline{D}_{\alpha_0-\delta}$ can be uniformly approximated on K by holomorphic $(p,0)$-forms defined on $D_{\alpha_0+\delta}$. If f is a holomorphic $(p,0)$-form on a neighbourhood of $\overline{D}_{\alpha_0-\delta}$ then

$$f(z) = \frac{(n-1)!}{(2i\pi)^n} \int_{\partial D_{\alpha_0-\delta}} f(\zeta) \frac{\omega'_\zeta(w_\rho(\zeta)) \wedge \omega(\zeta)}{\Phi^n(z,\zeta)}, \qquad \text{for any } z \in D_{\alpha_0-\delta}.$$

by Leray's formula. Since $\partial D_{\alpha_0-\delta}$ is compact, to construct the required approximating form it will be enough to prove that for any given point $\xi_0 \in \partial D_{\alpha_0-\delta}$ the function $1/\Phi(z,\zeta)$ can be uniformly approximated on $K \times V_{\xi_0}$ by functions which are holomorphic in z on $D_{\alpha_0+\delta}$, where V_{ξ_0} is some neighbourhood of ξ_0 in $\partial D_{\alpha_0-\delta}$. Fix a point $\xi_0 \in \partial D_{\alpha_0-\delta}$ and choose a finite number of points $\xi_1, \ldots, \xi_k$ such that

$$\alpha_0 - \delta = \rho(\xi_1) < \cdots < \rho(\xi_k) = \alpha_0 + \delta$$

and

$$\sup_{z\in K} \left|1 - \frac{\Phi(z,\xi_{j-1})}{\Phi(z,\xi_j)}\right| < 1 \quad \text{for any } j = 1, \ldots, k.$$

We note that if $\left|1 - \frac{\Phi(z,\xi_0)}{\Phi(z,\xi_1)}\right| < 1$ for any $z \in K$ then $\sup_{z\in K}\left|1 - \frac{\Phi(z,\zeta)}{\Phi(z,\xi_1)}\right| < r < 1$ for any ζ in some neighbourhood V_{ξ_0} of ξ_0. Then, for any $\zeta \in V_{\xi_0}$,

$$\frac{1}{\Phi(z,\zeta)} = \frac{1}{\Phi(z,\xi_1)} \sum_{s=0}^{\infty} \left[1 - \frac{\Phi(z,\zeta)}{\Phi(z,\xi_1)}\right]^s,$$

where the above sum converges uniformly on $K \times V_{\xi_0}$ and

$$\frac{1}{\Phi(z,\xi_{j-1})} = \frac{1}{\Phi(z,\xi_j)} \sum_{s=0}^{\infty} \left[1 - \frac{\Phi(z,\xi_{j-1})}{\Phi(z,\xi_j)}\right]^s,$$

where the above sum converges uniformly with respect to z on K. It follows that the function $1/\Phi(z,\zeta)$ can be uniformly approximated on $K \times V_{\xi_0}$ by polynomials in $\Phi(z,\zeta), \Phi(z,\xi_1), \ldots, \Phi(z,\xi_{k-1})$ and $1/\Phi(z,\xi_k)$, all of which are holomorphic functions on $D_{\alpha_0+\delta}$. After integrating over $\partial D_{\alpha_0-\delta}$ using a finite partition of unity subordinate to a finite cover of $\partial D_{\alpha_0-\delta}$ by open sets of the form V_{ξ_0}, we get a uniform approximation of f on K by holomorphic $(p,0)$-forms on $D_{\alpha_0+\delta}$. It follows that $\alpha_0 = +\infty$.

To complete the proof it remains to show that for any $\varepsilon > 0$ there is a sequence of holomorphic $(p,0)$-forms g_j defined on D_{j+1}, $j = 0,1,\dots$, such that $g_0 = h$, $D = D_0 \subset V$ and

$$\sup_{z\in D_j} |g_j(z) - g_{j+1}(z)| < \frac{\varepsilon}{2^{j+1}} \quad \text{for any } j = 0,1,\dots$$

Indeed, the sequence $(g_j)_{j\in\mathbb{N}}$ would then converge uniformly on all compact sets in $\mathbb{C}^n$ to a holomorphic $(p,0)$-form g on $\mathbb{C}^n$ such that $\sup_K |h-g| < \varepsilon$. To construct this sequence, assume $g_0,\dots,g_k$ already constructed (for $k = 0$ we have $g_0 = h$): since $\overline{D}_k$ is compact and convex the above method proves that there are holomorphic $(p,0)$-forms g_{k+1} defined on D_{k+2} such that $|g_k(z) - g_{k+1}(z)| < \varepsilon/2^{k+1}$ for any $z \in \overline{D}_k$. □

Theorem 2.2. *Let D be a bounded strictly convex open set with $\mathcal{C}^2$ boundary in $\mathbb{C}^n$. If K is a compact set in $\overline{D}$ then every $\overline{\partial}$-closed differential form on D which is continuous on $\overline{D}$ is a uniform limit on K of continuous $\overline{\partial}$-closed differential forms on $\mathbb{C}^n$.*

Proof. After translating if necessary, we can assume that $0 \in D$. Let $\lambda > 1$ be a real number and set $D_\lambda = \{z \in \mathbb{C}^n \mid z/\lambda \in D\}$. The set D_λ is then a bounded strictly convex open set with $\mathcal{C}^2$ boundary in $\mathbb{C}^n$ which contains $\overline{D}$. If f is a $\overline{\partial}$-closed differential (p,q)-form on D which is continuous on $\overline{D}$, where $0 \leqslant p,q \leqslant n$, then we set $f_\lambda = \varphi_\lambda^* f$, where φ_λ is the homothety of ratio $1/\lambda$. The function f_λ is then a $\overline{\partial}$-closed differential (p,q)-form on D_λ. Fix $\varepsilon > 0$ and choose $\lambda > 1$ such that

$$\sup_{\overline{D}} |f - f_\lambda| < \frac{\varepsilon}{2}. \tag{2.1}$$

Suppose initially that $q = 0$. The holomorphic $(p,0)$-form f_λ is defined on D_λ, which is a neighbourhood of the compact convex set $\overline{D}$. By Proposition 2.1, there is a holomorphic $(p,0)$-form g on $\mathbb{C}^n$ such that

$$\sup_{\overline{D}} |f_\lambda - g| < \frac{\varepsilon}{2}. \tag{2.2}$$

It then follows from (2.1) and (2.2) that $\sup_{\overline{D}} |f - g| < \varepsilon$.

If $q \geqslant 1$ then there is a form $u_\lambda \in \mathcal{C}^0_{p,q-1}(D_{(\lambda+1)/2})$ such that $\overline{\partial} u_\lambda = f_\lambda$ on $D_{(\lambda+1)/2}$ because $D_{(\lambda+1)/2}$ is a bounded strictly convex open set with $\mathcal{C}^2$ boundary contained in D_λ (cf. Th. 1.2). Since $D_{(\lambda+1)/2}$ contains $\overline{D}$ we can find a $\mathcal{C}^\infty$ function χ_λ with compact support in $D_{(\lambda+1)/2}$ which is equal to 1 on $\overline{D}$. We then set $g = \overline{\partial}\chi_\lambda u_\lambda$ and by (2.1) we have $\sup_{\overline{D}} |f - g| < \frac{\varepsilon}{2} < \varepsilon$. □

Corollary 2.3. *Let Ω be a bounded strictly pseudoconvex open set with $\mathcal{C}^2$ boundary in $\mathbb{C}^n$. For any $\xi \in \partial\Omega$ there is a neighbourhood U_ξ of ξ in $\mathbb{C}^n$ such that any continuous differential form on $\overline{U_\xi \cap \Omega}$ which is $\overline{\partial}$-closed on $U_\xi \cap \Omega$ is a uniform limit of continuous $\overline{\partial}$-closed differential forms on $\mathbb{C}^n$ on any compact set in $\overline{\Omega} \cap U_\xi$.*

Proof. This follows immediately from Theorem 2.2 and Corollary 3.24 of Chapter VI. □

3 Finiteness of Dolbeault cohomology groups of strictly pseudoconvex domains

The aim of this section is to deduce finiteness theorems for certain groups of $\overline{\partial}$-cohomology from local resolutions of $\overline{\partial}$ with Hölder estimates. To do this we will need some ideas and results from functional analysis which are summarised in Appendix C.

Let D be a bounded open set in $\mathbb{C}^n$. We consider the Banach spaces $\mathcal{C}^0_{p,q}(\overline{D})$ of continuous differential (p,q)-forms on $\overline{D}$ and the Banach spaces $\Lambda^{1/2}_{p,q}(D)$ of Hölder continuous differential (p,q)-forms of order $1/2$ on D (cf. Chap. III, § 2). We set $\Lambda^{1/2}_{p,-1}(D) = 0$ and if $q \geqslant 1$ we let $F^{1/2}_{p,q-1}(D)$ be the domain of definition of $\overline{\partial}$, considered as an operator from $\Lambda^{1/2}_{p,q-1}(D)$ to $\mathcal{C}^0_{p,q}(\overline{D})$, i.e. the subspace of differential forms $f \in \Lambda^{1/2}_{p,q-1}(D)$ such that the distribution $\overline{\partial} f$ is continuous on D and has a continuous extension to $\overline{D}$. We set $E^{1/2}_{p,q}(D) = \overline{\partial}(F^{1/2}_{p,q-1}(D))$. We let $Z^0_{p,q}(\overline{D})$ be the subspace of differential forms $f \in \mathcal{C}^0_{p,q}(\overline{D})$ such that $\overline{\partial} f = 0$ on D – this is a closed subspace of the Banach space $\mathcal{C}^0_{p,q}(\overline{D})$. And finally, we let $H^{p,q}_{0,1/2}(\overline{D})$ be the quotient space $Z^0_{p,q}(\overline{D})/E^{1/2}_{p,q}(D)$.

Proposition 3.1. *Let D be a bounded strictly pseudoconvex open set with $\mathcal{C}^2$ boundary in $\mathbb{C}^n$ and let p and q be integers such that $0 \leqslant p \leqslant n$ and $0 \leqslant q \leqslant n$. The following then hold:*

i) *there are continuous linear operators T^p_q from $\mathcal{C}^0_{p,q}(\overline{D})$ to $\Lambda^{1/2}_{p,q-1}(D)$ with the property that, for any $f \in \mathcal{C}^0_{p,q}(\overline{D})$ such that $\overline{\partial} f \in \mathcal{C}^0_{p,q}(\overline{D})$,*

$$f = \overline{\partial} T^p_q f + T^p_{q+1} \overline{\partial} f + K^p_q f, \tag{3.1}$$

where K^p_q is a compact operator from $\mathcal{C}^0_{p,q}(\overline{D})$ to itself.

ii) *if $q \geqslant 1$ then $\overline{\partial} T^p_q$ defines a continuous linear operator from $Z^0_{p,q}(\overline{D})$ to itself whose image has finite codimension.*

iii) *the space $E^{1/2}_{p,q}(D)$ is a closed subspace of finite codimension in $Z^0_{p,q}(\overline{D})$.*

Proof. By Corollary 3.24 of Chapter VI there is a finite cover of $\overline{D}$ by open sets $U_1, \dots, U_m$ in $\mathbb{C}^n$ such that for any $j = 1, \dots, m$ there is a biholomorphic map h_j defined on a neighbourhood of $\overline{U}_j$ such that $h_j(U_j \cap D)$ is a bounded strictly convex open set with $\mathcal{C}^2$ boundary. It then follows from Theorems 1.2 and 1.6 that we can find continuous linear operators $(T_q^p)_j$, $j = 1, \dots, m$, from $\mathcal{C}^0_{p,q}(\overline{D})$ to $\Lambda^{1/2}_{p,q-1}(U_j \cap D)$ such that, for any $f \in \mathcal{C}^0_{p,q}(\overline{D})$ and $\overline{\partial} f \in \mathcal{C}^0_{p,q+1}(\overline{D})$,

$$f = \overline{\partial}(T_q^p)_j f + (T_{q+1}^p)_j \overline{\partial} f \quad \text{on } U_j \cap D. \tag{3.2}$$

Choose $\mathcal{C}^\infty$ functions $(\varphi_j)_{j=1,\dots,m}$ with compact support on $\mathbb{C}^n$ such that $D \cap \operatorname{supp} \varphi_j \subset U_j$ and $\sum_{j=1}^m \varphi_j = 1$ on $\overline{D}$ and set $T_q^p f = \sum_{j=1}^m \varphi_j (T_q^p)_j f$ for any $f \in \mathcal{C}^0_{p,q}(\overline{D})$. This defines a continuous linear operator from $\mathcal{C}^0_{p,q}(\overline{D})$ to $\Lambda^{1/2}_{p,q-1}(D)$ such that

$$f = \overline{\partial} T_q^p f + T_{q+1}^p \overline{\partial} f - \sum_{j=1}^m \overline{\partial}\varphi_j \wedge (T_q^p)_j f \quad \text{on } D.$$

Set $K_q^p f = -\sum_{j=1}^m \overline{\partial}\varphi_j \wedge (T_q^p)_j f$ for any $f \in \mathcal{C}^0_{p,q}(\overline{D})$. The operator K_q^p thus defined is a continuous linear operator from $\mathcal{C}^0_{p,q}(\overline{D})$ to $\Lambda^{1/2}_{p,q}(D)$: by Ascoli's theorem it therefore defines a compact operator from $\mathcal{C}^0_{p,q}(\overline{D})$ to itself and i) follows. Moreover, K_q^p sends $Z^0_{p,q}(\overline{D})$ to itself. Indeed, if $f \in Z^0_{p,q}(\overline{D})$ then

$$\begin{aligned}
\overline{\partial} K_q^p f &= -\overline{\partial}\Big(\sum_{j=1}^m \overline{\partial}\varphi_j \wedge (T_q^p)_j f\Big) = \sum_{j=1}^m \overline{\partial}\varphi_j \wedge \overline{\partial}(T_q^p)_j f \\
&= \sum_{j=1}^m \overline{\partial}\varphi_j \wedge f \quad \text{by (3.2) since } f \in Z^0_{p,q}(\overline{D}) \\
&= 0 \quad \text{since } \textstyle\sum_{j=1}^m \varphi_j = 1 \text{ on } \overline{D}.
\end{aligned}$$

By (3.1) we know that $\overline{\partial} T_q^p = I - K_q^p$ on $Z^0_{p,q}(\overline{D})$ for any $q \geqslant 1$ and as K_q^p is compact the image of $I - K_q^p$ has finite codimension in $Z^0_{p,q}(\overline{D})$, which proves ii).

Condition iii) follows from Proposition 4 of Appendix C because $E^{1/2}_{p,q}(D)$, which is the image under $\overline{\partial}$ of $F^{1/2}_{p,q-1}(D)$, is of finite codimension because it contains the image of $\overline{\partial} T_q^p$. □

Corollary 3.2. *If D is a bounded strictly pseudoconvex open set with $\mathcal{C}^2$ boundary in $\mathbb{C}^n$ then*

$$\dim H^{p,q}_{0,1/2}(\overline{D}) < +\infty \quad \textit{whenever } 0 \leqslant p \leqslant n \textit{ and } 1 \leqslant q \leqslant n.$$

4 Invariance of Dolbeault cohomology under strictly pseudoconvex extensions

Using Grauert's Beulenmethode, we will now prove that for any bounded strictly pseudoconvex domain Ω in $\mathbb{C}^n$ the $\overline{\partial}$-cohomology group $H^{p,q}_{0,1/2}(\overline{\Omega})$ is isomorphic to the Dolbeault cohomology group $H^{p,q}(\Omega)$ defined in Chapter II. Corollary 3.2 then implies that the Dolbeault cohomology groups of a bounded strictly pseudoconvex domain in $\mathbb{C}^n$ are finite-dimensional.

Definition 4.1. By an *elementary strictly pseudoconvex extension* we mean an ordered pair $[\theta_1, \theta_2]$ of open sets in $\mathbb{C}^n$ with $\mathcal{C}^2$ boundary such that $\theta_1 \subset \theta_2$ and the following holds: there is an open set V containing $\overline{\theta_2 \setminus \theta_1}$, there are strictly pseudoconvex domains D_1 and D_2 such that $D_1 \subset D_2$, $\theta_2 = \theta_1 \cup D_2$, $\theta_1 \cap D_2 = D_1$ and $\overline{(\theta_1 \setminus D_2)} \cap \overline{(\theta_2 \setminus \theta_1)} = \varnothing$ and there is a biholomorphic map h defined on a neighbourhood of $\overline{V}$ such that for $j = 1, 2$ the set $h(D_j)$ is a bounded strictly convex domain with $\mathcal{C}^2$ boundary in $\mathbb{C}^n$.

Lemma 4.2. *Let $[\theta_1, \theta_2]$ be an elementary strictly pseudoconvex extension. For any (p, q) such that $0 \leqslant p \leqslant n$ and $0 \leqslant q \leqslant n$,*

i) *the restriction map $H^{p,q}_{0,1/2}(\overline{\theta}_2) \to H^{p,q}_{0,1/2}(\overline{\theta}_1)$ is surjective if $q \geqslant 1$,*
ii) *the restriction map $H^{p,q}_{0,1/2}(\overline{\theta}_2) \to H^{p,q}_{0,1/2}(\overline{\theta}_1)$ is injective if $q \geqslant 2$ or if $q = 1$ and θ_1 is bounded and strictly pseudoconvex,*
iii) *the restriction map $Z^0_{p,q}(\overline{\theta}_2) \to Z^0_{p,q}(\overline{\theta}_1)$ has dense image if θ_1 is bounded and strictly pseudoconvex.*

Proof. Since $\overline{(\theta_1 \setminus \theta_2)} \cap \overline{(\theta_2 \setminus \theta_1)} = \varnothing$ we can find neighbourhoods V' and V'' of $\overline{\theta_2 \setminus \theta_1}$ such that $V' \Subset V'' \Subset V$ and $V'' \cap \overline{(\theta_1 \setminus D_2)} = \varnothing$. Choose a $\mathcal{C}^\infty$ function χ on $\mathbb{C}^n$ such that $\chi = 1$ on V' and $\operatorname{supp} \chi \subset V''$.

Let us prove i). Consider an element $f_1 \in Z^0_{p,q}(\overline{\theta}_1)$, where $1 \leqslant q \leqslant n$: we seek differential forms $u_1 \in \Lambda^{1/2}_{p,q-1}(\theta_1)$ and $f_2 \in Z^0_{p,q}(\overline{\theta}_2)$ such that $f_2 = f_1 - \overline{\partial} u_1$ on θ_1. As D_1 is the image under a biholomorphic map defined in a neighbourhood of $\overline{D}_1$ of a bounded strictly convex domain with $\mathcal{C}^2$ boundary there is a form $u \in \Lambda^{1/2}(D_1)$ such that $f_1 = \overline{\partial} u$ on D_1 (cf. Theorem 1.6). Set

$$u_1 = \begin{cases} 0 & \text{on } \overline{\theta_1 \setminus D_2} \\ \chi u & \text{on } \overline{D}_1 \end{cases} \quad \text{and} \quad f_2 = \begin{cases} f_1 - \overline{\partial} u_1 & \text{on } \overline{\theta}_1 \\ 0 & \text{on } \overline{\theta_2 \setminus \theta_1}. \end{cases}$$

The differential forms u_1 and f_2 then satisfy the required conditions.

Let us prove ii). Let $f_2 \in Z^0_{p,q}(\overline{\theta}_2)$ and $u_1 \in \Lambda^{1/2}_{p,q-1}(\theta_1)$ be such that $\overline{\partial} u_1 = f_2$ on θ_1. We seek an element $u_2 \in \Lambda^{1/2}_{p,q-1}(\theta_2)$ such that $\overline{\partial} u_2 = f_2$ on θ_2. As D_2 is the image under a biholomorphic map defined in a neighbourhood of $\overline{D}_2$ of a bounded strictly convex domain with $\mathcal{C}^2$ boundary there is a function $u \in \Lambda^{1/2}_{p,q-1}(\overline{D}_2)$ such that $f_2 = \overline{\partial} u$ on D_2 and hence $u - u_1 \in Z^0_{p,q-1}(\overline{D}_1)$.

If $q \geqslant 2$ then there is a form $v \in \Lambda^{1/2}_{p,q-2}(D_1)$ such that $\overline{\partial} v = u - u_1$ on D_1. Therefore, $u_1 - \overline{\partial}(\chi v) = u$ on $V' \cap \theta_1$. Setting

$$u_2 = \begin{cases} u_1 - \overline{\partial}(\chi v) & \text{on } \theta_1 \\ u & \text{on } \theta_2 \cap V' \end{cases}$$

we get $u \in \Lambda^{1/2}_{p,q-1}(\theta_2)$ and $\overline{\partial} u_2 = f_2$ on θ_2.

Suppose that $q = 1$ and θ_1 is bounded and strictly pseudoconvex. It then follows from the definition of an elementary strictly pseudoconvex extension that θ_2 is also bounded and strictly pseudoconvex and hence $E^{1/2}_{p,1}(\theta_2)$ is a closed subspace of $Z^0_{p,1}(\overline{\theta}_2)$ by Proposition 3.1, iii). It is therefore enough to construct a sequence $(w_k)_{k\in\mathbb{N}}$ of Hölder continuous differential $(p,0)$-forms of order $1/2$ on θ_2 such that the forms $\overline{\partial} w_k$ are continuous on $\overline{\theta}_2$ and $\lim_{k\to\infty} |f_2 - \overline{\partial} w_k|_{0,\overline{D}_2} = 0$. The differential $(p,0)$-form $u - u_1$ is continuous on $\overline{D}_1$ and $\overline{\partial}$-closed on D_1. It therefore follows from Theorem 2.2 that there is a sequence $(v_k)_{k\in\mathbb{N}}$ of holomorphic $(p,0)$-forms on V such that

$$\lim_{k\to\infty} |v_k - (u - u_1)|_{0,\overline{D}_1\cap \operatorname{supp}\chi} = 0.$$

We now define a sequence $(w_k)_{k\in\mathbb{N}}$ by setting $w_k = (1-\chi)u_1 + \chi(u - v_k)$. Then $\overline{\partial} w_k = f_2 + (\overline{\partial}\chi)(u - u_1 - v_k)$ and the sequence $(w_k)_{k\in\mathbb{N}}$ therefore has the required properties.

We end the proof of the lemma by proving iii). Consider an element $f \in Z^0_{p,q}(\overline{\theta}_1)$. By Theorem 2.2, there is a sequence $(v_k)_{k\in\mathbb{N}}$ of $\overline{\partial}$-closed continuous (p,q)-forms on V such that $\lim_{k\to\infty} |f - v_k|_{0,\overline{\theta}_1\cap\operatorname{supp}\chi} = 0$. Then

$$\lim_{k\to\infty} |\overline{\partial}\chi \wedge (f - v_k)|_{0,\overline{\theta}_2} = 0.$$

Set $\widetilde{f}_k = (1-\chi)f + \chi v_k$: this is a sequence of continuous differential forms on $\overline{\theta}_2$ which converges uniformly to f on $\overline{\theta}_1$ such that $\overline{\partial}\widetilde{f}_k = \overline{\partial}\chi \wedge (f - v_k)$ converges uniformly to 0 on $\overline{\theta}_2$. By Proposition 5 of Appendix C, it follows from Proposition 3.1 that there is a continuous linear operator T from $Z^0_{p,q}(\overline{\theta}_2)$ to $C^0_{p,q-1}(\overline{\theta}_2)$ such that $\overline{\partial} T = I$ on $E^0_{p,q}(\theta_2)$ because θ_1 and therefore θ_2 are strictly pseudoconvex. We set $u_k = T(\overline{\partial}\chi \wedge (f - v_k))$. Then $\lim_{k\to\infty} |u_k|_{0,\overline{\theta}_2} = 0$ and $\overline{\partial} u_k = \overline{\partial}\widetilde{f}_k$ and it follows that if $f_k = \widetilde{f}_k - u_k$ then the sequence $(f_k)_{k\in\mathbb{N}}$ is contained in $Z^0_{p,q}(\overline{\theta}_2)$ and converges uniformly to f on $\overline{\theta}_1$. □

Lemma 4.3. *Let Ω be a bounded strictly pseudoconvex domain with $\mathcal{C}^2$ boundary in $\mathbb{C}^n$ and let ρ be a strictly plurisubharmonic defining function for Ω, i.e. $\Omega = \{z \in U_{\partial\Omega} \mid \rho(z) < 0\}$. Set $\Omega_\varepsilon = \{z \in U_{\partial\Omega} \mid \rho(z) < \varepsilon\} \cup \Omega$. For small enough $|\varepsilon|$, Ω_ε is a strictly pseudoconvex domain with $\mathcal{C}^2$ boundary. There is then a real number $\varepsilon_0 > 0$ such that whenever $-\varepsilon_0 \leqslant \alpha < 0 \leqslant \beta \leqslant \varepsilon_0$*

there is a finite set of domains $\theta_1, \dots, \theta_N$ *such that* $\Omega_\alpha = \theta_1 \subset \theta_2 \subset \cdots \subset \theta_N = \Omega_\beta$ *and* $[\theta_j, \theta_{j+1}]$ *is an elementary strictly pseudoconvex extension for any* $j = 1, \dots, N-1$.

Proof. Theorem 3.23 of Chapter VI says that for any $\xi \in \partial\Omega$ there is a neighbourhood U_ξ of ξ in $\mathbb{C}^n$ and a holomorphic function h_ξ defined in some neighbourhood of $\overline{U}_\xi$ such that $h(\overline{U_\xi \cap \Omega})$ is strictly convex and has $\mathcal{C}^2$ boundary. As $\partial\Omega$ is compact we can extract a finite subcover $(U_i)_{1 \leqslant i \leqslant n}$ from the cover $(U_\xi)_{\xi \in \partial\Omega}$. There is then a real number $\varepsilon_0 > 0$ such that $\overline{\Omega}_{\varepsilon_0} \setminus \Omega_{-\varepsilon_0} \subset \bigcup_{i=1}^N U_i$. Choose $\mathcal{C}^\infty$ functions $(\chi_j)_{j=1,\dots,N}$ with compact support in $\mathbb{C}^n$ such that $\operatorname{supp} \chi_j \subset U_j$, $j = 1, \dots, N$, and $\sum_{j=1}^N \chi_j = 1$ on $\overline{\Omega}_{\varepsilon_0} \setminus \Omega_{-\varepsilon_0}$. Given α and β such that $-\varepsilon_0 \leqslant \alpha < 0 < \beta \leqslant \varepsilon_0$ we set

$$\theta_k = \{z \in \mathbb{C}^n \mid \rho(z) - \alpha < (\beta - \alpha) \textstyle\sum_{j=1}^k \chi_j(z)\}.$$

Then $\Omega_\alpha = \theta_0 \subset \cdots \subset \theta_N = \Omega_\beta$ and for any small enough ε_0 the pair $[\theta_{j+1}, \theta_j]$ is then an elementary strictly pseudoconvex extension. □

Definition 4.4. Let $D \Subset \Omega \Subset \mathbb{C}^n$ be open sets in $\mathbb{C}^n$. We say that Ω is a *strictly pseudoconvex extension* of D if there is a neighbourhood U of $\overline{\Omega} \setminus D$ and a strictly plurisubharmonic $\mathcal{C}^2$ function ρ on U such that

$$D \cap U = \{z \in U \mid \rho(z) < 0\} \quad \text{and} \quad \Omega \cap U = \{z \in U \mid \rho(z) < 1\}.$$

We will say that the extension is *non-critical* if the function ρ has no critical points.

The results proved below are still valid if the non-critical hypothesis is dropped. They will be proved in full generality in Section 8.

Proposition 4.5. *Let* D *and* Ω *be open sets in* $\mathbb{C}^n$, $D \Subset \Omega \Subset \mathbb{C}^n$, *such that* Ω *is a non-critical strictly pseudoconvex extension of* D. *Then,*

i) *for any* (p,q) *such that* $0 \leqslant p \leqslant n$ *and* $0 \leqslant q \leqslant n$, *the restriction map* $H^{p,q}_{0,1/2}(\overline{\Omega}) \to H^{p,q}_{0,1/2}(\overline{D})$ *is an isomorphism,*
ii) *for any* (p,q) *such that* $0 \leqslant p \leqslant n$ *and* $0 \leqslant q \leqslant n$ *the restriction map* $Z^0_{p,q}(\overline{\Omega}) \to Z^0_{p,q}(\overline{D})$ *has dense image for the topology defined by the* $|\ |_{0,\overline{D}}$ *norm.*

Proof. By definition of a non-critical strictly pseudoconvex extension there is a neighbourhood U of $\overline{\Omega} \setminus D$ and a $\mathcal{C}^2$ strictly plurisubharmonic function ρ on U without critical points such that $D \cap U = \{z \in U \mid \rho(z) < 0\}$ and $\Omega \cap U = \{z \in U \mid \rho(z) < 1\}$. Set $\Omega_t = D \cup \{z \in U \mid \rho(z) < t\}$ for any $0 \leqslant t \leqslant 1$. Let T be the supremum of the set E of values $t \in [0,1]$ such that there is a finite sequence $\theta_1, \dots, \theta_k$ of open sets in $\mathbb{C}^n$ such that $D = \theta_0 \subset \cdots \subset \theta_k = \Omega_t$ and $[\theta_{j-1}, \theta_j]$, $j = 1, \dots, k$, is an elementary strictly pseudoconvex extension. Let us prove that $T = 1$. Indeed, if $T < 1$ then by Lemma 4.3 there is a $\varepsilon_0 > 0$ such that we can pass from $\Omega_{T-\varepsilon}$ to $\Omega_{T+\varepsilon}$ by a finite sequence of elementary strictly pseudoconvex extensions for every $\varepsilon < \varepsilon_0$. Choose a real number $\varepsilon < \varepsilon_0$ such that $T - \varepsilon \in E$. Then $T + \varepsilon$ is also in E, which contradicts the fact that T is the supremum of E. It follows that $T = 1$. Applying Lemma 4.3 to $\Omega = \Omega_1$ with $\beta = 0$ and $\alpha \in E$ we get in fact $1 \in E$.

We have therefore proved that there is a finite sequence $\theta_0, \ldots, \theta_N$ of open sets in $\mathbb{C}^n$ such that

$$D = \theta_0 \subset \cdots \subset \theta_N = \Omega$$

and that each $[\theta_{j-1}, \theta_j]$ is an elementary strictly pseudoconvex extension for $j = 1, \ldots, N$. Each of the sets θ_j, $j = 0, \ldots, N$, is then a bounded strictly pseudoconvex open set and by Lemma 4.2 all the restriction maps $H^{p,q}_{0,1/2}(\overline{\theta}_j) \xrightarrow{\varphi_j} H^{p,q}_{0,1/2}(\overline{\theta}_{j-1})$ are isomorphisms if $0 \leqslant p \leqslant n$ and $1 \leqslant q \leqslant n$. It follows that the restriction map $\varphi = \varphi_N \circ \cdots \circ \varphi_1 : H^{p,q}_{0,1/2}(\overline{\Omega}) \to H^{p,q}_{0,1/2}(\overline{D})$ is an isomorphism. Moreover, all the restriction maps $Z^0_{p,q}(\overline{\theta}_j) \xrightarrow{\psi_j} Z^0_{p,q}(\overline{\theta}_{j-1})$ have dense image if $0 \leqslant p \leqslant n$ and $0 \leqslant q \leqslant n$ and hence the restriction map $\psi = \psi_N \circ \cdots \circ \psi_1 : Z^0_{p,q}(\overline{\Omega}) \to Z^0_{p,q}(\overline{D})$ has dense image. □

Theorem 4.6. *Let Ω be a bounded strictly pseudoconvex domain with $\mathcal{C}^2$ boundary in $\mathbb{C}^n$. For any pair (p,q) such that $0 \leqslant p \leqslant n$ and $1 \leqslant q \leqslant n$ the restriction map*

$$H^{p,q}_{0,1/2}(\overline{\Omega}) \longrightarrow H^{p,q}(\Omega)$$

is then an isomorphism.

Proof. Let U be a neighbourhood of $\partial\Omega$. Choose $D \Subset \Omega$ such that $\Omega \setminus U \Subset D$ and Ω is a non-critical strictly pseudoconvex extension of D. We then have restriction maps

$$H^{p,q}_{0,1/2}(\overline{\Omega}) \xrightarrow{\varphi} H^{p,q}(\Omega) \xrightarrow{\psi} H^{p,q}_{0,1/2}(\overline{D}).$$

Indeed, by the Dolbeault isomorphism, $H^{p,q}(\Omega)$ can be identified with $Z^0_{p,q}(\Omega)/Z^0_{p,q}(\Omega) \cap \overline{\partial}\mathcal{C}^0_{p,q-1}(\Omega)$ and φ is therefore well defined since $E^{1/2}_{p,q}(\overline{\Omega}) \subset Z^0_{p,q}(\Omega) \cap \overline{\partial}\mathcal{C}^0_{p,q-1}(\Omega)$. The restriction map ψ is also well defined by the regularity of $\overline{\partial}$ (cf. Chap. III, Remark 3.6). By Proposition 4.5, $\psi \circ \varphi$ is an isomorphism so φ is injective and ψ is surjective. To prove the theorem it is enough to show that ψ is injective which follows from the following lemma. □

Lemma 4.7. *Let D and Ω be two open sets in $\mathbb{C}^n$ such that $D \Subset \Omega$. We suppose there is a neighbourhood U of $\Omega \setminus D$ and a $\mathcal{C}^2$ strictly plurisubharmonic function ρ without critical points such that $D \cap U = \{z \in U \mid \rho(z) < 0\}$ and $D \cup \{z \in U \mid \rho(z) \leqslant C\} \Subset \Omega$ for any $C > 0$. The restriction map*

$$H^{p,q}(\Omega) \longrightarrow H^{p,q}_{0,1/2}(\overline{D})$$

is then injective for any (p,q) such that $0 \leqslant p \leqslant n$ and $1 \leqslant q \leqslant n$.

Proof. We can construct a sequence

$$D = D_0 \subset D_1 \subset \cdots \subset D_k \subset \cdots$$

of bounded open sets in $\mathbb{C}^n$ such that $\Omega = \bigcup_{k \geqslant 0} D_k$ and the set D_{k+1} is a strictly pseudoconvex extension of D_k without critical points for any $k \in \mathbb{N}$. Let $f \in Z^0_{p,q}(\Omega)$ have the property that $f = \overline{\partial} w_0$ on D, where $w_0 \in \Lambda^{1/2}_{p,q-1}(D)$. By Proposition 4.5 i) there is a sequence $w_k \in \Lambda^{1/2}_{p,q-1}(D_k)$ such that $f = \overline{\partial} w_k$ on D_k, since every D_k is a non-critical strictly pseudoconvex extension of D_0. We will construct a sequence $(u_k)_{k \in \mathbb{N}}$ of differential forms $u_k \in \mathcal{C}^0_{p,q-1}(D_k)$ such that $\overline{\partial} u_k = f$ on D_k and $|u_{k+1} - u_k|_{0,\overline{D}_k} \leqslant \frac{1}{2^k}$. The sequence $(u_k)_{k \in \mathbb{N}}$ thus constructed then converges uniformly on any compact set in Ω to a differential form $u \in \mathcal{C}^0_{p,q-1}(\Omega)$ such that $\overline{\partial} u = f$ on D. Set $u_0 = w_0$ and assume $u_1, \ldots, u_k$ already constructed. The differential form $w_{k+1} - u_k$ is then an element of $Z^0_{p,q-1}(\overline{D}_k)$ and by Proposition 4.5 ii) there is an $\alpha_{k+1} \in Z^0_{p,q-1}(\overline{D}_{k+1})$ such that

$$|w_{k+1} - u_k - \alpha_{k+1}|_{0,\overline{D}_k} < 1/2^k.$$

We therefore set $u_{k+1} = w_{k+1} - \alpha_{k+1}$ and we have $\overline{\partial} u_{k+1} = f$ on D_{k+1} since $\overline{\partial} \alpha_{k+1} = 0$ on D_{k+1} and $|u_{k+1} - u_k|_{0,\overline{D}_k} < 1/2^k$. □

Corollary 4.8. *If Ω is a bounded strictly pseudoconvex open set with $\mathcal{C}^2$ boundary in $\mathbb{C}^n$ then*

$$\dim H^{p,q}(\Omega) < +\infty, \quad \textit{for any } (p,q) \textit{ such that } 0 \leqslant p \leqslant n \textit{ and } 1 \leqslant q \leqslant n.$$

Proof. This follows immediately from Corollary 3.2 and Theorem 4.6. □

5 Vanishing theorems for Dolbeault cohomology of strictly pseudoconvex domains in $\mathbb{C}^n$

We now present a result of Laufer's [Lau] which will enable us to deduce a vanishing theorem for Dolbeault cohomology from the finiteness theorems in Section 4. This result says that if Ω is an open set in $\mathbb{C}^n$ then the Dolbeault cohomology groups $H^{p,q}(\Omega)$, $0 \leqslant p \leqslant n$ and $1 \leqslant q \leqslant n$, are either 0 or infinite-dimensional.

Theorem 5.1. *Let Ω be an open set in $\mathbb{C}$ and let p and q be integers such that $0 \leqslant p \leqslant n$ and $1 \leqslant q \leqslant n-1$. If $E^\infty_{p,q}(\Omega) = \overline{\partial}\mathcal{C}^\infty_{p,q-1}(\Omega)$ is a finite-codimensional subspace in $Z^\infty_{p,q}(\Omega)$ then*

$$E^\infty_{p,q}(\Omega) = Z^\infty_{p,q}(\Omega).$$

Proof. Consider an element $f \in Z^\infty_{p,q}(\Omega) \setminus E^\infty_{p,q}(\Omega)$. Since $E^\infty_{p,q}(\Omega)$ has finite codimension in $Z^\infty_{p,q}(\Omega)$ there is an integer $N \in \mathbb{N}^*$ such that the subspace of $Z^\infty_{p,q}(\Omega)$ generated by $f, z_1 f, \dots, z_1^N f$ meets $E^\infty_{p,q}(\Omega)$. We can therefore find complex numbers $\alpha_0, \dots, \alpha_N$, not all zero, and a differential form $u \in \mathcal{C}^\infty_{p,q-1}(\Omega)$ such that

$$\alpha_0 f + \alpha_1 z_1 f + \cdots + \alpha_N z_1^N f = \overline{\partial} u \quad \text{on } \Omega.$$

This can be written as $P(z_1) f \in E^\infty_{p,q-1}(\Omega)$ where $P(z_1)$ is the polynomial $\alpha_0 + \alpha_1 z_1 + \cdots + \alpha_N z_1^N$. Denote by N_f the minimum of the degrees of the non-zero polynomials P in the variable z_1 such that $P(z_1) f \in E^\infty_{p,q}(\Omega)$ and let P_f be a polynomial of degree N_f such that $P_f(z_1) f \in E^\infty_{p,q}(\Omega)$. Let $\{f_1, \dots, f_k\} \subset Z^\infty_{p,q}(\Omega)$ be a basis for an algebraic complement of $E^\infty_{p,q}(\Omega)$ in $Z^\infty_{p,q}(\Omega)$ and set $P = P_{f_1} \cdots P_{f_k}$. The polynomial P then has the property that

$$P(z_1) f \in E^\infty_{p,q}(\Omega) \quad \text{for every } f \in Z^\infty_{p,q}(\Omega).$$

Let N_0 be the smallest integer such that there is a non-zero polynomial P in the variable z_1 of degree N such that $P(z_1) f \in E^\infty_{p,q}(\Omega)$ for every $f \in Z^\infty_{p,q}(\Omega)$. Assume that $N_0 \geqslant 1$. Let P_0 be a polynomial of degree N_0 in the variable z_1 such that $P_0(z_1) f \in E^\infty_{p,q}(\Omega)$ for every $f \in Z^\infty_{p,q}(\Omega)$. Fix an arbitrary differential form $f \in Z^\infty_{p,q}(\Omega)$: there is then a form $u \in \mathcal{C}^\infty_{p,q-1}(\Omega)$ such that

$$P_0(z_1) f = \overline{\partial} u \quad \text{on } \Omega. \tag{5.1}$$

Differentiating (5.1) with respect to z_1, we get

$$\frac{\partial P_0}{\partial z_1}(z_1) f + P_0(z_1) \frac{\partial f}{\partial z_1} = \overline{\partial}\left(\frac{\partial u}{\partial z_1}\right) \quad \text{on } \Omega,$$

where $\partial f / \partial z_1$ and $\partial u / \partial z_1$ are the differential forms obtained by differentiating the coefficients of the differential forms f and u with respect to z_1. The differential form $\partial f / \partial z_1$ is again an element of $Z^\infty_{p,q}(\Omega)$ and hence $P_0(z_1) \partial f / \partial z_1 \in E^\infty_{p,q}(\Omega)$. It follows that the polynomial $\widetilde{P}_0 = \partial P_0 / \partial z_1$ has the property that $\widetilde{P}_0(z_1) f \in E^\infty_{p,q}(\Omega)$ for any $f \in Z^\infty_{p,q}(\Omega)$. As $d^0 \widetilde{P}_0 = N_0 - 1$, this contradicts the minimality of N_0 and hence $N_0 = 0$. This proves that $E^\infty_{p,q}(\Omega) = Z^\infty_{p,q}(\Omega)$. □

Corollary 5.2. *If Ω is a bounded strictly pseudoconvex open set with $\mathcal{C}^2$ boundary in $\mathbb{C}^n$ then*

$$H^{p,q}_{0,1/2}(\overline{\Omega}) = H^{p,q}(\Omega) = 0, \quad \text{whenever } 0 \leqslant p \leqslant n \text{ and } 1 \leqslant q \leqslant n.$$

Proof. Since $H^{p,q}(\Omega) = Z^\infty_{p,q}(\Omega) / E^\infty_{p,q}(\Omega)$ it follows from Corollary 4.8 and Theorem 5.1 that $H^{p,q}(\Omega) = 0$. The corollary then follows from Theorem 4.6. □

Theorem 5.3. *Let Ω be a bounded strictly pseudoconvex open set with $\mathcal{C}^2$ boundary in $\mathbb{C}^n$ and let p and q be integers such that $0 \leqslant p \leqslant n$ and $1 \leqslant q \leqslant n$. There is then a continuous linear map T from $Z^0_{p,q}(\overline{\Omega})$ to $\Lambda^{1/2}_{p,q-1}(\Omega)$ such that $T(Z^0_{p,q}(\overline{\Omega})) \subset F^{1/2}_{p,q-1}(\Omega)$ and $\overline{\partial} \circ T = I$.*

Proof. Corollary 5.2 says that $Z^0_{p,q}(\overline{\Omega}) = E^{1/2}_{p,q}(\Omega)$. We then simply apply Proposition 5 of Appendix C to prove the existence of the operator T. □

We end this section with a lemma which will be useful in Section 6.

Lemma 5.4. *Let Ω be a bounded strictly pseudoconvex open set with $\mathcal{C}^2$ boundary in $\mathbb{C}^n$ and let $U_{\overline{\Omega}}$ be a neighbourhood of $\overline{\Omega}$. Then there is a continuous linear operator $T : Z^\infty_{0,1}(U_{\overline{\Omega}}) \to \mathcal{C}^\infty(\Omega)$ such that $\overline{\partial}Tf = f$ on Ω for any $f \in Z^\infty_{0,1}(U_{\overline{\Omega}})$.*

Proof. We will prove that the operator $T : Z^0_{0,1}(\overline{\Omega}) \to \Lambda^{1/2}(\Omega)$ whose existence is established in Theorem 5.3 has the required properties. Consider an element $f \in Z^\infty_{0,1}(U_{\overline{\Omega}})$. The form Tf is then defined and $\overline{\partial}Tf = f$ on Ω since $Z^\infty_{0,1}(U_{\overline{\Omega}}) \subset Z^0_{(0,1)}(\overline{\Omega})$. By Theorem 3.5 of Chapter III on the regularity of $\overline{\partial}$ we have $Tf \in \mathcal{C}^\infty(\Omega)$. The operator T therefore defines a linear operator from $Z^\infty_{0,1}(U_{\overline{\Omega}})$ to $\mathcal{C}^\infty(\Omega)$. The continuity of T as a linear operator between the Fréchet spaces $Z^\infty_{0,1}(U_{\overline{\Omega}})$ and $\mathcal{C}^\infty(\Omega)$ then follows from the closed graph theorem. Indeed, let $(f_n)_{n\in\mathbb{N}}$ be a sequence of elements of $Z^\infty_{0,1}(U_{\overline{\Omega}})$ such that the sequence (f_n, Tf_n) converges to (f, g) in $Z^\infty_{0,1}(U_{\overline{\Omega}}) \times \mathcal{C}^\infty(\Omega)$. The continuity of T on $Z^0_{0,1}(\overline{\Omega})$ implies that Tf_n converges uniformly to Tf on $\overline{\Omega}$ and as $\overline{\partial}Tf = f$ on Ω, the regularity of $\overline{\partial}$ implies that $Tf \in \mathcal{C}^\infty(\Omega)$. The uniqueness of this limit then implies that $(f, Tf) = (f, g)$, so (f, g) is in the graph of T. □

Remark. An alternative proof of this lemma, not using the operator T of Theorem 5.3, is given in [He/Le1] (cf. Lemma 2.4.1).

6 Integral formulae for solving $\overline{\partial}$ with Hölder estimates in strictly pseudoconvex domains

Let Ω be a bounded strictly pseudoconvex open set with $\mathcal{C}^2$ boundary in $\mathbb{C}^n$. We will construct a global support function $\Phi(z, \zeta)$ for Ω, i.e. a $\mathcal{C}^1$ function defined on $U_{\overline{\Omega}} \times U_{\partial\Omega}$, where $U_{\partial\Omega}$ is a neighbourhood of $\partial\Omega$ and $U_{\overline{\Omega}} = U_{\partial\Omega} \cup \Omega$, such that Φ depends holomorphically on z in $U_{\overline{\Omega}}$ and $\Phi(z, \zeta) \neq 0$ if $z \in \overline{\Omega}$, $\zeta \in U_{\partial\Omega}$ and $z \neq \zeta$. In the integral formulae below, this function will play the role of the function $\langle w(z, \zeta), \zeta - z\rangle$, where w is a Leray section for Ω.

Lemma 6.1. *Let θ be a bounded open set in $\mathbb{C}^n$ and let ρ be a strictly plurisubharmonic function defined on some neighbourhood of $\overline{\theta}$. Consider the function*

$$\beta = \frac{1}{3} \min_{\substack{\xi \in \mathbb{C}^n \\ |\xi|=1}} \sum_{j=1}^{n} \frac{\partial^2 \rho(\xi)}{\partial \zeta_j \partial \overline{\zeta}_k} \xi_j \xi_k. \tag{6.1}$$

Let the functions a_{jk} be $\mathcal{C}^1$ functions defined on some neighbourhood of $\overline{\theta}$ such that

$$\max_{\zeta \in \overline{\theta}} \left| a_{jk}(\zeta) - \frac{\partial^2 \rho}{\partial \zeta_j \partial \zeta_k}(\zeta) \right| < \frac{\beta}{n^2} \quad \text{for } j,k = 1, \dots, n \tag{6.2}$$

and choose a real number $\varepsilon > 0$ small enough that

$$\max_{\substack{\zeta, z \in \overline{\theta} \\ |\zeta - z| \leqslant \varepsilon}} \left| \frac{\partial^2 \rho}{\partial x_j \partial x_k}(\zeta) - \frac{\partial^2 \rho}{\partial x_j \partial x_k}(z) \right| < \frac{\beta}{2n^2} \quad \text{for } j,k = 1, \dots, 2n, \tag{6.3}$$

where the coordinates $x_j = x_j(\xi)$ are the real coordinates of $\xi \in \mathbb{C}^n$ such that $\xi_j = x_j(\xi) + i x_{j+n}(\xi)$. For any $z, \zeta \in \overline{\theta}$ we set

$$F(z,\zeta) = -\left[2 \sum_{j=1}^{n} \frac{\partial \rho}{\partial \zeta_j}(\zeta)(z_j - \zeta_j) + \sum_{j,k=1}^{n} a_{jk}(\zeta)(z_j - \zeta_j)(z_k - \zeta_k) \right].$$

The, for any $z, \zeta \in \overline{\theta}$ such that $|\zeta - z| \leqslant \varepsilon$,

$$\operatorname{Re} F(z,\zeta) \geqslant \rho(\zeta) - \rho(z) + \beta |\zeta - z|^2. \tag{6.5}$$

Remark. Since ρ is $\mathcal{C}^2$ and $\beta > 0$ (because ρ is strictly plurisubharmonic) we can find functions a_{jk} and a real number ε such that (6.2) and (6.3) hold.

Proof. Consider points ζ and z in $\overline{\theta}$ such that $|\zeta - z| \leqslant \varepsilon$. It then follows from Lemma 2.22 of Chapter VI and Taylor's formula that if $\widehat{F}_\rho$ is the Levi polynomial of ρ then

$$\operatorname{Re} \widehat{F}_\rho(z,\zeta) = \rho(\zeta) - \rho(z) + \sum_{j,k=1}^{n} \frac{\partial^2 \rho}{\partial \zeta_j \partial \overline{\zeta}_k}(\zeta)(z_j - \zeta_j)(\overline{z}_k - \overline{\zeta}_k) + R(z,\zeta),$$

where $|R(z,\zeta)| \leqslant \beta |\zeta - z|^2$ by (6.3). By (6.1),

$$\operatorname{Re} \widehat{F}_\rho(z,\zeta) \geqslant \rho(\zeta) - \rho(z) + 2\beta |\zeta - z|^2,$$

but then $|\widehat{F}_\rho(z,\zeta) - F(z,\zeta)| \leqslant \beta |\zeta - z|^2$ by (6.2) and hence $\operatorname{Re} F(z,\zeta) \geqslant \rho(\zeta) - \rho(z) + \beta |\zeta - z|^2$. □

Theorem 6.2. *Let Ω be a bounded strictly pseudoconvex domain with $\mathcal{C}^2$ boundary in $\mathbb{C}^n$, let θ be a neighbourhood of $\partial\Omega$ and let ρ be a $\mathcal{C}^2$ strictly plurisubharmonic function defined on a neighbourhood of $\overline{\theta}$ such that*

$$\Omega \cap \theta = \{z \in \theta \mid \rho(z) < 0\}.$$

Let ε, β and $F(z,\zeta)$ be chosen as in Lemma 6.1 with ε small enough that

$$\{z \in \mathbb{C}^n \mid \ |\zeta - z| \leqslant 2\varepsilon\} \subset \theta \quad \textit{for all } \zeta \in \partial\Omega.$$

There is then a $\mathcal{C}^1$ function $\Phi(z,\zeta)$ defined on $U_{\overline{\Omega}} \times U_{\partial\Omega}$, where $U_{\partial\Omega}$ is a neighbourhood of $\partial\Omega$ contained in θ and $U_{\overline{\Omega}} = U_{\partial\Omega} \cup \Omega$, such that

i) *$\Phi(z,\zeta)$ depends holomorphically on z in $U_{\overline{\Omega}}$,*
ii) *$\Phi(z,\zeta) \neq 0$ for any $z \in U_{\overline{\Omega}}$ and $\zeta \in U_{\partial\Omega}$ such that $|\zeta - z| \geqslant \varepsilon$,*
iii) *there is a $\mathcal{C}^1$ function, $M(z,\zeta)$, which is defined and non-zero for any $z \in U_{\overline{\Omega}}$ and $\zeta \in U_{\partial\Omega}$, such that $|\zeta - z| \leqslant \varepsilon$, and which has the property that*
$$\Phi(z,\zeta) = F(z,\zeta)M(z,\zeta)$$
for any $z \in U_{\overline{D}}$ and $\zeta \in U_{\partial D}$ such that $|\zeta - z| \leqslant \varepsilon$.

Proof. It follows from estimate (6.5) that $\operatorname{Re} F(z,\zeta) \geqslant \rho(\zeta) - \rho(z) + \beta\varepsilon^2$ for any $z,\zeta \in \theta$ such that $\varepsilon \leqslant |\zeta - z| \leqslant 2\varepsilon$. As $\rho = 0$ on $\partial\Omega$ our choice of ε implies that we can find a neighbourhood $V_{\partial\Omega}$ of $\partial\Omega$ contained in θ which is small enough that $|\rho| \leqslant \beta\varepsilon^2/3$ on $V_{\partial\Omega}$ and for any $\zeta \in V_{\partial\Omega}$ the ball with centre ζ and radius 2ε is contained in θ. Setting $V_{\overline{\Omega}} = \Omega \cup V_{\partial\Omega}$, we see that for any $(z,\zeta) \in V_{\overline{\Omega}} \times V_{\partial\Omega}$ such that $|\zeta - z| < 2\varepsilon$, z and ζ are in θ and moreover $\operatorname{Re} F(z,\zeta) \geqslant \beta\varepsilon^2/3$ for any $z \in V_{\overline{\Omega}}$ and $\zeta \in V_{\partial\Omega}$ such that $\varepsilon \leqslant |\zeta - z| \leqslant 2\varepsilon$. It follows that $\ell n\, F(z,\zeta)$, where ℓn is the principal branch of the logarithm, is defined for all $z \in V_{\overline{\Omega}}$ and $\zeta \in V_{\partial\Omega}$ such that $\varepsilon \leqslant |\zeta - z| \leqslant 2\varepsilon$. Choose a $\mathcal{C}^\infty$ function χ on $\mathbb{C}^n$ such that $\chi(\xi) = 1$ if $|\xi| \leqslant 5\varepsilon/4$ and $\chi(\xi) = 0$ whenever $|\xi| \geqslant 7\varepsilon/4$. For any $z \in V_{\overline{\Omega}}$ and $\zeta \in V_{\partial\Omega}$ we define f by

$$f(z,\zeta) = \begin{cases} \overline{\partial}_z[\chi(\zeta - z)\,\ell n\, F(z,\zeta)] & \text{if } \varepsilon \leqslant |\zeta - z| \leqslant 2\varepsilon, \\ 0 & \text{otherwise.} \end{cases}$$

The map $\zeta \mapsto f(\cdot,\zeta)$ is then $\mathcal{C}^1$ on $V_{\partial\Omega}$ and its image is contained in $Z^\infty_{0,1}(V_{\overline{\Omega}})$. Choose a neighbourhood $U_{\partial\Omega} \Subset V_{\partial\Omega}$ such that $U_{\overline{\Omega}} = \Omega \cup U_{\partial\Omega}$ is bounded and strictly pseudoconvex with $\mathcal{C}^2$ boundary. It follows from Lemma 5.4 that there is a continuous linear operator $T : Z^\infty_{0,1}(V_{\overline{\Omega}}) \to \mathcal{C}^\infty(U_{\overline{\Omega}})$ such that $\overline{\partial} T\varphi = \varphi$ on $U_{\overline{\Omega}}$ for any $\varphi \in Z^\infty_{0,1}(V_{\overline{\Omega}})$. For any $z \in U_{\overline{\Omega}}$ and $\zeta \in U_{\partial\Omega}$ we set

$$u(z,\zeta) = (T(f(\cdot,\zeta))(z), \quad M(z,\zeta) = \exp(-u(z,\zeta))$$

and

$$\Phi(z,\zeta) = \begin{cases} F(z,\zeta)M(z,\zeta) & \text{if } |\zeta - z| \leqslant \varepsilon, \\ \exp(\chi(\zeta - z)\,\ell n\, F(z,\zeta) - u(z,\zeta) & \text{if } |\zeta - z| \geqslant \varepsilon. \end{cases}$$

It is clear that Φ satisfies conditions i)–iii) of the theorem. □

Remarks.

1) Note that we have not used the fact that $\partial\Omega$ is $\mathcal{C}^2$. Theorem 6.2 remains true even if $d\rho$ vanishes on $\partial\Omega$ (cf. [He/Le1]).
2) We do not need the full force of Theorem 5.3 to prove Theorem 6.2 – we only need to know that if a differential $(0,1)$-form is $\overline{\partial}$-closed in a neighbourhood of a strictly pseudoconvex domain then it is exact on this domain and the solution of the Cauchy–Riemann equation is given by a linear operator.

Theorem 6.2 solves the Levi problem for bounded strictly pseudoconvex domains with $\mathcal{C}^2$ boundary. (The general case will be studied in § 7.)

Corollary 6.3. *Let Ω be a bounded strictly pseudoconvex open set with $\mathcal{C}^2$ boundary in $\mathbb{C}^n$. The open set Ω is then a domain of holomorphy.*

Proof. Consider the function Φ which Theorem 6.2 associates to the domain Ω and fix a point $\xi_0 \in \partial\Omega$. The function $f(z) = 1/\Phi(z,\zeta_0)$ is then holomorphic on Ω and $f(z)$ tends to infinity as z tends to ζ_0 in D, so the function $f(z)$ cannot be extended to a holomorphic function on any neighbourhood of ζ_0. □

In order to apply the Cauchy–Leray–Koppelman formula on a strictly pseudoconvex domain Ω we will now construct a Leray section for Ω associated to the support function Φ constructed in Theorem 6.2. Our aim is to find a $\mathcal{C}^1$ map w on $U_{\overline{\Omega}} \times U_{\partial\Omega}$, holomorphic with respect to z on $U_{\overline{\Omega}}$, such that $\langle w(z,\zeta), \zeta - z\rangle = \Phi(z,\zeta)$.

Lemma 6.4. *Let Ω be a bounded strictly pseudoconvex open set with $\mathcal{C}^2$ boundary in $\mathbb{C}^n$, set $M_1 = \{z \in \mathbb{C}^n \mid z_1 = 0\}$ and let $U_{\overline{\Omega}}$ be a neighbourhood of $\overline{\Omega}$. For any holomorphic function f on $M_1 \cap U_{\overline{\Omega}}$ there is a holomorphic function $\widetilde{f}$ defined on Ω such that $\widetilde{f} = f$ on $M_1 \cap \Omega$.*

Proof. If U_1 is a small enough neighbourhood of $M_1 \cap \overline{\Omega}$ then on setting $F(z) = f(0, z_2, \dots, z_n)$ we get a holomorphic extension of f to U_1. Choose neighbourhoods U_1' and U_1'' of $M_1 \cap \overline{\Omega}$ such that $U_1' \Subset U_1'' \Subset U_1$ and let χ be a $\mathcal{C}^\infty$ function on $\mathbb{C}^n$ such that $\chi = 1$ on U_1' and $\operatorname{supp}\chi \subset U_1''$. We then set

$$\varphi(z) = \begin{cases} \dfrac{F(z)\overline{\partial}\chi(z)}{z_1} & \text{if } z \in U_1, \\ 0 & \text{if } z \in \mathbb{C}^n \setminus U_1. \end{cases}$$

The form φ is then a differential $(0,1)$-form which is $\overline{\partial}$-closed and $\mathcal{C}^\infty$ on a neighbourhood U of $\overline{\Omega}$. By Corollary 5.2 applied to some strictly pseudoconvex domain with $\mathcal{C}^2$ boundary containing $\overline{\Omega}$ and contained in U there is a continuous function u on $\overline{\Omega}$ such that $\overline{\partial}u = \varphi$ on Ω, which implies that $\overline{\partial}(F\chi - z_1 u) = 0$ on Ω. It follows that $\widetilde{f} = F\chi - z_1 u$ is holomorphic on Ω and since $\chi \equiv 1$ on $\Omega \cap M_1$, $\widetilde{f} = F = f$ on $\Omega \cap M_1$. □

Lemma 6.5. *Let Ω be a bounded strictly pseudoconvex open set with C^2 boundary in $\mathbb{C}^n$ and set $M_k = \{z \in \mathbb{C}^n \mid z_1 = \cdots = z_k = 0\}$ for any k such that $1 \leqslant k \leqslant n$. Let $U_{\overline{\Omega}}$ be a neighbourhood of $\overline{\Omega}$. If f is a holomorphic function on $U_{\overline{\Omega}}$ such that $f = 0$ on $M_k \cap U_{\overline{\Omega}}$ then there are holomorphic functions $f_1, \ldots, f_k$ on Ω such that $f(z) = \sum_{j=1}^k z_j f_j(z)$ for any $z \in \Omega$.*

Proof. If $k = 1$ then we can set $f_1(z) = f(z)/z_1$. Arguing by induction, we assume that the lemma is proved for M_{k-1} in $\mathbb{C}^l$, whenever $k-1 \leqslant \ell$. Let f be a holomorphic function on $U_{\overline{\Omega}}$ such that $f(z) = 0$ whenever $z \in M_k \cap U_{\overline{\Omega}}$. Choose a strictly pseudoconvex open set $\widetilde{\Omega}$ such that $\Omega \Subset \widetilde{\Omega} \Subset U_{\overline{\Omega}}$. The set $\widetilde{\Omega} \cap M_1$ is then a bounded strictly pseudoconvex open set in $M_1 = \mathbb{C}^{n-1}$. By the induction hypothesis there are functions $\widetilde{f}_j(z_2, \ldots, z_n)$ for $j = 2, \ldots, k$, holomorphic on $\widetilde{\Omega} \cap M_1$, such that $f(z) = \sum_{j=2}^k z_j \widetilde{f}_j(z_2, \ldots, z_n)$ for any $z \in \widetilde{\Omega} \cap M_1$. Lemma 6.4 then says that there are holomorphic functions f_j on Ω for any $j = 2, \ldots, k$ such that $f_j(z) = \widetilde{f}_j(z_2, \ldots, z_n)$ for any $z = (0, z_2, \ldots, z_n) \in \Omega \cap M_1$. To complete the proof we then simply set $f_1(z) = \frac{1}{z_1}(f(z) - \sum_{j=2}^k z_j f_j(z))$ for any $z \in \Omega$. □

Theorem 6.6. *Let Ω be a bounded strictly pseudoconvex open set with C^2 boundary in $\mathbb{C}^n$ and let $U_{\overline{\Omega}}$ be a neighbourhood of $\overline{\Omega}$. If f is a holomorphic function on $U_{\overline{\Omega}}$ then there are holomorphic functions h_j defined on $\Omega \times \Omega$ such that, for any $(z, \zeta) \in \Omega \times \Omega$,*

$$f(\zeta) - f(z) = \sum_{j=1}^n h_j(z, \zeta)(\zeta_j - z_j).$$

Proof. Choose a bounded strictly pseudoconvex open set $\widetilde{\Omega}$ contained in $\mathbb{C}^{2n}$ such that $\Omega \times \Omega \Subset \widetilde{\Omega} \Subset U_{\overline{\Omega}} \times U_{\overline{\Omega}}$ and define new coordinates on $\mathbb{C}^{2n}$ by setting $u_j = \zeta_j - z_j$ and $u_{j+n} = z_j$ for any $j = 1, \ldots, n$. If we now define a function F by $F(z, \zeta) = f(\zeta) - f(z)$ then F is holomorphic on $U_{\overline{\Omega}} \times U_{\overline{\Omega}}$ and $F = 0$ on $U_{\overline{\Omega}} \times U_{\overline{\Omega}} \cap \{u \in \mathbb{C}^{2n} \mid u_1 = \cdots = u_n = 0\}$. We can therefore apply Lemma 6.5 to F, which proves the theorem on returning to the system of coordinates (z, ζ). □

To construct the Leray section w associated to Φ we need a version of Theorem 6.6 in which f depends once-continuously differentiably on some parameter and the f_j also depend once-continuously differentiably on this parameter.

Theorem 6.7. *Let X be a C^k manifold, $k \geqslant 1$, let Ω be a bounded strictly pseudoconvex open set with C^2 boundary in $\mathbb{C}^n$ and let $U_{\overline{\Omega}}$ be a neighbourhood of $\overline{\Omega}$. If f is a C^k function on X with image in $C^\infty(U_{\overline{\Omega}})$ such that $f(\cdot, x)$ is holomorphic on $U_{\overline{\Omega}}$ for any $x \in X$ then, for any j such that $1 \leqslant j \leqslant n$, there are C^∞ functions h_j defined on $(\Omega \times \Omega) \times X$ such that $h_j(\cdot, x)$ is holomorphic* *on $\Omega \times \Omega$ for any $x \in X$ and that, for all $x \in X$ and $(z, \zeta) \in \Omega \times \Omega$,*

$$f(\zeta, x) - f(z, x) = \sum_{j=1}^{n} h_j(\zeta, z, x)(\zeta_j - z_j).$$

The proof of Theorem 6.7 simply repeats the proof of Theorem 6.6. We now give a detailed proof of the paramterised version of Lemma 6.4, which is the only parameterised result we will prove in detail.

Lemma 6.8. *Let X be a $\mathcal{C}^k$ manifold, $k \geqslant 1$, let Ω be a bounded strictly pseudoconvex open set with $\mathcal{C}^2$ boundary in $\mathbb{C}^n$, set $M_1 = \{z \in \mathbb{C}^n \mid z_1 = 0\}$ and let $U_{\overline{\Omega}}$ be a neighbourhood of Ω. If f is a $\mathcal{C}^k$ function on X with image in $\mathcal{C}^\infty(U_{\overline{\Omega}} \cap M_1)$ such that $f(\cdot, x)$ is holomorphic on $U_{\overline{\Omega}} \cap M_1$ for any $x \in X$ then there is a $\mathcal{C}^k$ function $\widetilde{f}$ on X with image in $\mathcal{C}^\infty(\Omega)$ such that $\widetilde{f}(\cdot, x)$ is holomorphic on Ω for any $x \in X$ and that, for any $x \in X$ and $(0, z_2, \ldots, z_n) \in \Omega$,*

$$\widetilde{f}\big((0, z_2, \ldots, z_n), x\big) = f\big((0, z_2, \ldots, z_n), x\big).$$

Proof. The proof is similar to that of Lemma 6.4: we use the same notations as in the proof of this lemma. Set

$$\varphi(z, x) = \begin{cases} \dfrac{F(z, x)\overline{\partial}\chi(z)}{z_1} & \text{if } x \in X \text{ and } z \in U_1, \\ 0 & \text{if } x \in X \text{ and } z \in \mathbb{C}^n \setminus U_1. \end{cases}$$

The form φ is then a $\overline{\partial}_z$-closed $\mathcal{C}^k$ differential form on $X \times U_1$. Choose a strictly pseudoconvex open set $\widetilde{\Omega}$ with $\mathcal{C}^2$ boundary such that $\Omega \subset \widetilde{\Omega} \Subset U_1$. The map $x \mapsto \varphi(\cdot, x)$ is $\mathcal{C}^k$ on X and its image lies in $Z^\infty_{0,1}(U_1)$. By Lemma 5.4, there is therefore a continuous linear operator T from $Z^\infty_{0,1}(U_1)$ to $\mathcal{C}^\infty(\widetilde{\Omega})$ such that $\overline{\partial} \circ T = \mathrm{Id}$. Set $u(\cdot, x) = T(\varphi(\cdot, x))$. The function u is then a $\mathcal{C}^k$ function on X with image in $\mathcal{C}^\infty(\widetilde{\Omega})$ and the function $\widetilde{f}(z, x) = \chi(z)F(z, x) - z_1 u(z, x)$ is the function we seek. □

Corollary 6.9. *Let Ω be a bounded strictly pseudoconvex domain in $\mathbb{C}^n$ and let Φ be the support function associated to the domain Ω by Theorem 6.2. Suppose that the function Φ is defined on $U_{\overline{\Omega}} \times U_{\partial\Omega}$. There is then a neighbourhood $V_{\partial\Omega}$ of $\partial\Omega$ such that $V_{\partial\Omega} \Subset U_{\partial\Omega}$ and a $\mathcal{C}^1$ map w defined on $V_{\overline{\Omega}} \times V_{\partial\Omega}$ (here $V_{\overline{\Omega}} = V_{\partial\Omega} \cup \Omega$) which is holomorphic with respect to z on $V_{\overline{\Omega}}$ such that*

$$\Phi(z, \zeta) = \sum_{j=1}^{n} w_j(z, \zeta)(\zeta_j - z_j) = \langle w, \zeta - z \rangle$$

for any $z \in V_{\overline{D}}$ and $\zeta \in V_{\partial D}$. Moreover, w is a Leray section for Ω.

Proof. Applying Theorem 6.7 to Φ we get

$$\Phi(\xi,\zeta) - \Phi(z,\zeta) = \sum_{j=1}^{n} h_j(\xi,z,\zeta)(\xi_j - z_j)$$

and since $\Phi(\zeta,\zeta) = 0$ on taking $\xi = \zeta$ we get

$$\Phi(z,\zeta) = \sum_{j=1}^{n} w_j(z,\zeta)(\zeta_j - z_j),$$

where $w_j(z,\zeta) = -h_j(\zeta,z,\zeta)$ has the required properties. □

Remark. The Leray section w thus obtained can also be used for any strictly pseudoconvex domain of the form

$$\Omega \setminus V_{\partial\Omega} \cup \{z \in V_{\partial\Omega} \mid \rho(z) < -\delta\}$$

for small enough $\delta > 0$.

Using the global Leray section constructed above we can construct an integral operator which enables us to solve $\overline{\partial}$ with order 1/2 Hölder estimates.

Theorem 6.10. *Let Ω be a bounded strictly pseudoconvex domain with $\mathcal{C}^2$ boundary in $\mathbb{C}^n$ and let w be the Leray section associated to Ω by Corollary 6.9. There is then a constant C such that*

i) *for any continuous differential form f on $\overline{\Omega}$ the integral*

$$R^w_{\partial\Omega} f = \int_{(\zeta,\lambda)\in\partial\Omega\times[0,1]} f(\zeta) \wedge K^{\eta^w}(\cdot,\zeta,\lambda)$$

has the property that

$$|R^w_{\partial\Omega} f|_{1/2,\Omega} \leqslant C|f|_{0,r},$$

ii) *for any continuous differential (p,q)-form f on $\overline{\Omega}$ which is $\overline{\partial}$-closed on Ω such that $0 \leqslant p \leqslant n$ and $1 \leqslant q \leqslant n$ the form*

$$u = (-1)^{p+q}(R^w_{\partial\Omega} f + B_\Omega f)$$

is a solution of the equation $\overline{\partial} u = f$ on Ω and has the property that

$$|u|_{1/2,\Omega} \leqslant C|f|_{0,\Omega}.$$

Moreover, $u \in \mathcal{C}^\alpha_{p,q-1}(\Omega)$ for any $\alpha \in]0,1[$ and if f is $\mathcal{C}^k$ on Ω for some $k = 1,2,\ldots,\infty$ then $u \in \mathcal{C}^{k+\alpha}_{p,q-1}(\Omega)$ for any $\alpha \in]0,1[$.

Proof. By Theorem 3.3 of Chapter VI we only have to prove i). To do this it will be enough to show that the Leray section w satisfies the hypotheses of Lemma 1.3. As hypotheses i) and ii) follow from Corollary 6.9, it only remains to prove iii).

Fix a point $\xi \in \Omega$. By definition of w and Φ,

$$\langle w(z,\zeta), \zeta - z\rangle = F(z,\zeta)M(z,\zeta),$$

where F and M have the properties given in Lemma 6.1 and Theorem 6.2. We can then choose a neighbourhood U_ξ of ξ which is small enough that there is a $\delta_1 > 0$ such that

$$|\langle w(z,\zeta), \zeta - z\rangle| \geqslant \delta_1 |F(z,\zeta)| \tag{6.6}$$

for any $z, \zeta \in U_\xi$. Set $t_1(z,\zeta) = \operatorname{Im} F(z,\zeta)$ and let the coordinates $x_j = x_j(\zeta)$ be the real coordinates of $\zeta \in \mathbb{C}^n$ such that $\zeta_j = x_j(\zeta) + ix_{j+n}(\zeta)$. Then

$$t_1(z,\zeta) = \sum_{j=1}^n \Big[\frac{\partial\rho(\zeta)}{\partial x_j} x_{j+n}(\zeta - z) - \frac{\partial\rho}{\partial x_{j+n}}(\zeta) x_j(\zeta - z)\Big] - \operatorname{Im}\sum_{j,k=1}^n a_{jk}(\zeta)(\zeta_j - z_j)(\zeta_k - z_k)$$

and it follows that

$$d_\zeta t_1(z,\zeta)\big|_{\zeta=z} = \sum_{j=1}^n \Big[\frac{\partial\rho(z)}{\partial x_j} dx_{j+n}(z) - \frac{\partial\rho(z)}{\partial x_{j+n}} dx_j(z)\Big].$$

In particular, $n\|d_\zeta t_1(z,\zeta)\big|_{\zeta=z} \wedge d\rho(z)\| \geqslant \|d\rho(z)\|^2$ for every $z \in \partial\Omega$ because the coefficient of $dx_j(z) \wedge dx_{j+n}(z)$ in the differential form on the left-hand side of the above equation is $-\big[(\frac{\partial\rho}{\partial x_j}(z))^2 + (\frac{\partial\rho}{\partial x_{j+n}}(z))^2\big]$. Since $d\rho(z) \neq 0$ for any $z \in \partial\Omega$ because Ω has $\mathcal{C}^2$ boundary, on restricting U_ξ we can find $\mathcal{C}^1$ functions $\widetilde{t}_2, \dots, \widetilde{t}_{2n-1}$ defined on U_ξ such that for any $z \in U_\xi$ the functions $t_1(z,\cdot), \widetilde{t}_2, \dots, \widetilde{t}_{2n-1}$ are coordinates on $U_\xi \cap \partial\Omega$. We set $t_j(z,\zeta) = \widetilde{t}_j(\zeta) - \widetilde{t}_j(z)$ for any $z, \zeta \in U_\xi$ and any $j = 2, \dots, 2n-1$. Then for any $z \in U_\xi$ the functions $t_1(z,\cdot), \dots, t_{2n-1}(z,\cdot)$ are also coordinates on $\partial\Omega \cap U_\xi$. To complete the proof of the theorem we must prove that (1.2) holds for these coordinates. As $t_j(z,z) = 0$, on restricting U_ξ Taylor's formula implies there is a $\delta_2 > 0$ such that $|\zeta - z| \geqslant \delta_2 |t(z,\zeta)|$ for every $z, \zeta \in U_\xi$. It then follows from (6.5) of Lemma 6.1 that after again restricting U_ξ we can find a $\delta_3 > 0$ such that, for any $z, \zeta \in U_\xi$,

$$|F(z,\zeta)| \geqslant \delta_3\big(|t_1(z,\zeta)| + \rho(\zeta) - \rho(z) + |t(z,\zeta)|^2\big).$$

As $\rho(\zeta) = 0$ for any $\zeta \in \partial\Omega$ and as there is a $\delta_4 > 0$ such that $-\rho(z) \geqslant \delta_4 \operatorname{dist}(z, \partial\Omega)$ for any $z \in D \cap U_\xi$,

$$|F(z,\zeta)| \geqslant \delta_3\delta_4\big(|t_1(z,\zeta)| + |t(z,\zeta)|^2 + \operatorname{dist}(z,\partial\Omega)\big)$$

for any $z \in D \cap U_\xi$ and $\zeta \in \partial D \cap U_\xi$. The result now follows from (6.6). □

Remark. The advantage of Theorem 6.10 over the results proved in Section 5 is that the integral operator is much more explicit, which enables us to prove, amongst other things, that the constant C depends continuously on the domain Ω in a certain sense.

7 The Levi problem in $\mathbb{C}^n$

We start this section by proving the Oka–Weil approximation theorem. This theorem will enable us to solve the Levi problem and prove the vanishing of Dolbeault cohomology groups for pseudoconvex open sets in $\mathbb{C}^n$ in bidegree (p, q) for $0 \leqslant p \leqslant n$ and $1 \leqslant q \leqslant n$.

Theorem 7.1. *Let $\Omega \subset \mathbb{C}^n$ be a pseudoconvex open set and let $K \Subset \Omega$ be a compact set such that $K = \widehat{K}^p_\Omega$. For any $0 \leqslant p \leqslant n$ any holomorphic $(p, 0)$-form on a neighbourhood of K can be uniformly approximated on K by holomorphic functions on Ω.*

Proof. Let h be a holomorphic $(p, 0)$-form on a neighbourhood U_K of K. By Corollary 3.13 of Chapter VI, there is a $\mathcal{C}^\infty$ strictly plurisubharmonic function $\rho : \Omega \to \mathbb{R}$ such that

i) $\Omega_\alpha = \{z \in \Omega \mid \rho(z) < \alpha\} \Subset \Omega$ for any $\alpha \in \mathbb{R}$,
ii) the set $\mathrm{Crit}(\rho) = \{z \in \Omega \mid d_\rho(z) = 0\}$ is discrete in Ω,
iii) $\rho < 0$ on K and $\rho > 0$ on $\Omega \setminus U_K$.

Let α_0 be the supremum of the set of $\alpha \in \mathbb{R}$ such that h can be uniformly approximated on K by holomorphic $(p, 0)$-forms on Ω_α. Since $h \in \mathcal{C}_{p,0}(U_K)$ and $\overline{\partial} h = 0$ on U_K we have $\alpha_0 > 0$ because $\rho > 0$ on $\Omega \setminus U_K$. To prove that $\alpha_0 = +\infty$ and establish the theorem we simply copy the proof of Theorem 2.1, defining the function Φ and the associated Leray section w as follows. By Theorem 6.2, Corollary 6.9 and the estimate (6.5) there is a real number $\delta > 0$, a $\mathcal{C}^1$ function $\Phi(z, \zeta)$ and a $\mathcal{C}^1$ map to $\mathbb{C}^n$, $w(z, \zeta)$, defined for any z in a neighbourhood of $\overline{D}_{\alpha_0+\delta}$ and any ζ in a neighbourhood of $\overline{D}_{\alpha_0+\delta} \setminus D_{\alpha_0-\delta}$, such that

a) $\Phi(z, \zeta)$ and $w(z, \zeta)$ are holomorphic in z,
b) $\Phi(z, \zeta) = \langle w(z, \zeta), \zeta - z\rangle$,
c) $\Phi(z, \zeta) \neq 0$ whenever $\alpha_0 - \delta \leqslant \rho(\zeta) \leqslant \alpha_0 + \delta$ and $\rho(z) < \rho(\zeta)$,
d) $w(z, \zeta)$ is a Leray section for $D_{\alpha_0-\delta}$

By property ii) of ρ, we can assume that $d\rho(z) \neq 0$ for any $z \in \partial D_{\alpha_0-\delta}$ and then apply Leray's formula to the domain $D_{\alpha_0-\delta}$. □

Remark. Theorem 7.1 can be proved using Proposition 4.5 instead of § 6. This is the method of proof that will be used for ii) of Theorem 8.11.

Theorem 7.2. *Let $\Omega \subset \mathbb{C}^n$ be a pseudoconvex open set. Then, for any compact set K in Ω we have $\widehat{K}^p_\Omega = \widehat{K}_\Omega$.*

Proof. We already know that $\widehat{K}^p_\Omega \subset \widehat{K}_\Omega$ (cf. the remark following Definition 3.3 of Chapter VI). To prove the converse, choose a point $\xi \in \Omega \setminus \widehat{K}^p_\Omega$ and construct a holomorphic function h on Ω such that

$$|h(\xi)| > \max_{z\in K} |h(z)|. \tag{7.1}$$

It follows from Corollary 3.13 and Lemma 2.24 of Chapter VI that there is a strictly pseudoconvex open set $G \Subset \Omega$ such that $K \Subset G$, $\xi \in \partial G$ and $\widehat{\overline{G}}^p_\Omega = \overline{G}$ since G is defined by $\{z \in \Omega \mid \rho(z) < 0\}$ where ρ is some $\mathcal{C}^\infty$ plurisubharmonic function on Ω. By Theorem 6.2 and estimate (6.5) there is a $\mathcal{C}^1$ function $\Phi(z,\zeta)$ defined for any ζ in some neighbourhood $U_{\partial G}$ of ∂G and any z in some neighbourhood $U_{\overline{G}}$ of $\overline{G}$ such that $\Phi(z,\zeta)$ is holomorphic in z on $U_{\overline{G}}$, $\Phi(\zeta,\zeta) = 0$ and $\Phi(z,\zeta) \neq 0$ for any $\zeta \in U_{\partial G} \setminus \overline{G}$ and $z \in \overline{G}$. For any $\zeta_0 \in U_{\partial G} \setminus \overline{G}$, $1/\Phi(z,\zeta_0)$ is holomorphic in z on some neighbourhood of $\overline{G}$. As $K \Subset G$, $\xi \in \partial G$ and $\Phi(\xi,\xi) = 0$ we can choose $\zeta_0 \in U_{\partial G} \setminus \overline{G}$ close enough to ξ so that

$$\left|\frac{1}{\Phi(\xi,\zeta_0)}\right| \geqslant 1 + \max_{z\in K}\left|\frac{1}{\Phi(z,\zeta_0)}\right|. \tag{7.2}$$

As $\overline{G} = \widehat{\overline{G}}^p_\Omega$ it follows from Theorem 7.1 that there is a holomorphic function h defined on Ω such that $\left|h(z) - 1/\Phi(z,\zeta_0)\right| < 1/2$ for any $z \in \overline{G}$. By (7.2), Condition (7.1) then holds. □

Corollary 7.3. *An open set in $\mathbb{C}^n$ is a domain of holomorphy if and only if it is pseudoconvex.*

Proof. We saw in Chapter VI, Corollary 3.7 that any domain of holomorphy is pseudoconvex. The converse follows immediately from Theorem 7.2, the definition of pseudoconvex open sets (cf. Chap. VI, Def. 3.5) and the Cartan–Thullen theorem (cf. Chap. VI, Th. 1.13). □

Theorem 7.4. *Let $D \subset \mathbb{C}^n$ be a pseudoconvex open set, let p and q be integers such that $0 \leqslant p \leqslant n$ and $1 \leqslant q \leqslant n$ and consider an element k where $k \in \mathbb{N} \cup \{+\infty\}$. For any $\mathcal{C}^k$ differential (p,q)-form f on D such that $\overline{\partial} f = 0$ on D there is a solution u of the equation $\overline{\partial} u = f$ on D such that $u \in \mathcal{C}^{k+\alpha}_{p,q-1}(D)$ for any $\alpha \in]0,1[$.*

Proof. By Corollary 3.13 of Chapter VI there is a $\mathcal{C}^\infty$ strictly plurisubharmonic function ρ on D such that

i) $D_\alpha = \{z \in D \mid \rho(z) < \alpha\} \Subset D$ for any $\alpha \in \mathbb{R}$,
ii) the set $\mathrm{Crit}(\rho) = \{z \in D \mid d\rho(z) = 0\}$ is discrete in D.

By property ii) we can find a sequence $(\alpha_\nu)_{\nu\in\mathbb{N}}$ of real numbers tending to infinity such that $d\rho(z) \neq 0$ for any $z \in \partial D_{\alpha_\nu}$, the set D_{α_ν} is a strictly pseudoconvex domain with $\mathcal{C}^\infty$ boundary and $\bigcup_{\nu\in\mathbb{N}} D_{\alpha_\nu}$. By Theorem 6.10, for any $\nu \in \mathbb{N}$ there are differential forms $u_\nu \in \mathcal{C}^{k+\alpha}_{p,q-1}(D_{\alpha_\nu})$, $0 < \alpha < 1$, such that $\overline{\partial} u_\nu = f$ on D_{α_ν}. We will construct a solution u of $\overline{\partial} u = f$ on D using the forms u_ν.

Suppose first that $q \geqslant 2$. We will construct a sequence $(v_\nu)_{\nu\geqslant 3}$ of differential forms which are solutions of $\overline{\partial} v_\nu = f$ on D_{α_ν} such that $v_\nu \in \mathcal{C}^{k+\alpha}_{p,q-1}(D_{\alpha_\nu})$ for any $0 < \alpha < 1$ and $v_\nu = v_{\nu+1}$ on $D_{\alpha_{\nu-2}}$. The form $u = \lim_{\nu\to\infty} v_\nu$ will be the solution we seek. Set $v_3 = u_3$ and assume that $v_3, \dots, v_\ell$ have been constructed for some $\ell \geqslant 3$. The form $v_\ell - u_{\ell+1}$ is then $\overline{\partial}$-closed on D_{α_ℓ} and Theorem 6.10 says there is a solution to $\overline{\partial}\varphi = v_\ell - u_{\ell+1}$ in $D_{\alpha_{\ell-1}}$ such that $\varphi \in \mathcal{C}^{k+\alpha}_{p,q-1}(D_{\alpha_{\ell-1}})$ for any $0 < \alpha < 1$. Choose a $\mathcal{C}^\infty$ function χ with compact support on $D_{\alpha_{\ell-1}}$ which is equal to 1 on $D_{\alpha_{\ell-2}}$ and set $v_{\ell+1} = u_{\ell+1} + \overline{\partial}(\chi\varphi)$. The differential form $v_{\ell+1}$ then has the required properties.

Now consider the case where $q = 1$. We will construct a sequence $(v_\nu)_{\nu\geqslant 2}$ of solutions of $\overline{\partial} v_\nu = f$ on D_{α_ν} such that $v_\nu \in \mathcal{C}^{k+\alpha}_{p,0}(D_{\alpha_\nu})$ and $|v_\nu(z) - v_{\nu+1}(z)| < 2^{-\nu}$ for any $z \in D_{\alpha_{\nu-1}}$. Such a sequence converges uniformly on all compact sets in D to a $(p,0)$-form u which has the required properties since the differences $u - v_\nu$ are holomorphic on D_ν. Set $v_2 = u_2$ and assume that $v_2, \dots, v_\ell$ have been constructed for some $\ell \geqslant 2$. The difference $v_\ell - u_{\ell+1}$ is then holomorphic on D_{α_ℓ} and by Theorem 7.1 there is a $\overline{\partial}$-closed form $v \in \mathcal{C}_{p,0}(D)$ such that $|v_\ell(z) - u_{\ell+1}(z) - v(z)| < 2^{-\ell}$ for all $z \in D_{\alpha_{\ell-1}}$. We then simply set $v_{\ell+1} = u_{\ell+1} - v$. □

Corollary 7.5. *Let D be an open pseudoconvex set in $\mathbb{C}^n$. The cohomology group $H^{p,q}(D)$ then vanishes for any (p,q) such that $0 \leqslant p \leqslant n$ and $1 \leqslant q \leqslant n$.*

Theorem 7.6. *An open set D in $\mathbb{C}^n$ is a domain of holomorphy if and only if*

$$H^{0,q}(D) = 0 \quad \text{for } 1 \leqslant q \leqslant n-1.$$

Proof. It follows from Corollaries 7.3 and 7.5 that the condition is necessary.

We will prove that the condition is sufficient by induction on the complex dimension n. If $n = 1$ then the hypothesis is empty but the theorem holds because every open set in $\mathbb{C}$ is a domain of holomorphy. Suppose that the theorem holds for open sets in $\mathbb{C}^{n-1}$ for some $n \geqslant 2$ and consider an open set D in $\mathbb{C}^n$ such that $H^{0,q}(D) = 0$ whenever $1 \leqslant q \leqslant n-1$.

We start by showing that if L is an affine linear submanifold of $\mathbb{C}^n$ of complex dimension $n-1$ then any connected component of $D \cap L$, considered as an open set in $L \simeq \mathbb{C}^{n-1}$, is a domain of holomorphy. By the induction hypothesis, it is enough to prove that $H^{0,q}(D\cap L) = 0$ for any $1 \leqslant q \leqslant n-2$. Without loss of generality we can assume that $L = \{z \in \mathbb{C}^n \mid z_n = 0\}$. Let φ be a $\mathcal{C}^\infty$ and $\overline{\partial}$-closed differential $(0,q)$-form on $D\cap L$ such that $1 \leqslant q \leqslant n-2$; the form φ can then be extended to a neighbourhood U of $D \cap L$ in D as a $\mathcal{C}^\infty$ and $\overline{\partial}$-closed differential form Φ of type $(0,q)$ by considering φ as a form independent of z_n. Let χ be a $\mathcal{C}^\infty$ function on D which is equal to 1 on some neighbourhood of $D \cap L$ in D and which is supported on U. We set

$$\widetilde{\Phi}(z) = \begin{cases} \frac{1}{z^n}\big(\overline{\partial}\chi(z) \wedge \Phi(z)\big) & \text{if } z \in U \\ 0 & \text{if } z \in D \setminus U. \end{cases}$$

This defines a $\mathcal{C}^\infty$ and $\overline{\partial}$-closed differential $(0,q+1)$-form on the whole of D. Since $H^{0,q+1}(D) = 0$ by hypothesis, there is a $\mathcal{C}^\infty$ differential $(0,q)$-form $\widetilde{\Psi}$ on D such that $\overline{\partial}\widetilde{\Psi} = \widetilde{\Phi}$. Then

$$\overline{\partial}(\chi\Phi - z_n\widetilde{\Psi}) = \overline{\partial}\chi \wedge \Phi - z_n\overline{\partial}\widetilde{\Psi} = z_n(\widetilde{\Phi} - \overline{\partial}\widetilde{\Psi}) = 0$$

on D and as $H^{0,q}(D) = 0$ there is a $\mathcal{C}^\infty$ differential $(0,q-1)$-form Ψ on D such that $\chi\Phi - z_n\widetilde{\Psi} = \overline{\partial}\Psi$. Restricting this identity to L and setting $\theta = \Psi\big|_L$ we get a differential $(0,q-1)$-form θ on $D \cap L$ such that $\overline{\partial}\theta = \varphi$ which proves that $H^{0,q}(D \cap L) = 0$.

Suppose now that D is not a domain of holomorphy. There are then two open sets D_1 and D_2 such that $\varnothing \neq D_1 \subset D_2 \cap D$, D_2 is not contained in D and for any holomorphic function on D there is a function $g_2 \in \mathcal{O}(D_2)$ such that $g = g_2$ on D_1. Consider $\xi \in D_1$ and let L be a linear affine submanifold of $\mathbb{C}^n$ of complex dimension $n-1$ passing through ξ such that there is a point $\zeta \in \partial(D \cap L) \cap D_2$. Since we have just proved that the connected components of $D \cap L$ are domains of holomorphy in $L \simeq \mathbb{C}^{n-1}$ there is a holomorphic function f on $D \cap L$ which cannot be extended to a holomorphic function in a neighbourhood of ζ. As $H^{0,1}(D) = 0$ it follows from Lemma 6.4 that f is the restriction to $D \cap L$ of a holomorphic function g on D. (In the above, we proved Lemma 6.4 for a strictly pseudoconvex domain Ω, but in fact the proof only uses the fact that $H^{0,1}(\Omega) = 0$.) But this function g can now be extended to a holomorphic function in a neighbourhood of ζ, which contradicts the assumptions on f. It follows that D is a domain of holomorphy. □

We end this section with a result which gathers together Corollary 7.3, Theorem 7.6 and Theorem 1.13 of Chapter VI.

Corollary 7.7. *Let D be an open set in $\mathbb{C}^n$. The following are then equivalent:*

i) *D is a domain of holomorphy,*
ii) *D is holomorphically convex,*
iii) *D is pseudoconvex,*
iv) *$H^{0,q}(D) = 0$ if $1 \leqslant q \leqslant n-1$.*

8 The Levi problem for complex analytic manifolds

The aim of this section is to link holomorphic convexity to the existence of $\mathcal{C}^2$ plurisubharmonic exhaustion functions on complex analytic manifolds.

A. Solving $\overline{\partial}$ on complex analytic manifolds

Throughout this section X denotes a complex analytic manifold of dimension n.

Definition 8.1. A relatively compact open set D in X is said to be *strictly pseudoconvex* if there is a $\mathcal{C}^2$ strictly plurisubharmonic function ρ defined on a neighbourhood $U_{\partial D}$ of the boundary of D such that $D \cap U_{\partial D} = \{z \in U_{\partial D} \mid \rho(z) < 0\}$ and $d\rho(z) \neq 0$ for any $z \in \partial D$.

Remark. The set of (strictly) plurisubharmonic functions is well defined on a complex analytic manifold because (strict) plurisubharmonicity is preserved by holomorphic coordinate changes.

The definition of a strictly pseudoconvex open set given here coincides with that of Chapter VI when $X = \mathbb{C}^n$. Note further that any such domain has $\mathcal{C}^2$ boundary.

Unlike the $\mathbb{C}^n$ case, it is not always possible to solve $\overline{\partial}$ on strictly pseudoconvex open sets in arbitrary complex analytic manifolds. However, Proposition 3.1 and its proof are still valid in this more general setting.

Theorem 8.2. *Let D be a strictly pseudoconvex open set with $\mathcal{C}^2$ boundary, relatively compact in X, and let p and q be integers such that $0 \leqslant p \leqslant n$ and $0 \leqslant q \leqslant n$. Then:*

i) *there are continuous linear operators T_q^p from $\mathcal{C}^0_{p,q}(\overline{D})$ to $\Lambda^{1/2}_{p,q-1}(D)$ such that, if $f \in \mathcal{C}^0_{p,q}(\overline{D})$ has the property that $\overline{\partial} f \in \mathcal{C}^0_{p,q}(\overline{D})$, then*

$$f = \overline{\partial} T_q^p f + T_{q+1}^p \overline{\partial} f + K_q^p f, \tag{8.1}$$

where K_q^p is a compact operator from $\mathcal{C}^0_{p,q}(\overline{D})$ to itself,

ii) *if $q \geqslant 1$ then $\overline{\partial} T_q^p$ defines a continuous linear operator from $Z^0_{p,q}(\overline{D})$ to itself whose image is of finite codimension,*

iii) *the space $E^{1/2}_{p,q}(D)$ is a closed subspace of $Z^0_{p,q}(\overline{D})$ of finite codimension.*

We will now prove that if an open set D in X has a $\mathcal{C}^2$ strictly plurisubharmonic defining function defined not only on a neighbourhood of ∂D but on a neighbourhood of $\overline{D}$ then we can solve $\overline{\partial}$ on D. Note that any strictly pseudoconvex domain in $\mathbb{C}^n$ has this property (cf. Chap. VI, Th. 3.19), but in general this does not hold for a strictly pseudoconvex domain in an arbitrary complex analytic manifold.

Definition 8.3. A real-valued continuous function on X is said to be an *exhaustion* function if for every $\alpha \in \mathbb{R}$ the set $\{z \in X \mid \rho(z) < \alpha\}$ is relatively compact in X.

We will study those complex analytic manifolds which have $\mathcal{C}^2$ strictly plurisubharmonic exhaustion functions.

Proposition 8.4. *If X has a $\mathcal{C}^2$ strictly plurisubharmonic exhaustion function then it has a $\mathcal{C}^2$ strictly plurisubharmonic exhaustion function whose set of critical points is discrete.*

This proposition follows immediately from the following Morse lemma.

Lemma 8.5. *Let X be a complex analytic manifold and let ρ be a $\mathcal{C}^2$ strictly plurisubharmonic function on X. If K is a compact set in X such that $d\rho(z) \neq 0$ for any $z \in K$ then for any $\varepsilon > 0$ there is a $\mathcal{C}^2$ strictly plurisubharmonic function ρ_ε on X such that*

i) *the function $\rho - \rho_\varepsilon$ and its first and second derivatives are bounded by ε on X,*
ii) *the set $\mathrm{Crit}(\rho_\varepsilon) = \{z \in X \mid d\rho_\varepsilon(z) = 0\}$ is discrete in X,*
iii) *$\rho_\varepsilon = \rho$ on K.*

Proof. Let U_K be a neighbourhood of K such that $d\rho \neq 0$ on U_K. There are then two sequences of relatively compact open sets $(U_j)_{j\in\mathbb{N}^*}$ and $(V_j)_{j\in\mathbb{N}^*}$ in X such that

a) $X \setminus U_K \subset \bigcup_{j=1}^\infty U_j$,
b) for any j the open set V_j is a chart domain and $U_j \Subset V_j$,
c) $V_j \cap K = \varnothing$ for every j,
d) for any compact set L in X there are only a finite number of indices j such that $L \cap V_j \neq \varnothing$.

By the Lemmas 2.24 and 3.12 of Chapter VI, we can construct a sequence $(\chi_j)_{j\in\mathbb{N}}$ of $\mathcal{C}^\infty$ functions on X such that, for any j,

1) $\chi_j = 0$ on some neighbourhood of $X \setminus V_j$ and hence $\chi_j = 0$ on K,
2) the function $\rho + \chi_1 + \cdots + \chi_j$ is strictly plurisubharmonic on X and has only a finite number of critical points on $K \cup \overline{U}_1 \cup \cdots \cup \overline{U}_j$,
3) the function χ_j and its first and second derivatives are bounded by $\varepsilon/2^j$ on X.

The function $\rho_\varepsilon = \rho + \sum_{j=1}^\infty \chi_j$ is then the function we seek. □

In complex analytic manifolds we will have to replace elementary strictly pseudoconvex extensions by the more general strictly pseudoconvex extension elements in order to be able to cross critical points.

Definition 8.6. A *strictly pseudoconvex extension element* is an ordered pair $[\theta_1, \theta_2]$ of open sets in X with $\mathcal{C}^2$ boundary such that $\theta_1 \subset \theta_2$ and the following condition is satisfied: there is an open pseudoconvex set V contained in a chart domain of X containing $\overline{\theta_2 \setminus \theta_1}$, there are strictly pseudoconvex domains D_1 and D_2 such that

$$D_1 \subset D_2,\ \theta_2 = \theta_1 \cup D_2,\ \theta_1 \cap D_2 = D_1,\ (\overline{\theta_1 \setminus D_2}) \cap (\overline{\theta_2 \setminus \theta_1}) = \varnothing$$

and there is a biholomorphic map h defined on a neighbourhood of $\overline{V}$ with image in $\mathbb{C}^n$ such that $h(D_j)$, $j = 1, 2$, is a bounded strictly pseudoconvex domain with $\mathcal{C}^2$ boundary in $\mathbb{C}^n$.

Lemma 4.2 remains valid for strictly pseudoconvex extension elements: simply replace Theorem 1.6 by Theorem 5.3, Theorem 2.2 by Theorem 7.1 and Proposition 3.1 by Proposition 8.2 in the proof of Lemma 4.2. We will now extend Lemma 4.3 to the case where the function ρ has isolated critical points.

Lemma 8.7. *Let $\rho : X \to \mathbb{R}$ be a $\mathcal{C}^2$ strictly plurisubharmonic function such that* $\mathrm{Crit}(\rho)$ *is discrete. We set $D_\alpha = \{z \in X \mid \rho(z) < \alpha\}$ for any $\alpha \in \mathbb{R}$ and we assume that ∂D_0 is compact. There is an $\varepsilon > 0$ such that for any α, β satisfying $-\varepsilon \leqslant \alpha < 0 < \beta \leqslant \varepsilon$ there is a finite set of domains $\theta_1, \dots, \theta_N$ such that $D_\alpha = \theta_1 \subset \cdots \subset \theta_N \subset D_\beta$ and for any $j = 1, \dots, N-1$, $[\theta_j, \theta_{j+1}]$ is a strictly pseudoconvex extension element.*

Proof. As $\mathrm{Crit}\,\rho$ is discrete there is a $\varepsilon_0 > 0$ such that $\mathrm{Crit}(\rho) \cap (\overline{D}_{\varepsilon_0} \setminus D_{-\varepsilon_0})$ is finite and contained in ∂D_0. Let $\xi_1, \dots, \xi_M$ be the critical points of ρ contained in ∂D_0 and choose pairwise disjoint open sets $V_1, \dots, V_M$ in X such that $\xi_j \in V_j$ and V_j is contained in a chart domain of X for every $j = 1, \dots, M$. As ∂D_0 is compact there are open sets $V_{M+1}, \dots, V_N$, each contained in a chart domain of X, such that $\mathrm{Crit}(\rho) \cap \big(\bigcup_{j=M+1}^N V_j\big) = \varnothing$ and $\partial D_0 \setminus \big(\bigcup_{j=1}^M V_j\big) \subset \bigcup_{j=M+1}^N V_j$. Let ε_1 be such that $0 < \varepsilon_1 < \varepsilon_0$ and $\overline{D}_{\varepsilon_1} \setminus D_{-\varepsilon_1} \subset \bigcup_{j=1}^N V_j$. Choose $\mathcal{C}^\infty$ functions $(\chi_j)_{j=1,\dots,N}$ with compact support in $\mathbb{C}^n$ such that $\operatorname{supp}\chi_j \subset V_j$ for all $j = 1, \dots, N$ and $\sum_{j=1}^N \chi_j = 1$ on $\overline{D}_{\varepsilon_1} \setminus D_{-\varepsilon_1}$. Let ε be such that $0 < \varepsilon < \varepsilon_1$. Given α and β such that $-\varepsilon \leqslant \alpha < 0 < \beta \leqslant \varepsilon$ we set

$$\theta_k = \big\{z \in \mathbb{C}^n \mid \rho(z) - \alpha < (\beta - \alpha) \textstyle\sum_{j=1}^k \chi_j(z)\big\}.$$

Then $D_\alpha = \theta_1 \subset \cdots \subset \theta_N \subset D_\beta$. Let $(z_{1_j}, \dots, z_{n_j})$ be holomorphic coordinates on V_j and set

$$C = \sum_{j=1}^N \sup_{z \in V_j} \sum_{r_j, s_j}^n \Big| \frac{\partial^2 \chi_j(z)}{\partial z_{s_j} \partial \overline{z}_{r_j}} \Big|.$$

As ρ is strictly plurisubharmonic there is a $\gamma > 0$ such that, for every $\xi \in \mathbb{C}^n$ and $j = 1, \dots, N$,

$$\inf_{z \in V_j} \sum_{r_j, s_j}^N \frac{\partial^2 \rho}{\partial z_{s_j} \partial \overline{z}_{r_j}} \xi_s \overline{\xi}_r \geqslant \gamma |\xi|^2.$$

Take $\delta > 0$ such that $|d\rho(z)| \geqslant 3\delta \sum_{j=1}^N |d\chi_j(z)|$ for any $z \in \overline{D}_{\varepsilon_1} \setminus D_{-\varepsilon_1}$. If $\varepsilon < \min(\delta, \gamma/3C)$ for every $j = 1, \dots, N-1$ then $[\theta_j, \theta_{j+1}]$ is a strictly pseudoconvex extension element. □

Remark. If $\partial D_0 \cap \mathrm{Crit}(\rho) = \varnothing$ then we can take $\alpha = 0$ or $\beta = 0$ in Lemma 8.7.

We can define strictly pseudoconvex extensions on complex analytic manifolds as in $\mathbb{C}^n$.

Definition 8.8. Let $D \Subset \Omega \Subset X$ be open sets in X. We say that Ω is a *strictly pseudoconvex extension* of D if there is a neighbourhood U of $\overline{\Omega} \setminus D$ and a $\mathcal{C}^2$ strictly plurisubharmonic function ρ on U such that

$$D \cap U = \{z \in U \mid \rho(z) < 0\} \quad \text{and} \quad d\rho(z) \neq 0 \quad \text{if } z \in \partial D$$
$$\Omega \cap U = \{z \in U \mid \rho(z) < 1\} \quad \text{and} \quad d\rho(z) \neq 0 \quad \text{if } z \in \partial \Omega.$$

Remark. By Lemma 8.5 the function ρ can be chosen such that $\mathrm{Crit}(\rho)$ is discrete. Using Lemma 8.7 instead of Lemma 4.3, the proof of Proposition 4.5 can then be extended to strictly pseudoconvex extensions in complex analytic manifolds. This gives us the following proposition.

Proposition 8.9. *Let D and Ω be open sets in X, $D \Subset \Omega \Subset X$, such that Ω is a strictly pseudoconvex extension of D. Then:*

i) *the restriction map*

$$H^{p,q}_{0,1/2}(\overline{\Omega}) \longrightarrow H^{p,q}_{0,1/2}(\overline{D}), \quad 0 \leqslant p \leqslant n,\ 1 \leqslant q \leqslant n$$

is an isomorphism,

ii) *for any pair (p,q) such that $0 \leqslant p \leqslant n$ and $0 \leqslant q \leqslant n$ the restriction map $Z^0_{p,q}(\overline{\Omega}) \to Z^0_{p,q}(\overline{D})$ has dense image.*

Likewise, Lemma 4.7 still holds in the following form.

Proposition 8.10. *Let D and Ω be two open sets in X such that $D \Subset \Omega$. Suppose there is a neighbourhood U of $\Omega \setminus D$ and a $\mathcal{C}^2$ strictly plurisubharmonic function ρ such that $D \cap U = \{z \in U \mid \rho(z) < 0\}$, $d\rho(z) \neq 0$ for any $z \in \partial D$ and $D \cup \{z \in U \mid \rho(z) \leqslant C\} \Subset \Omega$ for any $C > 0$. The restriction map*

$$H^{p,q}(\Omega) \longrightarrow H^{p,q}_{0,1/2}(\overline{D})$$

is then injective for any (p,q) such that $0 \leqslant p \leqslant n$ and $1 \leqslant q \leqslant n$.

We can now prove the main result of this section.

Theorem 8.11. *Let X be a complex analytic manifold with a $\mathcal{C}^2$ strictly plurisubharmonic exhaustion function ρ and let D be an open set in X defined by $D = \{z \in X \mid \rho(z) < \alpha\}$ such that $d\rho(z) \neq 0$ if $z \in \partial D$. Then:*

i) *for any pair (p,q) such that $0 \leqslant p \leqslant n$ and $1 \leqslant q \leqslant n$*

$$H^{p,q}_{0,1/2}(\overline{D}) = 0.$$

More precisely, there is a continuous linear map T from the Banach space $Z^0_{p,q}(\overline{D})$ to the Banach space $\Lambda^{1/2}_{p,q-1}(D)$ such that

$$\overline{\partial} T f = f \text{ on } D \text{ for any } f \in Z^0_{p,q}(\overline{D}).$$

ii) *any continuous differential $(p,0)$-form on $\overline{D}$, holomorphic on D, can be uniformly approximated on $\overline{D}$ by holomorphic $(p,0)$-forms on X.*
iii) *for any pair (p,q) such that $0 \leqslant p \leqslant n$ and $1 \leqslant q \leqslant n$,*

$$H^{p,q}(X) = 0.$$

Proof. By Lemma 8.5 applied to $K = \partial D$ we can assume that $\mathrm{Crit}(\rho)$ is discrete. Set $\alpha_0 = \min_{z \in X} \rho(z)$. If $z \in X$ is such that $\rho(z) = \alpha_0$ then $z \in \mathrm{Crit}(\rho)$. As this set is discrete there are only a finite number of such points z in X. It follows that, for small enough $\varepsilon > 0$, $D_{\alpha_0+\varepsilon} = \{z \in X \mid \rho(z) < \alpha_0+\varepsilon\}$ is biholomorphic to a finite union of bounded strictly pseudoconvex domains with $\mathcal{C}^2$ boundary in $\mathbb{C}^n$ so by Corollary 5.2 if $0 \leqslant p < n$ and $1 \leqslant q \leqslant n$ then

$$H^{p,q}_{0,1/2}(\overline{D_{\alpha_0+\varepsilon}}) = 0. \tag{8.2}$$

But as D is a strictly pseudoconvex extension of $D_{\alpha_0+\varepsilon}$, Proposition 8.9 implies that if $0 \leqslant p \leqslant n$ and $1 \leqslant q \leqslant n$ then

$$H^{p,q}_{0,1/2}(\overline{D}) = 0.$$

The existence of the operator T then follows from Theorem 8.2 and Proposition 5 of Appendix C. We have therefore proved i). Since the pair $(D_{\alpha_0+\varepsilon}, X)$ satisfies the hypotheses of Proposition 8.10, iii) follows from (8.2).

We now complete the proof of the theorem by proving ii). Since $\mathrm{Crit}\,\rho$ is discrete, we can construct a sequence $(\beta_j)_{j\in\mathbb{N}}$ of real numbers $\beta_0 = \alpha < \beta_1 < \beta_2 < \cdots$ tending to infinity such that $D_{\beta_j} = \{z \in X \mid \rho(z) < \beta_j\}$ and $d\rho(z) \neq 0$ for any $z \in \partial D_{\beta_j}$. The open set $D_{\beta_{j+1}}$ is then a strictly pseudoconvex extension of D_{β_j} for any $j \in \mathbb{N}$. By ii) of Proposition 8.9, for any $\varepsilon > 0$ and any $f \in Z^0_{p,0}(\overline{D})$ we can construct a sequence $(f_j)_{j\in\mathbb{N}}$ of differential forms such that $f_j \in Z^0_{p,0}(\overline{D}_{\beta_j})$, $f_0 = f$ and $|f_{j+1} - f_j|_{0,D_{\beta_j}} < \varepsilon/2^{j+1}$. The sequence $(f_j)_{j\in\mathbb{N}}$ then converges uniformly on all compact sets to a holomorphic $(p,0)$-form $\widetilde{f}$ on X such that $|\widetilde{f} - f|_{0,D} < \varepsilon$. □

B. The Levi problem

Throughout this section X will be a complex analytic manifold of dimension n.

Definition 8.12. For any compact set K in X we define

$$\widehat{K}_X = \{z \in X \mid \forall f \in \mathcal{O}(X),\ |f(z)| \leqslant \sup_{\zeta\in K} |f(\zeta)|\}.$$

The set $\widehat{K}_X$ is called the *holomorphically convex hull* of K in X. If $K = \widehat{K}_X$ then K is said to be *$\mathcal{O}(X)$-convex.*

Definition 8.13. A complex analytic manifold X is *holomorphically convex* if for any compact set K in X the set $\widehat{K}_X$ is compact.

Definition 8.14. A complex analytic manifold X of dimension n is a *Stein manifold* if

i) X is holomorphically convex,
ii) for any point $z \in X$ there are n functions $f_1, \dots, f_n \in \mathcal{O}(X)$ which form a local system of coordinates on a neighbourhood of z. (In other words there is a neighbourhood U of z such that the map $F : \zeta \mapsto (f_1(\zeta), \dots, f_n(\zeta))$ is a biholomorphism between U and the open set $F(U)$ in $\mathbb{C}^n$.)

Example. By Theorem 1.13 of Chapter VI, any domain of holomorphy in $\mathbb{C}^n$ is a Stein manifold.

Definition 8.15. A subset V of a complex analytic manifold X of dimension n is an *analytic submanifold of dimension* $m < n$ if

i) V is closed,
ii) for any $z \in V$ there is a neighbourhood ω of z and local coordinates $(z_1, \dots, z_n)$ on X such that

$$\omega \cap V = \{\zeta \in \omega \mid z_{m+1}(\zeta) = \cdots = z_n(\zeta) = 0\}.$$

Note that if $(f_1, \dots, f_n)$ is a local system of coordinates on X in a neighbourhood of $z \in V$ then there are m of these functions which form a local system of coordinates on V in a neighbourhood of z. From this it is easy to deduce the following result.

Proposition 8.16. *Any complex analytic submanifold of a Stein manifold is a Stein manifold. In particular, any complex analytic submanifold of* $\mathbb{C}^n$ *is a Stein manifold.*

Remark. Although it will not be proved in this book, it can be shown that any Stein manifold is biholomorphic to a complex analytic submanifold of a certain $\mathbb{C}^N$ (cf. [Ho2, § 5.3] for example).

Theorem 8.17. *Let X be a Stein manifold, let K be a compact set in X and let U be a neighbourhood of $\widehat{K}_X$. There is then a $\mathcal{C}^\infty$ strictly plurisubharmonic function φ on X such that*

i) $\varphi < 0$ *on* K *and* $\varphi > 0$ *on* $X \smallsetminus U$,
ii) $\{z \in X \mid \varphi(z) < c\} \Subset X$ *for any* $c \in \mathbb{R}$.

Proof. As $\widehat{K}$ is $\mathcal{O}(X)$-convex and X is holomorphically convex we can find a sequence $(K_j)_{j \in \mathbb{N}^*}$ of $\mathcal{O}(X)$-convex compact sets such that $K_1 = \widehat{K}$, $K_j \subset \overset{\circ}{K}_{j+1}$ and $X = \bigcup_{j=1}^\infty K_j$. Set $U_1 = U$ and $U_j = \overset{\circ}{K}_{j+1}$ for any $j \geqslant 2$. For every j, choose functions $f_{jk} \in \mathcal{O}(X)$ for $k = 1, \dots, k_j$ such that $|f_{jk}| < 1$ on K_j and $\max_{1 \leqslant k \leqslant k_j} |f_{jk}(z)| > 1$ for any $z \in K_{j+2} \smallsetminus U$. (Such functions exist because $\widehat{K}_j = K_j$.) Moreover, by ii) of Definition 8.14, on adding functions we can assume that the rank of the matrix

$$\left(\frac{\partial f_{jk}}{\partial z_\ell}(z)\right)_{\substack{k=1,\dots,k_j\\ \ell=1,\dots,n}}$$

is n for all $z \in K_j$. Taking powers of f_{jk} we can even assume that

$$\sum_{k=1}^{k_j} |f_{jk}(z)|^2 < 2^{-j}, \quad \text{for any } z \in K_j \tag{8.3}$$

$$\sum_{k=1}^{k_j} |f_{jk}(z)|^2 > j, \quad \text{for any } z \in K_{j+2} \smallsetminus U_j. \tag{8.4}$$

By (8.3), the series $\sum_{j,k} f_{jk}(z)\overline{f_{jk}(\zeta)}$ converges uniformly on every compact set in $X \times X$ and defines a function which is holomorphic in z and antiholomorphic in ζ: it follows that the function $\varphi(z) = -1+\sum_{j=1}^{\infty}\left(\sum_{k=1}^{k_j} |f_{jk}(z)|^2\right)$ is $\mathcal{C}^\infty$ on X. It is clear that $\varphi(z) > j-1$ for any $z \in X \smallsetminus U_j$ by (8.4) and hence $\varphi > 0$ on $X \smallsetminus U$ and $\varphi < 0$ on $\widehat{K}$ which implies that $\varphi < 0$ on K by (8.3). Moreover, ii) holds by (8.4), and φ is plurisubharmonic because it is the supremum of a set of plurisubharmonic functions. It remains to show that φ is strictly plurisubharmonic. Consider a point $\zeta \in X$ and let $(z_1, \dots, z_n)$ be holomorphic local coordinates on some neighbourhood of ζ. Assume that, for every $\xi \in \mathbb{C}^n$,

$$\sum_{r,s=1}^{n} \frac{\partial^2 \varphi}{\partial z_s \partial \overline{z}_r}(\zeta)\xi_s\overline{\xi}_r = 0.$$

For every j we then have

$$\sum_{k=1}^{k_j}\left|\sum_{r=1}^{n} \frac{\partial f_{jk}}{\partial z_r}(\zeta)\xi_r\right|^2 = \left|\sum_{k=1}^{k_j}\sum_{r,s=1}^{n} \frac{\partial^2 |f_{jk}(\zeta)|^2}{\partial z_s \partial \overline{z}_r}\xi_s\overline{\xi}_r\right| = 0$$

which implies that $\xi = 0$ because the matrix $\left(\frac{\partial f_{jk}}{\partial z_\ell}(z)\right)_{\substack{k=1,\dots,kj\\ \ell=1,\dots,n}}$ has rank n if $z \in K_j$. □

To prove the converse of Theorem 8.17 we will need the following lemma.

Lemma 8.18. *Let X be a complex analytic manifold with a $\mathcal{C}^2$ strictly plurisubharmonic exhaustion function ρ. The following then hold.*

i) *Consider a point $\xi \in X$, set $\alpha = \rho(\xi)$ and set $D_\alpha = \{z \in X \mid \rho(z) < \alpha\}$. Assume that $d\rho(z) \neq 0$ for any $z \in \partial D_\alpha$. There is then a sequence of holomorphic functions $(f_k)_{k\in\mathbb{N}^*}$ on X and a constant C such that*
   a) *$f_k(\xi) = 1$ for any $k \in \mathbb{N}^*$,*

b) $|f_k|_{0,D_\alpha} \leqslant C$ *for any* $k \in \mathbb{N}^*$,

c) $\lim_{k\to\infty} |f_k|_{0,K} = 0$ *for any compact set* $K \subset \overline{D}_\alpha \setminus \{\xi\}$.

ii) *For any* $\xi \in X$ *and any holomorphic function* f *on a neighbourhood of* ξ *there is a sequence of holomorphic functions* $(f_k)_{k\in\mathbb{N}}$ *on* X *such that* $\lim_{k\to\infty} \partial f_k(\xi) = \partial f(\xi)$.

Proof.

i) Let $\lambda = (\lambda_1, \ldots, \lambda_n)$ be holomorphic coordinates on a neighbourhood V_ξ of ξ. Set

$$u(z) = 2\sum_{j=1}^{n} \frac{\partial \rho}{\partial \lambda_j}(\xi)(\lambda_j(z) - \lambda_j(\xi)) + \sum_{j,k=1}^{n} \frac{\partial^2 \rho}{\partial \lambda_j \partial \lambda_k}(\xi)(\lambda_j(z) - \lambda_j(\xi))(\lambda_k(z) - \lambda_k(\xi)).$$

The function u is then holomorphic on V_ξ, $u(\xi) = 0$ and by Lemma 2.22 of Chapter VI,

$$\operatorname{Re} u(z) = \rho(z) - \rho(\xi) - \sum_{j,k=1}^{n} \frac{\partial^2 \rho}{\partial \lambda_j \partial \overline{\lambda}_k}(\xi)(\lambda_j(z) - \lambda_j(\xi))(\overline{\lambda_k(z) - \lambda_k(\xi)}) + 0(|\lambda(\xi) - \lambda(z)|^2).$$

As ρ is strictly plurisubharmonic, on restricting V_ξ we can find $\beta > 0$ such that

$$\operatorname{Re} u(z) < \rho(z) - \rho(\xi) - \beta|\lambda(z) - \lambda(\xi)|^2 \quad \text{for any } z \in V_\xi. \tag{8.5}$$

Then $e^{u(\xi)} = 1$ and $|e^{u(z)}| < 1$ for any $z \in \overline{D}_\alpha \cap V_\xi \setminus \{\xi\}$. Choose a neighbourhood $W_\xi \Subset V_\xi$ of ξ and a $\mathcal{C}^\infty$ function χ on X such that $\chi = 1$ on W_ξ and $\operatorname{supp} \chi \Subset V_\xi$. The sequence $(e^{ku}\overline{\partial}\chi)_{k\in\mathbb{N}^*}$ is then a sequence of $\mathcal{C}^\infty$ and $\overline{\partial}$-closed differential $(0,1)$-forms on X such that

$$\lim_{k\to\infty} |e^{ku}\overline{\partial}\chi|_{0,D_\alpha} = 0.$$

Since $d\rho(z) \neq 0$ if $z \in \partial D_\alpha$ we can apply Theorem 8.11 i) to obtain a sequence $(v_k)_{k\mathbb{N}^*}$ of continuous functions on $\overline{D}_\alpha$ such that $\overline{\partial} v_k = e^{ku}\overline{\partial}\chi$ on D_α and $\lim_{k\to\infty} |v_k|_{0,D_\alpha} = 0$. Setting $\widetilde{f}_k = \chi e^{ku} - v_k + v_k(\xi)$, we get holomorphic functions $\widetilde{f}_k$ on D_α such that $\widetilde{f}_k(\xi) = 1$ for any $k \in \mathbb{N}^*$, $\sup_{k\in\mathbb{N}^*} |\widetilde{f}_k|_{0,D_\alpha} < +\infty$ and for any compact set $K \subset \overline{D}_\alpha \setminus \{\xi\}$, $\lim_{k\to\infty} |\widetilde{f}_k|_{0,K} = 0$. We then construct the required functions $f_k \in \mathcal{O}(X)$ using the approximation theorem 8.11 ii).

ii) Assume that $f(\xi) = 0$. Set $\alpha = \rho(\xi)$ and consider V_ξ, W_ξ, u and χ as in the proof of i). Restricting V_ξ if necessary, we can assume that f is holomorphic on V_ξ. Setting $\varphi_k = fe^{ku}\overline{\partial}\chi$ for any $k = 1, 2, \ldots$ we define a sequence of $\mathcal{C}^\infty$ and $\overline{\partial}$-closed $(0,1)$-forms on X such that $\operatorname{supp}\varphi_k \Subset V_\xi \setminus W_\xi$. It follows from (8.5) that there is a $\delta > 0$ such that $\lim_{k\to\infty} |\varphi_k|_{0,D_{\alpha+\delta}} = 0$. By Lemma 8.5, we can assume that $\mathrm{Crit}(\rho)$ is discrete in X and hence we can choose $0 < \varepsilon < \delta$ small enough that $d\rho(z) \neq 0$ if $z \in \partial D_{\alpha+\varepsilon}$. We can then apply 8.11 i) and find continuous functions v_k on $\overline{D}_{\alpha+\varepsilon}$ such that $\overline{\partial} v_k = \varphi_k$ on $D_{\alpha+\varepsilon}$ and $\lim_{k\to\infty} |v_k|_{0,D_{\alpha+\varepsilon}} = 0$. As $\varphi_k = 0$ on W_ξ, the functions v_k are holomorphic on W_ξ and by the Cauchy inequalities $\lim_{k\to\infty} \partial v_k(\xi) = 0$. The functions $\widetilde{f}_k$ defined by $\widetilde{f}_k = \chi f e^{ku} - v_k$ are continuous on $\overline{D}_{\alpha+\varepsilon}$ and holomorphic on $D_{\alpha+\varepsilon}$ and have the property that $\partial \widetilde{f}_k(\xi) = \partial f(\xi) - \partial v_k(\xi)$. Using Theorem 8.11 ii) we can find functions $f_k \in \mathcal{O}(X)$ such that $|f_k - \widetilde{f}_k|_{0,D_{\alpha+\varepsilon}} < 1/k$ and it follows that $\lim_{k\to\infty} |\partial f_k(\xi) - \partial \widetilde{f}_k(\xi)| = 0$ by the Cauchy inequalities. The functions f_k thus constructed have the required properties.

Theorem 8.19. *A complex analytic manifold X is Stein if and only if X has a $\mathcal{C}^2$ strictly plurisubharmonic exhaustion function ρ. For any $\alpha \in \mathbb{R}$ the sets $\{z \in X \mid \rho(z) \leqslant \alpha\}$ are then $\mathcal{O}(X)$-convex.*

Proof. Theorem 8.17 tells us that this condition is necessary.

Suppose that X has a $\mathcal{C}^2$ strictly plurisubharmonic exhaustion function ρ. Condition ii) of Definition 8.14 follows immediately from Lemma 8.18 ii) since for any finite family of holomorphic functions in a neighbourhood of a point ξ in X we can find holomorphic functions on X whose Jacobian at ξ is arbitrarily close to the Jacobian at ξ of the initial family.

Set $D_\alpha = \{z \in X \mid \rho(z) < \alpha\}, \alpha \in \mathbb{R}$. It is enough to prove that for any $\alpha \in \mathbb{R}$ the sets $\overline{D}_\alpha$ are $\mathcal{O}(X)$-convex. Consider a point $\xi \in X \setminus \overline{D}_\alpha$. By Lemma 8.5, there is a $\mathcal{C}^2$ strictly plurisubharmonic exhaustion function φ on X such that $\mathrm{Crit}(\varphi)$ is discrete and which is close enough to ρ that $D_\alpha \Subset \Omega_{\varphi(\xi)}$, where $\Omega_\beta = \{z \in X \mid \varphi(z) < \beta\}$ for any $\beta \in \mathbb{R}$. After adding a small constant to φ we can assume that $d\varphi(z) \neq 0$ for any $z \in \partial\Omega_{\varphi(\xi)}$. Applying Lemma 8.18 i) to $\Omega_{\varphi(\xi)}$ we can then find a function $f \in \mathcal{O}(X)$ such that $f(\xi) = 1$ and $|f| < 1$ on $\overline{D}_\alpha$. □

Corollary 8.20. *Let X be a Stein manifold and let z and ξ be two distinct points of X. There is then a function $f \in \mathcal{O}(X)$ such that $f(z) \neq f(\xi)$.*

Proof. By Theorem 8.19, X has a $\mathcal{C}^2$ strictly plurisubharmonic exhaustion function ρ. Lemma 8.5 implies there is a $\mathcal{C}^2$ strictly plurisubharmonic exhaustion function φ on X such that $\mathrm{Crit}(\varphi)$ is discrete. Without loss of generality we can assume that $\varphi(z) \leqslant \varphi(\xi)$. Set $D_{\varphi(\xi)} = \{\zeta \in X \mid \varphi(\zeta) < \varphi(\xi)\}$. After adding a small constant to φ, we can assume that $d\varphi(\zeta) \neq 0$ for any $\zeta \in \partial\Omega_{\varphi(\xi)}$. Applying Lemma 8.18 i) to $\Omega_{\varphi(\xi)}$ and $K = \{z\}$ we can find a function $f \in \mathcal{O}(X)$ such that $f(\xi) = 1$ and hence $|f(z)| < 1$ and $f(z) \neq f(\xi)$. □

Corollary 8.21. *If X is a Stein manifold of dimension n then*

$$H^{p,q}(X) = 0 \quad \textit{whenever } 0 \leqslant p \leqslant n \textit{ and } 1 \leqslant q \leqslant n.$$

Proof. This follows immediately from Theorem 8.11 iii) and Theorem 8.19. □

Remark. Unlike open sets in $\mathbb{C}^n$ (cf. Corollary 7.7), the vanishing of the cohomology groups $H^{0,q}(X)$, $1 \leqslant q \leqslant n-1$, does not characterise Stein manifolds amongst complex analytic manifolds; these groups also all vanish when X is a projective space, for example. A cohomological characterisation of Stein manifolds does exist, however (cf. [Gu], vol III).

Proposition 8.22. *Let X be a Stein manifold. Any strictly pseudoconvex open set $D \Subset X$ is a Stein manifold.*

Proof. Since D is strictly pseudoconvex there is a neighbourhood $U_{\partial D}$ of the boundary of D and a $\mathcal{C}^2$ strictly plurisubharmonic function ρ on $U_{\partial D}$ such that $D \cap U_{\partial D} = \{z \in U_{\partial D} \mid \rho(z) < 0\}$. By Theorem 8.19, X has a $\mathcal{C}^2$ strictly plurisubharmonic exhaustion function ρ_1. Fix a small enough $\varepsilon > 0$ that $\{z \in U_{\partial D} \mid -\varepsilon < \rho(z) < 0\} \Subset U_{\partial D}$ and choose a $\mathcal{C}^\infty$ function χ on $[-\infty, 0[$ such that $\chi(t) = 0$ for any $t < -\varepsilon$, $\chi(t) \to +\infty$ as $t \to 0$ and χ is strictly convex on $]-\varepsilon, 0[$. By Proposition 2.8 of Chapter VI $\chi \circ \rho$ is then strictly plurisubharmonic on $\{z \in U_{\partial D} \mid -\varepsilon < \rho(z) < 0\}$ and $\rho_1 + \chi \circ \rho$ is a $\mathcal{C}^2$ strictly plurisubharmonic exhaustion function for D. Theorem 8.19 then implies that D is Stein. □

C. Vanishing theorems for compactly supported cohomology and applications

It follows from Sections 8.1 and 8.2 that if X is a Stein manifold of dimension n then for any pair (p,q), where $0 \leqslant p \leqslant n$ and $1 \leqslant q \leqslant n$, we have

$$H^{p,q}(X) = 0.$$

We will now study the vanishing of the cohomology groups with compact support $H_c^{p,q}(X)$ on a complex analytic manifold X.

Proposition 8.23. *Let X be a complex analytic manifold of dimension n and let p and q be integers such that $0 \leqslant p \leqslant n$ and $1 \leqslant q \leqslant n$. We assume that the set*

$$E^0_{n-p,n-q+1}(X) = \{u \in \mathcal{C}^0_{n-p,n-q+1}(X) \mid u = \overline{\partial} v, v \in \mathcal{C}^0_{n-p,n-q}(X)\}$$

is closed in $\mathcal{C}^0_{n-p,n-q+1}(X)$ for the topology of uniform convergence on every compact set. If $f \in \mathcal{C}^0_{p,q}(X)$ is a $\overline{\partial}$-closed continuous differential form with compact support on X then the equation $\overline{\partial} g = f$ has a solution $g \in \mathcal{C}^0_{p,q-1}(X)$ with compact support in X if and only if $\int_X f \wedge \varphi = 0$ for any $\overline{\partial}$-closed form $\varphi \in \mathcal{C}^\infty_{n-p,n-q}(X)$ on X.

Proof. We start by proving the condition is necessary. Assume there is a compactly supported $g \in \mathcal{C}^0_{p,q-1}(X)$ such that $\overline{\partial} g = f$ on X. If $\varphi \in \mathcal{C}^\infty_{n-p,n-q}(X)$ is a $\overline{\partial}$-closed form on X then

$$\int_X f \wedge \varphi = \int_X \overline{\partial} g \wedge \varphi = \int_X \overline{\partial}(g \wedge \varphi) = 0$$

by Stokes' theorem because g has compact support.

Let us now prove that the condition is sufficient. Assume that $\int_X f \wedge \varphi = 0$ for any $\overline{\partial}$-closed form $\varphi \in \mathcal{C}^\infty_{n-p,n-q}(X)$. We define a linear form on $E^0_{n-p,n-q+1}(X)$ as follows: if $u \in E^0_{n-p,n-q+1}(X)$ then we choose $v \in \mathcal{C}^0_{n-p,n-q}(X)$ such that $\overline{\partial} v = u$ and we set

$$F(u) = \int_X f \wedge v.$$

The map F is well defined. Indeed, by the Dolbeault isomorphism if $v, w \in \mathcal{C}^0_{n-p,n-q}(X)$ are such that $\overline{\partial} v = \overline{\partial} w = h$ then there is a $\overline{\partial}$-closed $\varphi \in \mathcal{C}^\infty_{n-p,n-q}(X)$ such that $v - w - \varphi = \overline{\partial}\theta$ for some $\theta \in \mathcal{C}^0_{n-p,n-q-1}(X)$. It then follows by Stokes' formula that

$$\int_X f \wedge (v - w - \varphi) = 0$$

and by hypothesis $\int_X f \wedge v = \int_X f \wedge w$. As the space $E^0_{n-p,n-q+1}(X)$ is assumed closed, the open mapping theorem implies that there is a continuous linear map δ from $E^0_{n-p,n-q+1}(X)$ to $\mathcal{C}^0_{n-p,n-q}(X)$ such that $\overline{\partial}\delta = I$. The map F is then equal to $\Phi \circ \delta$ where

$$\Phi : \mathcal{C}^0_{n-p,n-q}(X) \longrightarrow \mathbb{C}$$

is the continuous linear form $v \mapsto \int_X f \wedge v$. It follows that F is a continuous linear form on $E^0_{n-p,n-q+1}(X)$. Applying the Hahn–Banach theorem we can extend F to a continuous linear form $\widetilde{F}$ on $\mathcal{C}^0_{n-p,n-q+1}(X)$. The form $\widetilde{F}$ therefore defines a compactly supported current such that

$$(-1)^{p+q-1}\overline{\partial}\widetilde{F}(h) = \widetilde{F}(\overline{\partial} h) = F(\overline{\partial} h) = \int_X f \wedge h = \langle T_f, h\rangle,$$

for any form $h \in \mathcal{D}_{n,n-q}(X)$, i.e. $(-1)^{p+q-1}\overline{\partial}\widetilde{F} = T_f$ where T_f is the current defined by f. The regularity of $\overline{\partial}$ (cf. Chap. V, Cor. 4.2) then implies that there is a $g \in \mathcal{C}^0_{p,q-1}(X)$ with compact support such that

$$(-1)^{p+q-1}\overline{\partial} g = f \quad \text{in } X. \qquad \square$$

Theorem 8.24. *Let X be a Stein manifold of dimension n. Then*

$$H^{p,q}_c(X) = 0 \quad \textit{whenever } 0 \leqslant p \leqslant n \textit{ and } 0 \leqslant q \leqslant n-1.$$

Proof. The principle of analytic continuation tells us that

$$H_c^{p,0}(X) = 0 \quad \text{for any } p,\ 0 \leqslant p \leqslant n,$$

since a Stein manifold does not have any compact connected components.

Suppose that $1 \leqslant q \leqslant n$. Since X is Stein the Dolbeault isomorphism tells us that

$$H_0^{p,q}(X) \simeq H^{p,q}(X) = 0.$$

It follows that $Z^0_{n-p,n-q+1}(X) = E^0_{n-p,n-q+1}(X)$ is closed in $\mathcal{C}^0_{n-p,n-q+1}(X)$ for $0 \leqslant p \leqslant n$ and $1 \leqslant q \leqslant n$. We can therefore apply Proposition 8.23. If $f \in \mathcal{C}^0_{p,q}(X)$ is a $\overline{\partial}$-closed differential form with compact support on X we will calculate $\int_X f \wedge \varphi$ for any $\overline{\partial}$-closed $\varphi \in \mathcal{C}^\infty_{n-p,n-q}(X)$. If $1 \leqslant q \leqslant n-1$ then $H^{n-p,n-q}(X) = 0$ and hence $\varphi = \overline{\partial}\Psi$ for some $\Psi \in \mathcal{C}^\infty_{n-p,n-q-1}(X)$. Applying Stokes' theorem we get

$$\int_X f \wedge \varphi = \int_X f \wedge \overline{\partial}\Psi = \int_X (-1)^{p+q}\overline{\partial}(f \wedge \Psi) = 0$$

since f has compact support. There is therefore a compactly supported $g \in \mathcal{C}^0_{p,q-1}(X)$ such that $\overline{\partial} g = f$ and by Dolbeault's isomorphism this implies that $H_c^{p,q}(X) = 0$ (cf. Chap. VI, Cor. 4.3). □

This vanishing theorem allows us to give sufficient geometric conditions for the Hartogs–Bochner phenomenon and the extension of CR functions to hold using the results of Chapter V.

We deduce a Hartogs–Bochner theorem for Stein manifolds from Corollaries 1.4 and 5.2 of Chapter V.

Theorem 8.25. *Let X be a Stein manifold of dimension n where $n \geqslant 2$. Hartogs' phenomenon then holds for X. More precisely, for any relatively compact domain D with $\mathcal{C}^k$ boundary ($k \geqslant 1$) in X such that $X \smallsetminus \overline{D}$ is connected and for any* CR *function f of class $\mathcal{C}^s$ on ∂D where $0 \leqslant s \leqslant k$ there is a $\mathcal{C}^s$ function F on $\overline{D}$ which is holomorphic on D such that $F|_{\partial D} = f$.*

Theorem 8.26. *Let X be a Stein manifold of dimension n, $n \geqslant 2$, and let K be a $\mathcal{O}(X)$-convex compact set in X. Then for any relatively compact domain D in X such that*

1) *$\partial D \smallsetminus K$ is a $\mathcal{C}^k$ submanifold, $k \geqslant 1$, of $X \smallsetminus K$,*
2) *$D \smallsetminus K = \mathrm{Int}(\overline{D} \smallsetminus K)$,*
3) *$X \smallsetminus (\overline{D} \cup K)$ is connected,*

and for any $\mathcal{C}^s$ CR *function f on $\partial D \smallsetminus K$ there is a $\mathcal{C}^s$ function F on $\overline{D} \smallsetminus K$ which is holomorphic on $D \smallsetminus K$ such that $F|_{\partial D \smallsetminus K} = f$.*

Proof. By Theorem 5.1 of Chapter V, it is enough to prove that $H_\Phi^{0,1}(X \smallsetminus K) = 0$.

If X is Stein and of dimension $n \geqslant 2$ then by Theorem 8.24 $H_c^{0,1}(X) = 0$. If K is $\mathcal{O}(X)$-convex then, by Theorem 8.17, K has a decreasing sequence $(U_p)_{p\in\mathbb{N}}$ of Stein neighbourhoods such that $\bigcap_{p\in\mathbb{N}} U_p = K$. Moreover, $H_c^{0,2}(U_p) = 0$ for $n \geqslant 3$ and the hypotheses of Theorem 2.5 of Chapter V hold. It follows that $H_\Phi^{0,1}(X \setminus K) = 0$.

Consider now the case where $n = 2$. Let us prove that hypothesis ii) of Theorem 2.5 of Chapter V is again satisfied for the neighbourhoods $(U_p)_{p\in\mathbb{N}}$ of K given by Theorem 8.17. These neighbourhoods are of the form $\{\rho_p < c\}$ where ρ_p is a strictly plurisubharmonic exhaustion function on X. By Theorem 8.11 ii), any holomorphic $(r,0)$-form on U_p is therefore a uniform limit on any compact set in U_p of holomorphic $(r,0)$-forms on X. Let $f \in \mathcal{C}_{0,2}^\infty(U_p)$ be a $\overline{\partial}$-closed form with compact support on U_p which can be written as $f = \overline{\partial} g$ where the form $g \in \mathcal{C}_{0,1}^\infty(X)$ has compact support on X. By Stokes' formula, such a form has the property that

$$\int_X f \wedge \varphi = 0$$

for any holomorphic $(2,0)$-form φ on X. By Proposition 8.23, to prove that $f = \overline{\partial} g_0$ for some compactly supported $g_0 \in \mathcal{C}_{0,1}^\infty(U_p)$ on U_p, it is enough to check firstly that $E_{2,1}^0(U_p)$ is closed, which is the case because U_p is Stein (and hence $H_0^{2,1}(U_p) = 0$) and secondly that

$$\int_{U_p} f \wedge \psi = 0$$

for any holomorphic $(2,0)$-form ψ on U_p. But by definition of U_p, $\psi = \lim_{n\to\infty} \varphi_n$, where the φ_n are holomorphic $(2,0)$-forms on X and hence $\int_{U_p} f \wedge \psi = \lim_{n\to\infty} \int_{U_p} f \wedge \varphi_n = \lim_{n\to\infty} \int_X f \wedge \varphi_n = 0$. □

Remark. For $n \geqslant 3$ we have not used the full force of the hypothesis that K is $\mathcal{O}(X)$-convex, we have simply used the fact that K has a decreasing sequence $(U_p)_{p\in\mathbb{N}}$ of Stein neighbourhoods such that $K = \bigcap_{p=0}^\infty U_p$. A compact set in a complex analytic manifold which has this property is called a *Stein compact.* This notion will be useful in Chapter VIII. In particular, it follows from Proposition 8.22 that if D is a relatively compact strictly pseudoconvex domain in a complex analytic manifold then $\overline{D}$ is a Stein compact.

Comments

The Levi problem for domains in $\mathbb{C}^n$ was solved by Oka [Ok] in 1942 for $n = 2$, and at the beginning of the 1950s for arbitrary n by Oka [Ok], H. Bremermann [Br1] and F. Norguet [No]. In 1958, H. Grauert [Gr] solved the Levi problem in Stein manifolds using the theory of coherent sheaves. The first proof of the solvability of $\overline{\partial}$ in pseudoconvex domains which does not use the solution of the Levi problem is due to L. Hörmander [Ho1]. This proof which appeared in 1965 is based on L^2 bounds for the $\overline{\partial}$-Neumann problem. The first integral operators for solving $\overline{\partial}$ in strictly pseudoconvex domains in $\mathbb{C}^n$ were constructed at the start of the 1970s by H. Grauert and I. Lieb [Gr/Li] and G.M. Henkin [He2]. Their construction uses an integral formula proved independently by G.M. Henkin [He1] and E. Ramirez [Ram].

The methods used in Sections 1 to 4 can be used to study $\overline{\partial}$ in q-convex domains à la Andreotti–Grauert: for more information, we recommend [He/Le2] to the interested. The proof of Theorem 5.1 is due to Laufer [Lau]. The resolution of the Levi problem by the integral representation method can be found in [He/Le1] and [Ra]: for more information on an alternative theory, namely Hörmander's L^2 theory, see [Ho2]. The cohomological characterisation of domains of holomorphy can be found in [Gu]. Theorem 8.26 is proved in [L-T2] using a generalisation of the Bochner–Martinelli kernel to Stein manifolds.

VIII

Characterisation of removable singularities of CR functions on a strictly pseudoconvex boundary

We start this chapter by giving various characterisations of the compact sets K in the boundary of a strictly pseudoconvex domain D in a Stein manifold of dimension n which have the following property: any continuous CR function on $\partial D \setminus K$ can be extended holomorphically to the whole of D. We will obtain a geometric characterisation of such sets for $n = 2$ and a cohomological characterisation of such sets for $n \geqslant 3$. Amongst other things, we prove that the sufficient cohomological condition given in Theorem 5.1 of Chapter V is necessary if the ambient manifold is Stein and the domain D is assumed strictly pseudoconvex. We end the section with a geometric characterisation of the compact sets K such that any continuous CR function defined on $\partial D \setminus K$ which is orthogonal to the set of $\overline{\partial}$-closed $(n, n-1)$-forms whose support does not meet K can be extended holomorphically to the whole of D. When K is empty this condition is just the hypothesis of Theorem 3.2 of Chapter IV.

1 Reduction to continuous functions

Let X be a complex analytic manifold of dimension n, let D be a relatively compact domain in X and let K be a compact subset of ∂D such that $\partial D \setminus K$ is a $\mathcal{C}^k$ submanifold of $X \setminus K$ for some $k \geqslant 1$. The compact set K is said to be a *removable singularity* for CR functions on ∂D of class $\mathcal{C}^s$ for some integer $0 \leqslant s \leqslant k$ if any CR function of class $\mathcal{C}^s$ defined on $\partial D \setminus K$ can be extended to a holomorphic function on D which is $\mathcal{C}^k$ on $\overline{D} \setminus K$.

In previous chapters we gave cohomological conditions (cf. Chap. V, Th. 5.1 and Th. 2.5 and 2.6) and geometric conditions (cf. Chap. VII, Th. 8.26) under which a compact set K in ∂D is a removable singularity for CR functions on ∂D of class $\mathcal{C}^s$. In this chapter we will study the special case where D is a strictly pseudoconvex domain with $\mathcal{C}^k$ boundary for some $k \geqslant 2$, and we will prove various characterisations of removable singularities of CR functions in this case.

Note first that if D is a strictly pseudoconvex domain with $\mathcal{C}^k$ boundary in X for some $k \geqslant 2$ and K is a compact set in ∂D then there is an open

C. Laurent-Thiébaut, *Holomorphic Function Theory in Several Variables: An Introduction*, Universitext, DOI 10.1007/978-0-85729-030-4_8, © Springer-Verlag London Limited 2011

set Ω in D such that $(\partial D \setminus K) \cup \Omega$ is a neighbourhood of $\partial D \setminus K$ in $\overline{D} \setminus K$ and for any integer $0 \leqslant s \leqslant k$ any CR function of class $\mathcal{C}^s$ on $\partial D \setminus K$ can be extended to a holomorphic function on Ω which is $\mathcal{C}^s$ on $(\partial D \setminus K) \cup \Omega$. This follows immediately from the following proposition.

Proposition 1.1. *Let D be a strictly pseudoconvex domain with $\mathcal{C}^k$ boundary in X for some $k \geqslant 2$, and let x be a point of ∂D. There is then a neighbourhood V_x of x in X such that any* CR *function of class $\mathcal{C}^s$ on $V_x \cap \partial D$ can be extended to a holomorphic function on $V_x \cap D$ which is $\mathcal{C}^s$ on $V_x \cap \overline{D}$.*

Proof. By Theorem 3.23 of Chapter VI there is a neighbourhood U_x of x and a choice of holomorphic coordinates on this neighbourhood in which ∂D is convex. Suppose that these coordinates are chosen such that $x = 0$ and the hyperplane tangent to ∂D at x has equation $\operatorname{Re} z_n = 0$. For small enough $\varepsilon \neq 0$ we then have $x \in \{z \in U_x \mid \operatorname{Re} z_n > \varepsilon\} \cap \overline{D} \subset U_x$. We set $\Gamma_x = \partial D \cap \{z \in U_x \mid \operatorname{Re} z_n > \varepsilon\}$. By the results of Section 6 of Chapter IV any CR function of class $\mathcal{C}^s$ on Γ_x can be extended to a holomorphic function on $D \cap \{z \in U_x \mid \operatorname{Re} z_n > \varepsilon\}$ which is $\mathcal{C}^s$ on $\overline{D} \cap \{z \in U_x \mid \operatorname{Re} z_n > \varepsilon\}$. It follows that the open set $V_x = \{z \in U_x \mid \operatorname{Re} z_n > \varepsilon\}$ has the properties we seek. □

This remark implies that when studying removable singularities of CR functions on a strictly pseudoconvex boundary we can restrict ourselves to continuous CR functions.

2 The two-dimensional case

Let X be a complex analytic manifold and let D be a relatively compact domain in X. Let $\mathcal{O}(\overline{D})$ be the vector space of holomorphic functions defined on some neighbourhood of $\overline{D}$.

Definition 2.1. A subset E in $\overline{D}$ is $\mathcal{O}(\overline{D})$*-convex* if and only if

$$E = \widehat{E}_{\mathcal{O}(\overline{D})} = \{z \in \overline{D} \mid \ |f(z)| \leqslant \sup_{x \in E} |f(x)|, f \in \mathcal{O}(\overline{D})\}.$$

Remark. If D is strictly pseudoconvex then the condition that a subset E in $\overline{D}$ be $\mathcal{O}(\overline{D})$-convex is equivalent to the condition that E be $\mathcal{O}(U)$-convex for some neighbourhood U of $\overline{D}$, which can be chosen to be Stein if X is Stein. This follows from ii) of Theorem 8.11 of Chapter VII, which in this case implies that the holomorphic functions on some neighbourhood U of D are dense in $\mathcal{O}(\overline{D})$. If X is Stein this neighbourhood can be chosen to be Stein and to have the property that $\overline{D}$ is $\mathcal{O}(U)$-convex. If D has a strictly plurisubharmonic defining function ρ defined on a neighbourhood $V_{\partial D}$ of ∂D, i.e. $D \cap V_{\partial D} = \{z \in V_{\partial D} \mid \rho(z) < 0\}$ then we simply set $U = D \cup \{z \in V_{\partial D} \mid \rho(z) < \varepsilon\}$ for small enough ε (cf. Chap. VII, Th. 8.19 and Prop. 8.22).

Theorem 2.2. *Let D be a relatively compact, strictly pseudoconvex domain with $\mathcal{C}^k$ boundary in a Stein manifold X of dimension 2 where $k \geqslant 2$. For any compact set K contained in ∂D the following are then equivalent.*

i) *K is a removable singularity for* CR *functions on ∂D,*
ii) *K is $\mathcal{O}(\overline{D})$-convex.*

Proof. Let us prove that ii) implies i). Let U be a Stein neighbourhood of $\overline{D}$ such that K is $\mathcal{O}(U)$-convex: by the above remark, some such neighbourhood exists because D is strictly pseudoconvex and X is Stein. Theorem 8.26 of Chapter VII applied to the triple (U, D, K) (which clearly satisfies the hypotheses of this theorem) then implies that K is a removable singularity for CR functions on ∂D.

We now prove the converse. Since D is strictly pseudoconvex, Lemma 8.18 of Chapter VII implies that $\widehat{K}_{\mathcal{O}(\overline{D})} \cap \partial D = K$. Let D' be a strictly pseudoconvex domain in X such that $D \subset D'$, $\overline{D} \cap \partial D' = K$ and $\widehat{K}_{\mathcal{O}(\overline{D}')} = \widehat{K}_{\mathcal{O}(\overline{D})}$. (To construct such a domain D' it is enough to take a small $\mathcal{C}^2$ perturbation of ∂D leaving K pointwise fixed: we can then pass from D to D' by a countable sequence of strictly pseudoconvex extension elements, and this proves that $\mathcal{O}(\overline{D}')$ is dense in $\mathcal{O}(\overline{D})$.) By a theorem of Slodkowski's which is proved below for $X = \mathbb{C}^2$ the set $D' \setminus \widehat{K}_{\mathcal{O}(\overline{D}')}$ is pseudoconvex and this set is therefore a domain of holomorphy. There is therefore a holomorphic function f on $D' \setminus \widehat{K}_{\mathcal{O}(\overline{D}')}$ which cannot be extended to any open set containing $D' \setminus \widehat{K}_{\mathcal{O}(\overline{D}')}$. The function f is holomorphic on some neighbourhood of $\partial D \setminus K$ and since K is a removable singularity for CR functions on ∂D it can be extended holomorphically to D. It follows that $D \subset D' \setminus \widehat{K}_{\mathcal{O}(\overline{D})}$, and hence $\widehat{K}_{\mathcal{O}(\overline{D})} = K$. □

Corollary 2.3. *Let X be a Stein manifold of dimension 2 and let $D \Subset X$ be a strictly pseudoconvex domain with $\mathcal{C}^k$ boundary for some $k \geqslant 2$ such that $\overline{D}$ is $\mathcal{O}(X)$-convex. For any compact set K contained in ∂D the following are then equivalent.*

i) *K is a removable singularity for* CR *functions on ∂D,*
ii) *K is $\mathcal{O}(X)$-convex.*

Example. If $X = \mathbb{C}^2$ and $D = B$ is the unit ball in $\mathbb{C}^2$ then a compact set K in the unit sphere is a removable singularity for CR functions if and only if K is polynomially convex.

We now prove the result of Z. Slodkowski's [Sl] which is used in the proof of Theorem 2.2.

Lemma 2.4. *If K is a compact subset in $\mathbb{C}^2$ and Ω is a pseudoconvex domain in $\mathbb{C}^2$ such that $K \cap \Omega = \varnothing$ then $\Omega \setminus \widehat{K}$ is pseudoconvex.*

More generally, if D is a strictly pseudoconvex domain and K is a compact set in ∂D then $D \setminus \widehat{K}_{\mathcal{O}(\overline{D})}$ is pseudoconvex.

This lemma follows from the "Kontinuitätsatz" (cf. Chap. VI, Th. 3.4) and Rossi's local maximum principle.

The local maximum principle. *Let X be a compact set in an open set U in $\mathbb{C}^n$ and let S be a subset of $\widehat{X}_U$. Then $S \subset (\widehat{\partial S \cup (S \cap X)})_U$, where ∂S is the boundary of S in $\widehat{X}_U$.*

We will not prove this result. The interested reader will find more details in [Ros], [Sto] or [St].

Proof of Lemma 2.4. We argue by contradiction. Assume that $\Omega \setminus \widehat{K}$ is not pseudoconvex and let Δ be the unit disc in $\mathbb{C}$. By the "Kontinuitätsatz", on changing holomorphic coordinates on $\mathbb{C}^2$ we may assume that the bidisc Δ^2 does not meet K and there is a sequence $(\varphi_j)_{j\in\mathbb{N}}$ of holomorphic functions defined on a neighbourhood of $\overline{\Delta}$ in $\mathbb{C}$ such that $|\varphi_j| < 1/2$ on Δ and if we set $\Phi_j(s) = (s, \varphi_j(s))$ then the map Φ_j maps $\overline{\Delta}$ to $\Omega \setminus \widehat{K}$ in such a way that

i) $\mathrm{dist}(\Phi_j(e^{i\theta}), \widehat{K}) > \delta > 0$ for all $\theta \in \mathbb{R}$ and
ii) $\Phi_j(0)$ tends to a point $p_0 \in \widehat{K}$ as j tends to infinity.

Set $f_j(z_1, z_2) = 1/(z_2 - \varphi_j(z_1))$: the function f_j thus defined is holomorphic on some neighbourhood of $\widehat{K} \cap \overline{\Delta}^2$. By Theorem 7.1 of Chapter VII we can approximate f_j uniformly on $\widehat{K} \cap \overline{\Delta}^2$ by holomorphic functions on $\mathbb{C}^2$. As $|f_j| \leqslant \max(1/\delta, 2)$ on $\widehat{K} \cap \partial\Delta^2$ and the supremum of the functions $|f_j|$ on $\widehat{K} \cap \overline{\Delta}^2$ tends to infinity with j we can therefore construct a holomorphic function f on $\mathbb{C}^2$ whose supremum on $\widehat{K} \cap \overline{\Delta}^2$ is strictly greater than its supremum on $\widehat{K} \cap \partial\Delta^2$, which contradicts the local maximum principle.

The second part of the lemma is proved by simply replacing $\mathbb{C}^2$ by a pseudoconvex open set U in $\mathbb{C}^2$ such that the holomorphic functions on U are dense in $\mathcal{O}(\overline{D})$ and $\overline{D}$ is $\mathcal{O}(U)$-convex in the above proof. □

3 A cohomological characterisation in dimension $n \geqslant 3$

When the dimension of the ambient manifold is greater than 3 we only get a cohomological characterisation of removable singularities, not a geometric characterisation.

Theorem 3.1. *Let D be a relatively compact strictly pseudoconvex domain with $\mathcal{C}^k$ boundary for some $k \geqslant 2$ in a Stein manifold X of dimension $n \geqslant 3$. For any compact set K contained in ∂D the following are then equivalent.*

i) *K is a removable singularity for* CR *functions on ∂D,*
ii) $H^{0,1}(X \setminus K) = 0$,
iii) $H^{0,1}_{\Phi}(X \setminus K) = 0$.

Proof. The fact that iii) $\Rightarrow$ i) follows from Theorem 5.1 of Chapter V: the hypotheses 2) and 3) of this theorem are satisfied because D is strictly pseudoconvex and $K \subset \partial D$. The fact that ii) implies iii) follows from Theorem 2.7 of Chapter V – since X is a Stein manifold we know that $H_c^{0,1}(X) = 0$ (cf. Chap. VII, Th. 8.24). It only remains to prove that i) implies ii). We will use the following lemma.

Lemma 3.2. *Let X be a Stein manifold of dimension $n \geqslant 3$ and let K be a Stein compact in X. Then*

$$H^{0,1}(X \setminus K) = 0.$$

Proof. As K is a Stein compact by definition there is a decreasing sequence $(U_p)_{p\in\mathbb{N}}$ of Stein neighbourhoods of K such that the closure of U_{p+1} is contained in U_p and $K = \bigcap_{p\in\mathbb{N}} U_p$. Let f be a $\overline{\partial}$-closed differential $(0,1)$-form of class $\mathcal{C}^\infty$ on $X \setminus K$. Choose a sequence $(\chi_p)_{p\in\mathbb{N}}$ of elements of $\mathcal{D}(X)$ such that $\chi_p = 1$ on some neighbourhood of $\overline{U}_{p+1}$ and $\chi_p = 0$ on some neighbourhood of $X \setminus U_p$ and extend $(1-\chi_p)f$ by 0 on $\overline{U}_{p+1}$. The form $\overline{\partial}(1-\chi_p)f = \overline{\partial}\chi_p \wedge f$ is then a $\overline{\partial}$-closed $(0,2)$-form with compact support on U_p. Since U_p is a Stein open set and $n \geqslant 3$ there is a $\mathcal{C}^\infty$ differential $(0,1)$-form g_p with compact support on U_p such that $\overline{\partial}(1-\chi_p)f = \overline{\partial}g_p$. It follows that $\overline{\partial}(g_p + (1-\chi_p)f) = 0$ on X and as X is Stein there is a $\gamma_p \in \mathcal{C}^\infty(X)$ such that $g_p + (1-\chi_p)f = \overline{\partial}\gamma_p$ on X. Note that $\gamma_{p+1} - \gamma_p$ is holomorphic on $X \setminus \overline{U}_p$ and $X \setminus \overline{U}_p$ is connected because U_p is pseudoconvex. It then follows from Hartogs' phenomenon for Stein manifolds (cf. Chap. VII, Th. 8.25) that there is a holomorphic function h_p on X such that $h_p|_{X\setminus\overline{U}_p} = \gamma_{p+1} - \gamma_p$. Set $h = \gamma_1 + \sum_{j=1}^\infty(\gamma_{j+1} - \gamma_j - h_j)$. This sum is locally finite on $X \setminus K$ and therefore defines a $\mathcal{C}^\infty$ function h on $X \setminus K$. Moreover,

$$\overline{\partial}h = (1-\chi_1)f + g_1 + \sum_{j=1}^\infty (\chi_j - \chi_{j+1})f + g_{j+1} - g_j = f + \lim_{p\to\infty} g_p = f. \quad \square$$

End of the proof of Theorem 3.1. As D is strictly pseudoconvex there are strictly pseudoconvex domains D' and D'' such that $D'' \subset D \subset D'$ and $\overline{D}'' \cap \partial D = K = \overline{D} \cap \partial D'$. We will prove that for any $\overline{\partial}$-closed differential $(0,1)$-form f of class $\mathcal{C}^\infty$ on $X \setminus K$ there is a $\mathcal{C}^\infty$ function g on $X \setminus K$ such that $\overline{\partial}g = f$. Since D' is a Stein open set there is a function $u_1 \in \mathcal{C}^\infty(D')$ such that $\overline{\partial}u_1 = f$ on D' (cf. Chap. VII, Cor. 8.21 and Prop. 8.22). As $\overline{D}''$ is a Stein compact because D'' is strictly pseudoconvex it follows from Lemma 3.2 that we can find a function $u_2 \in \mathcal{C}^\infty(X \setminus \overline{D}'')$ such that $\overline{\partial}u_2 = f$ on $X \setminus \overline{D}''$. The function $v = u_1 - u_2$ is then defined and holomorphic on $D' \setminus \overline{D}''$, which is a neighbourhood of $\partial D \setminus K$. As K is a removable singularity for CR functions on ∂D there is a function $V \in \mathcal{O}(D)$ extending v to D. Set

$$\begin{aligned} g &= u_1 - V \text{ on } D' \\ &= u_2 \qquad \text{on } X \setminus \overline{D}''. \end{aligned}$$

This defines a $\mathcal{C}^\infty$ function on $X \setminus K$ such that $\overline{\partial} g = f$ on $X \setminus K$. □

Remark. It follows from Theorem 3.1 that for $n \geqslant 3$ the fact that K is or is not a removable singularity for CR functions on the boundary of a strictly pseudoconvex domain D does not depend on D but only on K itself and the inclusion of K in the manifold X.

4 Characterisation of weakly removable singularities

Let X be a complex analytic manifold of dimension n, let D be a relatively compact domain in X and let K be a compact subset of ∂D such that $\Gamma = \partial D \setminus K$ is a $\mathcal{C}^k$ submanifold of $X \setminus K$ for some $k \geqslant 1$. The compact set K is a *weakly removable singularity* for CR functions on ∂D if for any function $f \in \mathcal{C}(\Gamma)$ which is orthogonal to any $\overline{\partial}$-closed $\mathcal{C}^\infty$ form φ of bidegree $(n, n-1)$ defined in some neighbourhood of D with support disjoint from K, there is a holomorphic function F on D which is continuous on $\overline{D} \setminus K$ such that $F\big|_\Gamma = f$. (Here, when we say that two forms are orthogonal we mean that $\int_\Gamma f\beta = 0$). Note that any such function is CR because all $\overline{\partial}$-exact forms are $\overline{\partial}$-closed. We proved in Section 3 of Chapter IV that if $X = \mathbb{C}^n$ then $K = \varnothing$ is a weakly removable singularity for CR functions on ∂D and we noted that CR functions are not always orthogonal to the set of $\overline{\partial}$-closed $(n, n-1)$-forms, particularly if Γ is not connected.

We start with cohomological characterisations of these weakly removable singularities.

Theorem 4.1. *Let D be a relatively compact strictly pseudoconvex domain with $\mathcal{C}^2$ boundary in a Stein manifold of dimension $n \geqslant 2$. For any compact set K contained in ∂D the following are then equivalent.*

i) *K is a weakly removable singularity for* CR *functions on ∂D,*
ii) *$H^{0,1}(X \setminus K)$ is Hausdorff,*
iii) *$H^{0,1}_\Phi(X \setminus K)$ is Hausdorff.*

Proof. We start by proving that ii) implies iii). Let f be a $\mathcal{C}^\infty$ differential $(0,1)$-form on $X \setminus K$ contained in the closure of $\overline{\partial}\mathcal{E}^{0,0}_\Phi(X \setminus K)$ whose support is contained in some compact set L in X. Since $H^{0,1}(X \setminus K)$ is Hausdorff there is a $g \in \mathcal{C}^\infty(X \setminus K)$ such that $f = \overline{\partial} g$. The function g is holomorphic on $X \setminus (L \cup K)$ and since X is a Stein manifold of dimension $n \geqslant 2$ it follows from Hartogs' phenomenon that there is a function $\widetilde{g} \in \mathcal{O}(X)$ equal to g outside a compact set in X. If we set $g_0 = g - \widetilde{g}$ then $f = \overline{\partial} g_0$ and the support of g_0 is relatively compact in X. It follows that $H^{0,1}_\Phi(X \setminus K)$ is Hausdorff.

Let us now prove that iii) implies i). Let f be a continuous CR function on Γ which is orthogonal to all $\overline{\partial}$-closed differential $(n, n-1)$-forms of class $\mathcal{C}^\infty$ whose support is disjoint from K. By Proposition 1.1, f has a holomorphic extension $\widetilde{f}$ to a neighbourhood U of Γ in $\overline{D}$. Let χ be a $\mathcal{C}^\infty$ function on

$X \setminus K$ equal to 1 on some neighbourhood of $X \setminus (D \cup K)$ and equal to 0 on $D \setminus U$. The support of $\overline{\partial}\chi$ is then relatively compact in $\overline{D}$. Let E be the closure of the set of $\overline{\partial}$-exact differential $(0,1)$-forms in $(\mathcal{C}^\infty_{0,1}(X \setminus K))_\Phi$. We will prove that $\widetilde{f}\overline{\partial}\chi \in E$. We argue by contradiction. If $\widetilde{f}\overline{\partial}\chi \notin E$ then there is a continuous linear form T on $(\mathcal{C}^\infty_{0,1}(X \setminus K))_\Phi$ such that $T\big|_E = 0$ and $\langle T, \widetilde{f}\overline{\partial}\chi\rangle \neq 0$. We can consider T as an $(n, n-1)$-current on $X \setminus K$ with closed support in X, and $\langle T, \overline{\partial} g\rangle = 0$ for any function $g \in \mathcal{C}^\infty(X \setminus K)$ with relatively compact support in X, which implies that T is $\overline{\partial}$-closed. By Corollary 4.2 i) of Chapter V, there is a $\overline{\partial}$-closed differential form φ of class $\mathcal{C}^\infty$ on $X \setminus K$ and a current S on $X \setminus K$, both with closed support in X, such that $T = \overline{\partial} S + \varphi$. As $\widetilde{f}\overline{\partial}\chi$ is $\overline{\partial}$-closed in $X \setminus K$ we have

$$\langle T, \widetilde{f}\overline{\partial}\chi\rangle = \langle \overline{\partial} S, \widetilde{f}\overline{\partial}\chi\rangle + \int_D \varphi \wedge \widetilde{f}\overline{\partial}\chi = \int_\Gamma f\varphi = 0$$

by Stokes' formula and the hypothesis on f. This contradicts our choice of T, so $\widetilde{f}\overline{\partial}\chi \in E$. By hypothesis $H^{0,1}_\Phi(X \setminus K)$ is Hausdorff so $\widetilde{f}\overline{\partial}\chi = \overline{\partial} g$, for some $\mathcal{C}^\infty$ function g on $X \setminus K$ with relatively compact support on X. The function g is then holomorphic on some neighbourhood of $X \setminus (D \cup K)$ and since X is Stein of dimension $n \geqslant 2$ it follows from Hartogs' phenomenon that there is a function $\widetilde{g} \in \mathcal{O}(X)$ such that $g = \widetilde{g}$ on some neighbourhood $X \setminus (D \cup K)$. Setting $F = \chi\widetilde{f} - g + \widetilde{g}$ we get a holomorphic function on $X \setminus K$ which is continuous on $\overline{D} \setminus K$ and equal to $\widetilde{f}$ on some neighbourhood of Γ in $\overline{D}$. This function is therefore equal to f on Γ.

We end by proving that i) implies ii). Since D is strictly pseudoconvex there are strictly pseudoconvex domains D' and D'' such that $D'' \subset D \subset D'$ and $\overline{D}'' \cap \partial D = K = \overline{D} \cap \partial D'$. Let f be a $\overline{\partial}$-closed differential $(0,1)$-form of class $\mathcal{C}^\infty$ on $X \setminus K$ such that f is the limit in $\mathcal{C}^\infty_{0,1}(X \setminus K)$ of a sequence $(\overline{\partial} g_j)_{j\in\mathbb{N}}$ where each function g_j is contained in $\mathcal{C}^\infty(X \setminus K)$. Since D' is a Stein open set there is a $u_1 \in \mathcal{C}^\infty(D')$ such that $f = \overline{\partial} u_1$ on D'.

Suppose initially that $n \geqslant 3$. As $\overline{D}''$ is a Stein compact, it follows from Lemma 3.2 that $H^{0,1}(X \setminus \overline{D}'') = 0$. There is then a $u_2 \in \mathcal{C}^\infty(X \setminus \overline{D}'')$ such that $f = \overline{\partial} u_2$ on $X \setminus \overline{D}''$. The function $v = u_1 - u_2$ is holomorphic on $D' \setminus \overline{D}''$ which is a neighbourhood of Γ and if φ is a $\overline{\partial}$-closed $(n, n-1)$-form of class $\mathcal{C}^\infty$ with compact support in $X \setminus K$ then

$$\begin{aligned}\int_\Gamma v\varphi = \int_\Gamma u_1\varphi - \int_\Gamma u_2\varphi &= \int_D f \wedge \varphi + \int_{X \setminus D} f \wedge \varphi \\ &= \int_X f \wedge \varphi = \lim_{j\to\infty} \int_X \overline{\partial} g_j \wedge \varphi = 0\end{aligned}$$

by applying Stokes' theorem twice.

Let φ now be an arbitrary $\overline{\partial}$-closed differential $(n, n-1)$-form of class $\mathcal{C}^\infty$ whose support does not meet K. Since D is strictly pseudoconvex $\overline{D}$ has a basis of Stein neighbourhoods and there is therefore an $(n, n-2)$-form ψ of class $\mathcal{C}^\infty$ on some relatively compact neighbourhood V of $\overline{D}$ such that $\varphi = \overline{\partial}\psi$ on V. Let χ be a $\mathcal{C}^\infty$ function on X which vanishes on some neighbourhood of $\overline{D}$ and is equal to 1 on $X \setminus V$ and set $\varphi_0 = \varphi - \overline{\partial}(\chi\psi) = (1-\chi)\varphi - \overline{\partial}\chi \wedge \psi$. The form φ_0 can be extended by 0 on $X \setminus K$ to a $\overline{\partial}$-closed form with compact support in $X \setminus K$. Then

$$\int_\Gamma v\varphi = \int_\Gamma v\varphi_0 + \int_\Gamma v\overline{\partial}(\chi\psi) = \int_\Gamma v\overline{\partial}(\chi\psi) = 0$$

since χ vanishes on some neighbourhood of $\overline{D}$. Since K is a weakly removable singularity there is a holomorphic function V on D' such that $V = v$ on $D' \setminus \overline{D}''$. If we set

$$\begin{aligned} g &= u_1 - V \text{ on } D' \\ &= u_2 \qquad \text{on } X \setminus \overline{D}'', \end{aligned}$$

then we get a $\mathcal{C}^\infty$ function g on $X \setminus K$ such that $\overline{\partial}g = f$, which proves that $H^{0,1}(X \setminus K)$ is Hausdorff.

The proof for $n = 2$ is similar but we need to prove that $H^{0,1}(X \setminus \overline{D}'')$ is Hausdorff. (Of course, this group vanishes for $n \geqslant 3$.) We will now prove this fact. □

Proposition 4.2. *Let X be a Stein manifold of dimension 2 and let K be a Stein compact in X. For any $p \geqslant 0$ the group $H^{p,1}(X \setminus K)$ is then a Hausdorff topological vector space.*

Proof. We start by proving that for any compact set L in $X \setminus K$ the space $\mathcal{D}_L^{p,2}(X \setminus K) \cap \overline{\partial}\mathcal{D}^{p,1}(X \setminus K)$ is a closed subspace of $\mathcal{D}_L^{p,2}(X \setminus K)$. It is enough to show that if a function $f \in \mathcal{D}^{p,2}(X \setminus K)$ supported in a compact set L is the limit of a sequence of elements $(f_j)_{j \in \mathbb{N}}$ in $\mathcal{D}^{p,2}(X \setminus K)$ which are supported on L such that $f_j = \overline{\partial}g_j$ for some $g_j \in \mathcal{D}^{p,1}(X \setminus K)$ then $f = \overline{\partial}g$ for some $g \in \mathcal{D}^{p,1}(X \setminus K)$. We extend f to X by zero. If $\varphi \in \mathcal{C}^\infty_{n-p,0}(X)$ is a $\overline{\partial}$-closed form then, by Stokes' theorem,

$$\int_X f \wedge \varphi = \lim_{j \to \infty} \int_X \overline{\partial}g_j \wedge \varphi = 0.$$

As the manifold X is Stein $H_0^{n-p,1}(X) = 0$ and it follows that $E^0_{n-p,1}(X)$ is closed in $\mathcal{C}^0_{n-p,1}(X)$. Proposition 8.23 of Chapter VII and the regularity of $\overline{\partial}$ then imply that $f = \overline{\partial}h$ for some $h \in \mathcal{D}^{p,1}(X)$. Since K is a Stein compact there is a Stein open set U in X such that $K \subset U \subset X \setminus L$. The form h is $\overline{\partial}$-closed on U and $h = \overline{\partial}u$ on U for some $u \in \mathcal{C}^\infty_{p,0}$. Let $\chi \in \mathcal{D}(X)$ be a function which is identically equal to 1 in some neighbourhood of K and which is supported in U. We set $g = h - \overline{\partial}(\chi u)$: the form g is then a $\mathcal{C}^\infty$ differential $(p, 1)$-form on X with compact support on $X \setminus K$ such that $\overline{\partial}g = f$.

The proposition then follows from the following general result. □

Theorem 4.3. *Let M be a complex analytic manifold of dimension n and let p and q be integers such that $p \geqslant 0$ and $1 \leqslant q \leqslant n$. We assume that*

$F_{p,q}(M)$: For any compact set L in M, $\mathcal{D}_L^{p,q}(M) \cap \overline{\partial}\mathcal{D}^{p,q-1}(M)$ is a closed subspace of $\mathcal{D}_L^{p,q}(M)$.

The group $H^{n-p,n-q+1}(M)$ is then Hausdorff. In particular, $H_c^{p,q}(M)$ is Hausdorff if and only if $H^{n-p,n-q+1}(M)$ is Hausdorff.

Lemma 4.3.1. *Let L be a compact subset of M such that $\mathcal{D}_L^{p,q}(M) \cap \overline{\partial}\mathcal{D}^{p,q-1}(M)$ is a closed subspace of $\mathcal{D}_L^{p,q}(M)$. There is then a compact subset L_0 in M such that $L \subset L_0$ and*

$$\mathcal{D}_L^{p,q}(M) \cap \overline{\partial}\mathcal{D}^{p,q-1}(M) = \mathcal{D}_L^{p,q}(M) \cap \overline{\partial}\mathcal{D}_{L_0}^{p,q-1}(M)$$

Proof. Let $(K_j)_{j\in\mathbb{N}}$ be an exhaustion of M by compact sets. Then

$$\mathcal{D}_L^{p,q}(M) \cap \overline{\partial}\mathcal{D}^{p,q-1}(M) = \cup_{j\in\mathbb{N}}\mathcal{D}_L^{p,q}(M) \cap \overline{\partial}\mathcal{D}_{K_j}^{p,q-1}(M).$$

Since $\mathcal{D}_L^{p,q}(M) \cap \overline{\partial}\mathcal{D}^{p,q-1}(M)$ is a Fréchet space by hypothesis there is an index j_0 such that $\mathcal{D}_L^{p,q}(M) \cap \overline{\partial}\mathcal{D}_{K_{j_0}}^{p,q-1}(M)$ is of the second Baire category and by the open mapping theorem $L_0 = K_{j_0}$ is then the compact set we seek. □

A pair (L, L_0) satisfying the conclusion of Lemma 4.3.1 is called a (p,q)-*admissible* pair.

Lemma 4.3.2. *If M is such that $F_{p,q}(M)$ holds, U is an open subset in M and L is a compact subset of M such that $(\overline{U}, L)$ is (p,q)-admissible then*

i) *For any neighbourhood U_0 of L and any differential form $f \in \mathcal{C}_{n-p,n-q+1}^{\infty}(U_0)$ such that $\int_{U_0} f \wedge \varphi = 0$ for any $\varphi \in \mathcal{D}_L^{p,q-1}(M) \cap \operatorname{Ker}\overline{\partial}$ there is a $g \in \mathcal{C}_{n-p,n-q}^{\infty}(U)$ such that $\overline{\partial}g = f_{|U}$.*
ii) *For any neighbourhood U_0 of L the vector space $\mathcal{C}_{n-p,n-q}^{\infty}(M) \cap \operatorname{Ker}\overline{\partial}$ is dense in $\mathcal{C}_{n-p,n-q}^{\infty}(U_0) \cap \operatorname{Ker}\overline{\partial}$ for the Fréchet vector space topology on $\mathcal{C}_{n-p,n-q}^{\infty}(U)$.*

Proof. Let us prove i). The hypotheses imply that the operator $\overline{\partial}$ from the set $\{\varphi \in \mathcal{D}_L^{p,q-1}(M) \mid \operatorname{supp}(\overline{\partial}\varphi) \subset \overline{U}\}$ to $\mathcal{D}_{\overline{U}}^{p,q}(M) \cap \overline{\partial}\mathcal{D}^{p,q-1}(M)$ is a continuous surjective linear map between Fréchet spaces. It is therefore an open map. It follows that the linear form F on $\mathcal{D}_{\overline{U}}^{p,q}(M) \cap \overline{\partial}\mathcal{D}^{p,q-1}(M)$ defined by $F(\overline{\partial}\varphi) = \int_{U_0} f \wedge \varphi$ for any $\varphi \in \mathcal{D}_L^{p,q-1}(M)$ is continuous and therefore by the Hahn–Banach theorem it has a continuous extension to $\mathcal{D}_{\overline{U}}^{p,q}(M)$. Let G be the $(n-p,n-q)$-current on U defined by this extension. Then $\overline{\partial}G = f_{|U}$. The Dolbeault isomorphism then completes the proof of i).

To prove ii), let us first prove that if $U_0 \subset \widetilde{U}_0$ are two neighbourhoods of L in M then the vector space $\mathcal{C}_{n-p,n-q}^{\infty}(\widetilde{U}_0) \cap \operatorname{Ker}\overline{\partial}$ is dense in $\mathcal{C}_{n-p,n-q}^{\infty}(U_0) \cap$

$\operatorname{Ker}\overline{\partial}$ for the Fréchet space topology on $\mathcal{C}^\infty_{n-p,n-q}(U)$. We use the Hahn–Banach theorem. Let T be a continuous linear form on $\mathcal{C}^\infty_{n-p,n-q}(U)$ – or in other words, let T be a (p,q)-current on M compactly supported in U – such that $T(\varphi) = 0$ for any differential form $\varphi \in \mathcal{C}^\infty_{n-p,n-q}(\widetilde{U}_0) \cap \operatorname{Ker}\overline{\partial}$. The current T is then $\overline{\partial}$-closed and by the Dolbeault isomorphism there is a differential form $f \in \mathcal{D}^{p,q}_{\overline{U}}(M)$ and a $(p,q-1)$-current S on M supported in $\overline{U}$ such that $f = T + \overline{\partial}S$. We deduce that, for any differential form $\varphi \in \mathcal{C}^\infty_{n-p,n-q}(\widetilde{U}_0) \cap \operatorname{Ker}\overline{\partial}$,

$$\int_{\widetilde{U}_0} f \wedge \varphi = T(\varphi) \pm S(\overline{\partial}\varphi) = 0.$$

An argument similar to that used for i) then proves there is a differential form g with compact support such that $f = \overline{\partial}g$ and since $(\overline{U}, L)$ is admissible we can assume that g is supported in L. It remains to show that $T(\varphi) = 0$ for any differential form $\varphi \in \mathcal{C}^\infty_{n-p,n-q}(U_0) \cap \operatorname{Ker}\overline{\partial}$. But

$$T(\varphi) = \int_{\widetilde{U}_0} f \wedge \varphi \pm S(\overline{\partial}\varphi) = \int_{\widetilde{U}_0} \overline{\partial}g \wedge \varphi = 0$$

by Stokes' formula.

Set $U_{-1} = U$ and $K_{-1} = L$. It follows from Lemma 4.3.1 that we can find an increasing sequence $(U_j)_{j\in\mathbb{N}}$ of open sets $U_j \Subset M$ and an increasing sequence $(K_j)_{j\in\mathbb{N}}$ of compact sets in M such that for any $j \geqslant -1$, $U_j \subset K_j \subset U_{j+1}$ and the pair $(\overline{U}_j, K_j)$ is (p,q)-admissible. For every $j \geqslant -1$, fix a metric ρ_j on $\mathcal{C}^\infty_{n-p,n-q}(U_j)$ defining its Fréchet space topology. Consider a form $f_0 \in \mathcal{C}^\infty_{n-p,n-q}(U_0)) \cap \operatorname{Ker}\overline{\partial}$ and a real number $\varepsilon > 0$: we can recursively construct a sequence $(f_j)_{j\in\mathbb{N}}$ of elements in $\mathcal{C}^\infty_{n-p,n-q}(U_j)) \cap \operatorname{Ker}\overline{\partial}$ such that

$$\rho_k(f_j, f_{j+1}) \leqslant \frac{\varepsilon}{2^{j+1}}, \quad \text{for any } k \text{ such that } -1 \leqslant k < j,$$

and it follows that

$$\rho_k(f_i, f_j) \leqslant \sum_{l=i}^{j-1} \rho_k(f_l, f_{l+1}) \leqslant \sum_{l=i}^{j-1} \frac{\varepsilon}{2^{l+1}} \leqslant \frac{\varepsilon}{2^i},$$

for any k such that $-1 \leqslant k < i < j$. The sequence $(f_j)_{j=k+1}^\infty$ is therefore Cauchy for any $k \geqslant -1$ and converges to a form $F_k \in \mathcal{C}^\infty_{n-p,n-q}(U_k) \cap \operatorname{Ker}\overline{\partial}$ such that $\rho_k(f_{k+1}, F_k) \leqslant \varepsilon/2^{k+1}$. It follows that $F_k = F_{k+1|_{U_k}}$ for any k. We can therefore define a form $F \in \mathcal{C}^\infty_{n-p,n-q}(M) \cap \operatorname{Ker}\overline{\partial}$ by setting $F_{|_{U_k}} = F_k$. This form has the property that $\rho_k(f_{k+1}, F) \leqslant \varepsilon/2^{k+1}$ for any $k \geqslant -1$ and in particular $\rho_{-1}(f_0, F) \leqslant \varepsilon$. □

We can now prove Theorem 4.3. We will prove that under the hypothesis of the theorem the space

$$E^\infty_{n-p,n-q+1}(M) = \{u \in \mathcal{C}^\infty_{n-p,n-q+1}(M) \mid u = \overline{\partial}v, v \in \mathcal{C}^\infty_{n-p,n-q}(M)\}$$

is equal to the closed subspace $\widetilde{Z}^{\infty}_{n-p,n-q+1}(M)$ in $Z^{\infty}_{n-p,n-q+1}(M)$ consisting of differential forms f such that $\int_M f \wedge \varphi = 0$ for all $\varphi \in \mathcal{D}^{p,q-1}(M)$ such that $\overline{\partial}\varphi = 0$ in M. Stokes' theorem implies that $E^{\infty}_{n-p,n-q+1}(M)$ is contained in $\widetilde{Z}^{\infty}_{n-p,n-q+1}(M)$. We now consider the inverse inclusion. Fix a form $f \in \widetilde{Z}^{\infty}_{n-p,n-q+1}(M)$. It follows from Lemma 4.3.1 that we can find an increasing sequence $(U_j)_{j\in\mathbb{N}}$ of open subsets $U_j \Subset M$ and an increasing sequence $(K_j)_{j\in\mathbb{N}}$ of compact subsets of M such that $U_j \subset K_j \subset U_{j+1}$ and the pair $(\overline{U}_j, K_j)$ is (p,q)-admissible for any $j \in \mathbb{N}$. Applying i) of Lemma 4.3.2 to each triple $U_{j-1} \subset K_{j-1} \subset U_j$ we get a sequence $(g_j)_{j\in\mathbb{N}}$ of elements of $\mathcal{C}^{\infty}_{n-p,n-q}(U_j)$ such that $\overline{\partial} g_j = f_{|U_j}$ for all j. Now for any $j \geqslant 0$, fix a metric ρ_j on $\mathcal{C}^{\infty}_{n-p,n-q}(U_j)$ defining its Fréchet space topology. Claim ii) of Lemma 4.3.2 enables us to recursively construct a sequence $(h_j)_{j\in\mathbb{N}}$ of elements of $\mathcal{C}^{\infty}_{n-p,n-q}(M) \cap \operatorname{Ker}\overline{\partial}$ such that, for any $j \geqslant 2$,

$$\rho_i(g_{j-1} + h_{j-1}, g_j + h_j) \leqslant \frac{1}{2^j} \quad \text{for any } i \text{ such that } 0 \leqslant i \leqslant j-2.$$

This proves that there is a differential form $g \in \mathcal{C}^{\infty}_{n-p,n-q}(M)$ such that $\lim_{\nu\to\infty} \rho_j(g, g_\nu + h_\nu) = 0$ and $\overline{\partial} g = f$.

For the second part of the theorem, the fact that the condition is sufficient follows from Proposition 8.23 of Chapter VII and the Dolbeault isomorphism. Conversely, if the cohomology group $H^{p,q}_c(M)$ is Hausdorff then for any compact set L in M the subspace $\mathcal{D}^{p,q}_L(M) \cap \overline{\partial}\mathcal{D}^{p,q-1}(M)$ is closed in $\mathcal{D}^{p,q}_L(M)$, so this condition is necessary by the first part of the theorem. □

We end this section with a geometric characterisation and an intrinsic cohomological characterisation of weakly removable singularities.

If F is a closed set in X then we let $\mathcal{C}^{\infty}_{p,q}(F)$ be the space of germs of differential (p,q)-forms on F, i.e. the inductive limit over open sets U in X containing F of the family $\mathcal{C}^{\infty}_{p,q}(U)$, where the map $i_{UV} : \mathcal{C}^{\infty}_{p,q}(U) \to \mathcal{C}^{\infty}_{p,q}(V)$ for any $V \subset U$ is simply the restriction map. The operator $\overline{\partial} : \mathcal{C}^{\infty}_{p,q}(F) \to \mathcal{C}^{\infty}_{p,q+1}(F)$ is well defined. It is also a continuous map such that $\overline{\partial}\circ\overline{\partial} = 0$. We can therefore consider the cohomology group

$$H^{p,q}(F) = Z^{\infty}_{p,q}(F)/\overline{\partial}\mathcal{C}^{\infty}_{p,q-1}(F),$$

where $Z^{\infty}_{p,q}(F) = \{u \in \mathcal{C}^{\infty}_{p,q}(F) \mid \overline{\partial} u = 0\}$. The group $H^{p,q}(F)$ is equipped with a natural quotient topology.

Theorem 4.4. *Let D be a relatively compact strictly pseudoconvex domain with $\mathcal{C}^2$ boundary in a Stein manifold of dimension $n \geqslant 2$. For any compact set K contained in ∂D the following are then equivalent.*

i) *K is a weakly removable singularity for* CR *functions on ∂D,*
ii) *$D \subset \widehat{\Gamma}_{\mathcal{O}(\overline{D})}$, where $\widehat{\Gamma}_{\mathcal{O}(\overline{D})} = \cup\{\widehat{E}_{\mathcal{O}(\overline{D})} \mid E \subset \Gamma$ compact $\}$.*

iii) *For any compact set* $L \subset \overline{D} \setminus K$ *there is a compact set* $\Gamma_L \subset \Gamma$ *such that* $L \subset (\widehat{\Gamma}_L)_{\mathcal{O}(\overline{D})}$.
iv) $H^{n,n-1}(K) = 0$.

Proof. We first prove that iv) implies i). By the equivalence between i) and ii) of Theorem 4.1, it is enough to show that if the group $H^{n,n-1}(K)$ is zero then $H^{0,1}(X \setminus K)$ is Hausdorff. We will prove that for any compact subset L in $X \setminus K$, $\mathcal{D}_L^{n,n}(X \setminus K) \cap \overline{\partial}\mathcal{D}^{n,n-1}(X \setminus K)$is a closed subset of $\mathcal{D}_L^{n,n}(X \setminus K)$ if iv) holds and by Theorem 4.3 this will prove this first implication. Let f be a $\mathcal{C}^\infty$-smooth, $\overline{\partial}$-closed differential form of bidegree (n, n) with support in L such that $f = \lim_{j\to 0} \overline{\partial} g_j$, where $(g_j)_{j\in\mathbb{N}}$ is a sequence of $\mathcal{C}^\infty$-smooth differential forms of bidegree $(n, n-1)$ with compact support in $X \setminus K$. From Stokes' formula it follows that

$$\int_X f \wedge \varphi = \lim_{j\to 0} \int_X \overline{\partial} g_j \wedge \varphi = 0$$

for any $\mathcal{C}^\infty$-smooth, holomorphic function φ on X. Since X is a Stein manifold, Proposition 8.23 then implies that there exists a $\mathcal{C}^\infty$-smooth differential form g of bidegree $(n, n-1)$ with compact support in X such that $\overline{\partial} g = f$. In particular g is $\overline{\partial}$-closed on a neighbourhood U of K and by iv) there exist a neighbourhood $V \subset U$ of K and a $\mathcal{C}^\infty$-smooth differential form h of bidegree $(n, n-2)$ on V such that $\overline{\partial} h = g$ on V. Let χ be a function of class $\mathcal{C}^\infty$ with compact support in V identically equal to 1 in a neighbourhood of K, we define χh on X by extending it by zero. The differential form $u = g - \overline{\partial}(\chi h)$ is then of class $\mathcal{C}^\infty$ on X, vanishes in a neighbourhood of K and its restriction to $X \setminus K$ is a differential form of bidegree $(n, n-1)$ with compact support in $X \setminus K$ which satisfies $\overline{\partial} u_{|X\setminus K} = f$.

Let us now prove that i) implies ii). Suppose that K is a weakly removable singularity for CR functions on ∂D and let $\mathcal{A}(D \cup \Gamma)$ be the algebra of continuous functions on $\Gamma \cup D$ which are holomorphic on D. As K is a weakly removable singularity, $\mathcal{A}(D\cup\Gamma)\cap\mathcal{C}(\Gamma)$ is a closed subalgebra of $\mathcal{C}(\Gamma)$. (In fact it is the intersection of the closed subspaces $\mathcal{F}_\varphi = \{f \in \mathcal{C}(\Gamma) \mid \int_\Gamma f\varphi = 0\}$, where φ runs over the set of $\overline{\partial}$-closed differential $(n, n-1)$-forms of class $\mathcal{C}^\infty$ defined on some neighbourhood of $\overline{D}$ whose support does not meet K.) It follows that the restriction map ρ from $\mathcal{A}(D\cup\Gamma)$ to $\mathcal{A}(D\cup\Gamma)\cap\mathcal{C}(\Gamma)$ is a topological isomorphism: we denote its inverse by χ. If z is a point of D then the Hahn–Banach theorem implies there is a continuous linear form $\psi : \mathcal{C}(\Gamma) \to \mathbb{C}$ equal to the map $f\,|_\Gamma \mapsto \chi(f\,|_\Gamma)(z) = f(z)$ on $\mathcal{A}(D\cup\Gamma)\cap\mathcal{C}(\Gamma)$. By the Riesz representation theorem there is therefore a measure μ_z of finite mass with compact support on Γ such that

$$f(z) = \int_\Gamma f d\mu_z \quad \text{for any } f \in \mathcal{A}(D \cup \Gamma).$$

Since $\mathcal{A}(D \cap \Gamma)$ is an algebra,

$$f^k(z) = \int_\Gamma f^k d\mu_z \quad \text{for any } f \in \mathcal{A}(D \cup \Gamma)$$

and any $k \in \mathbb{N}$. It follows that, for any $k \in \mathbb{N}$,

$$|f(z)| \leqslant \sup\{|f(\zeta)| \mid \zeta \in \operatorname{supp} \mu_z\} \|\mu_z\|^{1/k}.$$

If we let k tend to infinity we get

$$|f(z)| \leqslant \sup\{|f(\zeta)| \mid \zeta \in \operatorname{supp} \mu_z\} \quad \text{for any } f \in \mathcal{A}(D \cup \Gamma),$$

which proves that $z \in (\widehat{\operatorname{supp} \mu_z})_{\mathcal{O}(\overline{D})}$ and hence $D \subset \widehat{\Gamma}_{\mathcal{O}(\overline{D})}$ by definition of $\widehat{\Gamma}_{\mathcal{O}(\overline{D})}$.

We now prove that ii) $\Rightarrow$ iii). We will need two lemmas.

Lemma 4.5. *Let D be a relatively compact strictly pseudoconvex domain with $\mathcal{C}^2$ boundary in a Stein manifold X of dimension $n \geqslant 2$ and let K be a compact set contained in ∂D. For any compact set $E \subset \partial D \setminus K$ we can find a pair (D', E') such that*

i) *D' is a relatively compact strictly pseudoconvex domain with $\mathcal{C}^2$ boundary containing D such that $\partial D' \cap \overline{D} = K$ and $\mathcal{O}(\overline{D}') \mid_{\overline{D}}$ is dense in $\mathcal{O}(\overline{D})$*

ii) *E' is a compact subset of $\partial D' \setminus K$ such that $E \subset \widehat{E'}_{\mathcal{O}(\overline{D}')}$.*

Proof. We get D' by making a small $\mathcal{C}^2$ perturbation of ∂D leaving K pointwise fixed. We can then pass from D to D' by a sequence of pseudoconvex extension elements, which implies that $\mathcal{O}(\overline{D}') \mid_{\overline{D}}$ is dense in $\mathcal{O}(\overline{D})$. For any $x \in E$ consider the neighbourhood V_x of x in X constructed in Proposition 1.1. We can assume that $\overline{V}_x$ does not meet K. As E is compact, E is covered by a finite number $V_1, \ldots, V_p$ of such sets V_x and we set $E' = \bigcup_{i=1}^p (\overline{V}_i \cap \partial D')$. If D' is close enough to D then $\overline{V}_i \cap D'$ is a domain of the type studied in Section 6 of Chapter V for any $i = 1, \ldots, p$ and hence $E \subset \widehat{E'}_{\mathcal{O}(\overline{D}')}$. □

Lemma 4.6. *Let Y be a Stein open set in a Stein manifold of dimension $n \geqslant 2$ and let E be a compact set in Y. We set $E_\varepsilon = \{z \in X \mid \operatorname{dist}(z, E) \leqslant \varepsilon\}$ and we choose $\varepsilon > 0$ small enough that E_ε is again a compact set in Y. Then*

$$(\widehat{E}_Y)_\varepsilon = (\widehat{E}_\varepsilon)_Y.$$

Proof. Set $K = (\widehat{E}_\varepsilon)_Y$: this is a $\mathcal{O}(Y)$-convex compact set and it therefore has a basis $\mathcal{U}$ of Stein neighbourhoods such that for any $U \in \mathcal{U}$ the set $\mathcal{O}(Y) \mid_U$ is dense in $\mathcal{O}(U)$. (This follows immediately from Theorem 8.17 and Proposition 8.9 of Chapter VII.) Then

$$\operatorname{dist}(\widehat{E}_Y, \partial K) = \inf_{U \in \mathcal{U}} \operatorname{dist}(\widehat{E}_Y, \partial U) = \inf_{U \in \mathcal{U}} \operatorname{dist}(\widehat{E}_U, \partial U)$$

since $\widehat{E}_Y = \widehat{E}_U$ because $\mathcal{O}(Y) \mid_U$ is dense in $\mathcal{O}(U)$. As any element U in $\mathcal{U}$ is a domain of holomorphy because it is Stein it follows from Theorem 1.13 of Chapter VI that

$$\operatorname{dist}(\widehat{E}_Y, \partial K) = \inf_{U \in \mathcal{U}} \operatorname{dist}(E, \partial U).$$

But if $U \in \mathcal{U}$ then U contains E_ε and hence $\operatorname{dist}(\widehat{E}_Y, \partial K) \geqslant \varepsilon$ from which it follows that $(\widehat{E}_Y)_\varepsilon \subset K$. □

End of the proof of Theorem 4.4. Suppose that $D \subset \widehat{\Gamma}_{\mathcal{O}(\overline{D})}$. For any $a \in \overline{D} \setminus K$ there is then a compact set E_a in Γ such that $a \in (\widehat{E}_a)_{\mathcal{O}(\overline{D})}$. By Lemma 4.5 we can find a strictly pseudoconvex domain D' containing D such that $\partial D' \cap \overline{D} = K$ and $\mathcal{O}(\overline{D}') \mid_{\overline{D}}$ is dense in $\mathcal{O}(\overline{D})$, and a compact set $E'_a \subset \Gamma' = \partial D' \setminus K$ such that $E_a \subset (\widehat{E'}_a)_{\mathcal{O}(\overline{D}')}$. Then

$$a \in (\widehat{E}_a)_{\mathcal{O}(\overline{D})} = (\widehat{E}_a)_{\mathcal{O}(\overline{D}')} \subset (\widehat{E'}_a)_{\mathcal{O}(\overline{D}')}.$$

As mentioned in Section 1 of this chapter, we can find a Stein neighbourhood U of $\overline{D}'$ such that $\widehat{C}_{\mathcal{O}(\overline{D}')} = \widehat{C}_U$ for any compact set C in $\overline{D}'$. With the notations of Lemma 4.6 we choose ε small enough that $(E'_a)_\varepsilon \cap \overline{D} = \varnothing$ and $\big(\widehat{(E'_a)_\varepsilon}\big)_U \cap K = \varnothing$, which is possible because E'_a is a compact set in $\mathbb{C}^n \setminus \overline{D}$ and the points of K are peak points of $\mathcal{O}(\overline{D}')$. Lemma 4.6 then implies that

$$B(a,\varepsilon) \subset \big((\widehat{E'_a})_U\big)_\varepsilon \subset \big(\widehat{(E'_a)_\varepsilon}\big)_U$$

and by the local maximum principle $\overline{D} \cap B(a,\varepsilon) \subset (\widehat{\Gamma}_a)_U$, where $\Gamma_a = \big(\widehat{(E'_a)_\varepsilon}\big)_U \cap \partial D$ is a compact set in Γ. If L is a compact set in $\partial D \setminus K$ then L is covered by a finite number of open sets V_a in $\overline{D}$ of the form $\overline{D} \cap B(a,\varepsilon)$ and it follows that there is a compact set $\Gamma_L \subset \Gamma$ such that $L \subset (\widehat{\Gamma}_L)_{\mathcal{O}(\overline{D})}$.

We end by proving that iii) implies iv). Consider a $\overline{\partial}$-closed $\mathcal{C}^\infty$ differential $(n,n-1)$-form φ on a neighbourhood U of K and let λ be a $\mathcal{C}^\infty$ function with compact support on U which is equal to 1 on some neighbourhood V of K. The map $f \mapsto \int_D f\overline{\partial}\lambda \wedge \varphi$ is a continuous linear form on $\mathcal{O}(\overline{D})$. We set $L = \overline{D} \cap \operatorname{supp} \overline{\partial}\lambda$: the set L is a compact set disjoint from K and by iii) there is a compact set $\Gamma_L \subset \Gamma$ such that

$$\Big| \int_D f\overline{\partial}\lambda \wedge \varphi \Big| \leqslant C \sup_{z \in L} |f(z)| \leqslant C \sup_{z \in \Gamma_L} |f(z)|.$$

It follows from the Hahn–Banach theorem and the Riesz representation theorem that there is a measure μ_L on Γ_L such that $\langle \mu_L, f\rangle = \int_D f\overline{\partial}\lambda \wedge \varphi$ for any $f \in \mathcal{O}(\overline{D})$. If χ_D is the characteristic function of D then $\sigma = \chi_D\overline{\partial}\lambda \wedge \varphi - \mu_L$ is a measure on $D \cup \Gamma$ such that $\langle \sigma, f\rangle = 0$ for every $f \in \mathcal{O}(\overline{D})$. As the open set D is strictly pseudoconvex there is a strictly pseudoconvex domain G' such that $\operatorname{supp}\sigma \subset G'$, $K \cap \overline{G}' = \varnothing$ and $\mathcal{O}(\overline{D})$ is dense in $\mathcal{O}(G') \mid_{G' \cap (D \cup \Gamma)}$. (To get G', we simply push the boundary of D slightly towards the interior of D in a neighbourhood of K and push it slightly towards the exterior of D near the support of σ.) The measure σ then defines an (n,n)-current T_σ with compact support in G' such that $\langle T_\sigma, f\rangle = 0$ for any $f \in \mathcal{O}(G')$. As G' is Stein there is an $(n,n-1)$-current S with compact support in G' such that $T_\sigma = \overline{\partial}S$. But now T_σ is a $\mathcal{C}^\infty$ differential form on $D \cap G'$, so it follows from the regularity of $\overline{\partial}$ (cf. Chap. VI, Corollary 4.2 ii)) that there is a $\mathcal{C}^\infty$ differential form φ_0 on $D \cap G'$ which vanishes on some neighbourhood of K such that $\overline{\partial}\varphi_0 = T_\sigma$ on $D \cap G'$. Now consider a strictly pseudoconvex domain G'' such that $K \subset G'' \subset D \cup V$. Restricting the relation $\overline{\partial}\varphi_0 = T_\sigma$ to G'' we get $\overline{\partial}\lambda \wedge \varphi = \overline{\partial}\varphi_0$ or alternatively $\overline{\partial}(\lambda\varphi - \varphi_0) = 0$ on G'' (here we implicitly define $\lambda\varphi$ on the whole of G'' by extending it by 0). Since G'' is Stein, $H^{n,n-2}(G'') = 0$ and there is therefore a $\mathcal{C}^\infty$ differential $(n, n-2)$-form ψ on G'' such that $\lambda\varphi - \varphi_0 = \overline{\partial}\psi$ on G''. As $\lambda = 1$ and $\varphi_0 = 0$ on some neighbourhood of K we have $\overline{\partial}\psi = \varphi$ on some neighbourhood of K, which proves that $H^{n,n-1}(K) = 0$. □

Comments

The study of removable singularities for CR functions defined on the boundary of a domain started with the work of G. Lupacciolu and G. Tomassini [Lu/To] in 1984 and developed quickly in the following years. The most important results on the subject were proved by G. Lupacciolu [Lu1, Lu2]. A panorama of all known results on the subject is presented in [Ci/St].

Appendix

A

Differentiable manifolds and differential forms

This appendix contains some of the basic tools of differential geometry used in this book. After introducing the concept of a differentiable manifold we define the algebra of differential forms and prove Stokes' theorem, which is the main result of this appendix.

1 Differentiable manifolds

Definition 1.1. A *chart* h on a topological space X is a homeomorphism from an open set U of X to an open set of $\mathbb{R}^n$ for some integer n. The open set U is the *domain of the chart* h. We sometimes denote the chart h by the pair (U, h).

If V is an open set in X and $V \subset U$ then $h\big|_V$ is a chart whose domain is V.

Definition 1.2.

a) Two charts on X, h and h', which have the same domain U are said to be *q-compatible* for some integer $q \in \mathbb{N}^* \cup \{\infty\}$ if the two inverse homeomorphisms
$$h' \circ h^{-1} : h(U) \longrightarrow h'(U)$$
$$h \circ h'^{-1} : h'(U) \longrightarrow h(U)$$
are C^q as maps from an open set in $\mathbb{R}^p$ to an open set in $\mathbb{R}^{p'}$.

b) Two charts (U, h) and (U', h') are said to be q-compatible if $U \cap U' = \varnothing$ or $h\big|_{U\cap U'}$ and $h'\big|_{U\cap U'}$ are q-compatible as in a).

Remark. If n and n' are the integers associated to compatible charts h and h' then $n = n'$. Indeed as $q \geqslant 1$ the differential map $d(h' \circ h^{-1})(x)$ is a linear bijection from $\mathbb{R}^n$ to $\mathbb{R}^{n'}$ for any $x \in h(U)$ so $n = n'$.

Definition 1.3. A $\mathcal{C}^q$ *atlas* on X is a set of charts which are pairwise q-compatible and whose domains form an open cover of X. Two $\mathcal{C}^q$ atlases are said to be *compatible* if their union is a $\mathcal{C}^q$ atlas.

Remark. Compatibility is an equivalence relation on the set of $\mathcal{C}^q$ atlases on X.

Definition 1.4. A $\mathcal{C}^q$ *differentiable manifold* is a Hausdorff topological space which is a countable union of compact subsets equipped with an equivalence class of $\mathcal{C}^q$ atlases.

Examples.

1) Any non-empty open set D in $\mathbb{R}^n$ has a natural $\mathcal{C}^\infty$ manifold structure. The chart $\{(D, \mathrm{Id}_D)\}$ is an atlas for this structure.
2) The sphere in $\mathbb{R}^3$ is a $\mathcal{C}^\infty$ manifold with an atlas containing two charts.
3) Suppose that D is a relatively compact open set in $\mathbb{R}^n$ such that for any $P \in \partial D$ there is a neighbourhood W_p of P such that $\partial D \cap W_p$ can be written in the form $\partial D \cap W_p = \{x \in W_p \mid r_p(x) = 0\}$ where r_p is a $\mathcal{C}^k$ function $(1 \leqslant k \leqslant \infty)$ on W_p and $dr_p \neq 0$ on $\partial D \cap W_p$. The set ∂D is then a $\mathcal{C}^k$ manifold. (This follows from the implicit function theorem.)

If X is a differentiable manifold then for any $x \in X$ there is a chart (U, h) such that $x \in U$ and h is a map from U to $\mathbb{R}^n$. This integer n only depends on x: we call it the dimension of X at x. It is clear that n is constant on any connected component of X.

Definition 1.5. A differentiable manifold all of whose connected components are of dimension n is said to be of *dimension* n.

Definition 1.6. Let X and Y be two $\mathcal{C}^q$ differentiable manifolds. A map $f : X \to Y$ is said to be $\mathcal{C}^p$ for some integer $p \leqslant q$ if it is continuous and the following condition holds: for any pair of charts (U, h) and (V, k) of X and Y such that $f(U) \subset V$ the map

$$k \circ (f|_U) \circ h^{-1} : h(U) \longrightarrow k(V)$$

is $\mathcal{C}^p$ as a map from an open set of $\mathbb{R}^n$ to an open set of $\mathbb{R}^m$.

We note that it is enough to check these properties on every chart in some atlas.

If $Y = \mathbb{R}$ or $\mathbb{C}$ then a $\mathcal{C}^p$ map from X to Y is called a $\mathcal{C}^p$ function on X.

Definition 1.7. Let (U, h) be a $\mathcal{C}^q$ chart on a differentiable manifold X of class $\mathcal{C}^q$. The map h is then a $\mathcal{C}^q$ map from U to $\mathbb{R}^n$

$$\begin{aligned} U &\longrightarrow h(U) \subset \mathbb{R}^n \\ x &\longmapsto h(x) = (x_1(x), \ldots, x_n(x)), \end{aligned}$$

where each $x_j : U \to \mathbb{R}$ $(j = 1, \ldots n)$ is a $\mathcal{C}^q$ function on U. The functions $x_1, \ldots, x_n$ are called the *local coordinates* on U defined by the chart (U, h).

2 Partitions of unity

In this section, we construct a tool – the partition of unity – which enables us to localise problems on a manifold. In particular, it enables us to restrict to chart domains.

Lemma 2.1. *Let A be a compact set in $\mathbb{R}^n$ and let U be an open set containing A. There is a real $\mathcal{C}^\infty$ function with compact support in U whose image is contained in $[0,1]$ and which is equal to 1 on A.*

Proof. Consider the function defined on $\mathbb{R}$ by

$$\theta(t) = \begin{cases} Ce^{-1/(1-t^2)} & \text{if } |t| < 1, \\ 0 & \text{if } |t| \geqslant 1, \end{cases}$$

where C is a constant such that $\int_{\mathbb{R}} \theta(t)\, dt = 1$. This function is $\mathcal{C}^\infty$ on $\mathbb{R}$ and is supported on the interval $[-1,1]$. For any real number $\varepsilon > 0$ the function $x \mapsto \theta_\varepsilon(x) = \theta(|x|/\varepsilon)$ (where $|x| = (x_1^2 + \cdots + x_n^2)^{1/2}$) is $\mathcal{C}^\infty$ on $\mathbb{R}^n$. Its support is the closed ball of centre 0 and radius ε and its image is contained in $[0,1]$.

Let B be a relatively compact open set in $\mathbb{R}^n$ such that $A \Subset B \Subset U$. Set

$$\psi = \chi_B * \theta_\varepsilon,$$

where χ_B is the characteristic function of B. The function ψ is $\mathcal{C}^\infty$ on $\mathbb{R}$ and its image is contained in $[0,1]$. If $\varepsilon < \frac{1}{2} \min\big(\mathrm{dist}(\overline{B}, \complement U), \mathrm{dist}(A, \complement \overline{B})\big)$ then the support of ψ is a compact set contained in U and ψ is constant and equal to 1 on A. □

Lemma 2.2. *Let $\mathcal{U}$ be a locally finite open cover of a differentiable manifold X. We can choose open sets U' for every $U \in \mathcal{U}$ such that $\overline{U}' \subset U$ and the family of open sets U' is again an open cover of X.*

Proof. We assume without loss of generality that X is connected. Passing to a subsequence if necessary, we may assume that $\mathcal{U}$ is countable since X is a countable union of compact sets. (We take $U' = \varnothing$ for any U we have suppressed.) We write $\mathcal{U} = \{U_1, U_2, U_3, \ldots\}$.

Set $C_1 = U_1 \smallsetminus (\bigcup_{k \geqslant 2} U_k)$: C_1 is a closed subset of U_1 and $X = C_1 \cup U_2 \cup U_3 \cup \cdots$. Let U_1' be an open set such that $C_1 \subset U_1' \subset \overline{U}_1' \subset U_1$.

Set $C_2 = U_2 \smallsetminus (U_1' \cup (\bigcup_{k \geqslant 3} U_k))$: C_2 is a closed subset of U_2 and $X = U_1' \cup C_2 \cup U_3 \cup \cdots$. Let U_2' be an open set such that $C_2 \subset U_2' \subset \overline{U}_2' \subset U_2$.

Iterating this construction we get a family $(U_k')_{k \in \mathbb{N}}$ of open sets in X such that $\overline{U}_k' \subset U_k$. It remains to prove that this family is an open cover of X.

For any $x \in X$, there is a largest integer n such that $x \in U_n$, since $\mathcal{U}$ is locally finite. By construction of the sets U_k',

$$x \in U_1' \cup \cdots U_n' \cup \Big(\bigcup_{k \geqslant n+1} U_k\Big).$$

It follows that $x \in \bigcup_{k\geqslant 1} U'_k$ because x cannot disappear when we replace U_k by U'_k for any $k \geqslant n+1$ – by definition of n we know that $x \notin (\bigcup_{k\geqslant n+1} U_k)$. □

Theorem 2.3. *Let X be a differentiable manifold. For any open covering $(\Omega_i)_{i\in I}$ of X we can find functions $(\alpha_i)_{i\in I}$ such that*

1) *α_i is $\mathcal{C}^\infty$ and has compact support in Ω_i,*
2) *only a finite number of the functions α_i are not identically zero on any compact set in X,*
3) *$\alpha_i(x) \geqslant 0$ and $\sum_{i\in I} \alpha_i(x) = 1$ for any $x \in X$.*

Definition 2.4. A family of functions $(\alpha_i)_{i\in I}$ satisfying the conditions of Theorem 2.3 is called a *locally finite partition of unity subordinate to the open cover $(\Omega_i)_{i\in I}$*.

Proof (of Theorem 2.3). Suppose initially that the open cover $(\Omega_i)_{i\in I}$ is locally finite (i.e. every compact set only meets a finite number of sets Ω_i), all the sets Ω_i are relatively compact in X and the closure of each Ω_i is contained in a chart domain. Applying Lemma 2.2, we can find a new open cover $(\Omega'_i)_{i\in I}$ indexed by the same set such that $\overline{\Omega}'_i \subset \Omega_i$. By Lemma 2.1, there are $\mathcal{C}^\infty$ functions $(\varphi_i)_{i\in I}$ with image contained in $[0,1]$ such that the support of φ_i is contained in Ω_i and φ_i is equal to 1 on Ω'_i. The sum $\varphi = \sum_{i\in I} \varphi_i$ is well defined on X since for any $x \in X$ only a finite number of the values $\varphi_i(x)$ are non-zero. This function is everywhere $\geqslant 1$. The functions $\alpha_i = \frac{\varphi_i}{\varphi}$ satisfy the conditions of the theorem.

Suppose now that the open cover $(\Omega_i)_{i\in I}$ is arbitrary. As X is a countable union of compact sets we can find a locally finite refinement of $(\Omega_i)_{i\in I}$, $(G_j)_{j\in J}$, indexed by a set J and a map from J to I, $j \mapsto i(j)$ such that each G_j is relatively compact in X, the set $\overline{G}_j$ is contained in a chart domain and $G_j \subset \Omega_{i(j)}$ for any $j \in J$. By the first part of the proof there is a partition of unity $(\gamma_j)_{j\in J}$ subordinate to the cover $(G_j)_{j\in J}$. Set $\alpha_i = \sum_{i(j)=i} \gamma_j$ for every $i \in I$. The family $(\alpha_i)_{i\in I}$ then satisfies the conditions of the theorem. □

3 Cotangent space at a point and differential forms of degree 1

Let X be a $\mathcal{C}^q$ differentiable manifold of dimension n and consider a point $x \in X$. We consider the set of pairs (U, f) such that U is an open set in X containing x and f is a function such that $f \in \mathcal{C}^p(U)$ for some $p \leqslant q$. We define an equivalence relation on this set by setting $(U, f) \sim (U', f')$ if and only if there is an open set $W \subset U \cap U'$ such that $x \in W$ and $f|_W = f'|_{W'}$. A *germ* of a $\mathcal{C}^p$ function at x is an equivalence class for the above relation; when there is no risk of confusion we will often identify a germ with one of its representatives.

Definition 3.1. For any $p \geqslant 1$ a $\mathcal{C}^p$ function f defined on a neighbourhood W of a point $x \in X$ is said to be *stationary* at x if there is a chart (U, h) such that $x \in U \subset W$ and all the first-order partial derivatives of $f \circ h^{-1}$ vanish at $h(x)$. A germ of a $\mathcal{C}^p$ function ($p \geqslant 1$) at x is stationary if and only if any of its representatives is stationary at x. (Note that if one representative is stationary then all the others are as well.)

Let $\mathcal{C}^p_x(X)$ be the set of germs of $\mathcal{C}^p$ functions at x and let $\mathcal{S}^p_x(X)$ be the subset of $\mathcal{C}^p_x(X)$ consisting of stationary germs. The set $\mathcal{C}^p_x(X)$ is an $\mathbb{R}$-vector space and $\mathcal{S}^p_x(X)$ is a subspace of $\mathcal{C}^p_x(X)$. We note that $\mathcal{C}^p_x(X)/\mathcal{S}^p_x(X) \simeq \mathcal{C}^1_x(X)/\mathcal{S}^1_x(X)$ for any $p \geqslant 1$.

Definition 3.2. Let X be a $\mathcal{C}^q$ manifold for some $q \geqslant 1$. The vector space $T^*_x(X) = \mathcal{C}^1_x(X)/\mathcal{S}^1_x(X)$ is called the *cotangent space to X at x*. If $f \in \mathcal{C}^1_x(X)$ then we denote its image in $T^*_x(X)$ by $(df)_x$. Elements of $T^*_x(X)$ are called *differentials* at x.

Let (U, h) be a chart such that $x \in U$ and consider the map

$$\begin{aligned} \theta_{h,x} : T^*_x(X) &\longrightarrow \mathcal{L}(\mathbb{R}^n, \mathbb{R}) \\ (df)_x &\longmapsto d(f \circ h^{-1})(h(x)). \end{aligned}$$

The map $\theta_{h,x}$ is linear and injective. If $(x_1, \ldots, x_n)$ are the local coordinates at x defined by the chart (U, h) then the family of forms $d(x_j \circ h^{-1})(h(x))$ is a basis for $\mathcal{L}(\mathbb{R}^n, \mathbb{R})$. This means that $\theta_{h,x}$ is also surjective, so it is an isomorphism. The vector space $T^*_x(X)$ has dimension n and $(dx_1)_x, \ldots, (dx_n)_x$ is a basis of $T^*_x(X)$.

We now express an element $(df)_x$ of $T^*_x(X)$ in terms of this basis:

$$d(f \circ h^{-1})(h(x)) = \sum_{j=1}^{n} \frac{\partial f \circ h^{-1}}{\partial x_j}(h(x)) d(x_j \circ h^{-1})(h(x)),$$

so $(df)_x = \sum_{j=1}^n \frac{\partial f \circ h^{-1}}{\partial x_j}(h(x))(dx_j)_x$.

Definition 3.3. We define $T^*(X)$ to be the disjoint union of the spaces $T^*_x(X)$ for all $x \in X$. This union is called the *cotangent space* of X. We denote the natural projection from $T^*(X)$ to X by p.

Definition 3.4. Let X be a $\mathcal{C}^q$ differentiable manifold for some $q \geqslant 1$. Let A be an open set in X. A *differential form* of degree 1 on A is a map $\omega : A \to T^*(X)$ such that $p \circ A = \mathrm{Id}$.

Let (U, h) be a chart on X and let $(x_1, \ldots, x_n)$ be the associated local coordinates. If ω is a differential form of degree 1 on A then, for any $x \in A \cap U$,

$$\omega\big|_{A \cap U}(x) = \sum_{j=1}^{n} a_j(x)(dx_j)_x.$$

Let dx_j be the differential form of degree 1 defined on U by $x \mapsto (dx_j)_x$. We can then write

$$\omega\big|_{A\cap U} = \sum_{j=1}^{n} a_j dx_j,$$

where the functions a_j are defined on $A \cap U$.

The form ω is said to be $\mathcal{C}^\ell$ for some $\ell < q$ if the functions a_j are $\mathcal{C}^\ell$ on $A \cap U$. A straightforward calculation shows that this is independent of the choice of coordinates.

Example 3.5. If f is a $\mathcal{C}^1$ function on an open set A in X then the map $x \mapsto (df)_x$ defines a differential form of degree 1 which is continuous on A. This form is called the differential of the function f. If A is an open set in $\mathbb{R}^n$ then the differential df thus defined is simply the differential map of f. (Here we have identified T_x^*X with $\mathcal{L}(\mathbb{R}^n, \mathbb{R})$ for all $x \in A$.)

4 The tangent space at a point and vector fields

Let X be a $\mathcal{C}^q$ differentiable manifold of dimension n for some $q \geqslant 1$ and let x be a point in X.

A curve in X passing through $x \in X$ is a $\mathcal{C}^1$ map α from an open interval I in $\mathbb{R}$ to X such that $0 \in I$ and $\alpha(0) = x$. We define an equivalence relation on the set of curves in X passing through x by $\alpha \sim \beta$ if and only if for any $\mathcal{C}^1$ function f defined in a neighbourhood of x,

$$\frac{d}{dt}(f \circ \alpha)(0) = \frac{d}{dt}(f \circ \beta)(0).$$

We leave it to the reader to check that if X is an open set in $\mathbb{R}^n$ then $\alpha \sim \beta$ if and only if $\alpha'(0) = \beta'(0)$.

Definition 4.1. The set of equivalence classes of curves in X passing through x for the above relationship is called the *tangent space* to X at x. We denote it by $T_x(X)$.

If $\nu \in T_x(X)$ and f is a $\mathcal{C}^1$ function defined in a neighbourhood of x then we set

$$\nu(f) = \frac{d}{dt}(f \circ \alpha)(0),$$

for any representative α of ν. This can be interpreted as a directional derivative. When $X = \mathbb{R}^n$, it is easy to check that

$$\nu(f) = \lim_{t \to 0} \frac{f(x + t\alpha'(0)) - f(x)}{t}.$$

Note that $\nu(f) = \nu(g)$ if f and g are functions in $\mathcal{C}^1_x(X)$ and $f - g \in \mathcal{S}^1_x(X)$. If $\nu \in T_x(X)$ we can therefore define a map, also denoted by ν, from $T^*_x(X)$ to $\mathbb{R}$ by

$$\nu : (df)_x \longmapsto \frac{d}{dt}(f \circ \alpha)(0),$$

where f represents the element $(df)_x$ in $T^*_x(X)$ and α represents ν. Let us check that if ν_1 and ν_2 give rise to the same map from $T^*_x(X)$ to $\mathbb{R}$ then they are equal in $T_x(X)$. Let α_i be a representative of ν_i for $i = 1, 2$. If ν_1 and ν_2 are equal as maps from $T^*_x(X)$ to $\mathbb{R}$ then, for any $\mathcal{C}^1$ function f defined in a neighbourhood of x,

$$\frac{d}{dt}(f \circ \alpha_1)(0) = \frac{d}{dt}(f \circ \alpha_2)(0)$$

but this implies that α_1 and α_2 are in the same class and hence $\nu_1 = \nu_2$ in T_xX. A tangent vector to X at x can therefore be thought of as a linear form on $T^*_x(X)$.

Proposition 4.2. *Consider a tangent vector $\nu \in T_x(X)$ and let f and g be $\mathcal{C}^1$ functions defined in a neighbourhood of x. Then, for any $a, b \in \mathbb{R}$,*

$$\begin{aligned} \nu(af + bg) &= a\nu(f) + b\nu(g) \\ \nu(fg) &= \nu(f)g(x) + f(x)\nu(g). \end{aligned}$$

(A map with these properties is called a derivation at x.)

Proof. This is a straightforward application of the definitions. □

Example. Let (U, h) be a chart at $x \in X$ such that $h(x) = 0$ and let $(x_1, \dots, x_n)$ be the local coordinates defined by this chart. Let $(\frac{\partial}{\partial x_j})_x$ be the class of the curve $t \mapsto h^{-1}(0, \dots, 0, \underset{rgj}{t}, 0, \dots, 0)$ for all $j = 1, \dots, n$. If f is a $\mathcal{C}^1$ function on some neighbourhood of x then, for any $j = 1, \dots, n$,

$$\left(\frac{\partial}{\partial x_j}\right)_x (f) = \frac{\partial}{\partial x_j}(f \circ h^{-1})(0).$$

Proposition 4.3. *The vectors $(\frac{\partial}{\partial x_1})_x, \dots, (\frac{\partial}{\partial x_n})_x$ form a basis for $(T^*_x(X))^*$, the dual of $T^*_x(X)$.*

Proof. It is enough to check that the vectors $(\frac{\partial}{\partial x_1})_x, \dots, (\frac{\partial}{\partial x_n})_x$ are dual to the basis $((dx_j)_x)_{j=1,\dots,n}$ of $T^*_x(X)$. To do this we calculate $(\frac{\partial}{\partial x_j})_x(x_k)$:

$$\left(\frac{\partial}{\partial x_j}\right)_x (x_k) = 0 \text{ if } j \neq k \quad \text{and} \quad \left(\frac{\partial}{\partial x_j}\right)_x (x_k) = 1 \text{ if } j = k$$

which proves the proposition. □

Theorem 4.4. *The tangent space to X at x can be identified with the dual of the cotangent space to X at x. In other words, $T_x X = (T_x^*(X))^*$.*

Proof. By the remark following Definition 4.1 and the first part of Proposition 4.2 we know that $T_x(X)$ injects into $(T_x^*(X))^*$. It remains to prove the opposite inclusion. Let (U, h) be a chart of X such that $x \in U, h(x) = 0$ and $h(U)$ is the open set $\{x \in \mathbb{R}^n \mid \sup_{j=1,\ldots,n} |x_j| < 1\}$. By Proposition 4.3, for any $L \in (T_x^*(X))^*$ there are real numbers $a_1, \ldots, a_n$ such that $L = \sum_{j=1}^n a_j (\frac{\partial}{\partial x_j})_x$. Consider a set of functions γ_j, $j = 1, \ldots, n$, defined on an open interval I of $\mathbb{R}$ containing 0 such that $|\gamma_j(t)| < 1$ for any $t \in I$ and $\gamma_j(t) = a_j t$ in some neighbourhood of 0. Let Γ be the curve in X passing through x defined by $\Gamma = h^{-1} \circ \gamma$, where $\gamma(t) = (\gamma_1(t), \ldots, \gamma_n(t))$. The form L is then equal to the class of Γ and it follows that $(T_x^*(X))^* \subset T_x(X)$. □

Remark. The space $T_x^*(X)$ is therefore the dual of $T_x(X)$ and the bases $((dx_j)_x)_{j=1,\ldots,n}$ and $((\frac{\partial}{\partial x_j})_x)_{j=1,\ldots,n}$ are dual bases.

It follows that, for any $(df)_x \in T_x^*(X)$,

$$(df)_x = \sum_{j=1}^n \left(\frac{\partial}{\partial x_j}\right)_x (df)_x (dx_j)_x = \sum_{j=1}^n \left(\frac{\partial}{\partial x_j}\right)_x f (dx_j)_x.$$

We can define the $\mathbb{C}$-vector space of complex differential 1-forms at the point $x \in X$ by considering complex-valued functions. We denote this space by $\mathbb{C}T_x^*X$: it can be identified with the space of $\mathbb{R}$-linear maps from $T_x X$ to $\mathbb{C}$. The vector space $\mathbb{C}T_x^*X$ is in fact the complexification $\mathbb{C} \otimes_R T_x^*X$ of the real vector space T_x^*X. It is a vector space of complex dimension n. We can also consider the complexification $\mathbb{C}T_x X$ of the real vector space $T_x X$. An element $\nu \in \mathbb{C}T_x X$ can be uniquely written in the form $\nu = \nu_1 + i\nu_2$ where $\nu_1, \nu_2 \in T_x X$. A complex-valued differential 1-form ω can be naturally extended to a $\mathbb{C}$-linear map $\omega^{\mathbb{C}} : \mathbb{C}T_x X \to \mathbb{C}$ on setting $\omega^{\mathbb{C}}(\nu_1 + i\nu_2) = \omega(\nu_1) + i\omega(\nu_2)$. It is easy to show that $\mathbb{C}T_x^*X$ and $\mathbb{C}T_x X$ are naturally dual as $\mathbb{C}$-vector spaces.

Definition 4.5. We define $T(X)$ to be the disjoint union of the spaces $T_x(X)$ for all $x \in X$. It is called the *tangent space* of X and we denote the natural projection from $T(X)$ to X by p.

Definition 4.6. Let X be a $\mathcal{C}^q$ differentiable manifold for some $q \geqslant 1$, and let A be an open set in X. A *vector field* on A is a map $V : A \to T(X)$ such that $p \circ V = \mathrm{Id}$.

Let (U, h) be a chart on X and let $(x_1, \ldots, x_n)$ be the associated local coordinates. If V is a vector field on A and x is a point in $A \cap U$ then

$$V|_{A \cap U}(x) = \sum_{j=1}^n a_j(x) \left(\frac{\partial}{\partial x_j}\right)_x.$$

Let $\frac{\partial}{\partial x_j}$ be the vector field defined on U by $x \mapsto (\frac{\partial}{\partial x_j})_x$. We can then write

$$V\big|_{A\cap U} = \sum_{j=1}^{n} a_j \frac{\partial}{\partial x_j},$$

where the functions a_j are defined on $A \cap U$.

The vector field V is said to be $\mathcal{C}^\ell$ for some $\ell < q$ if the functions a_j are all $\mathcal{C}^\ell$ on $A \cap U$.

5 The algebra of differential forms

Let X be a $\mathcal{C}^q$ differentiable manifold for some $q \geqslant 1$, and let x be a point of X.

We consider the rth exterior product over $\mathbb{R}$ of the cotangent space $T_x^*(X)$ to X at x, which we denote by $\Lambda^r T_x^*(X)$. The reader will find the definitions and main properties of the rth exterior product of a vector space in [Lan]. In our setting we will use the following interpretation of the rth exterior power, which is the most concrete one.

By definition $\Lambda^0 T_x^*(X) = \mathbb{R}$ and for any $r \geqslant 1$ we identify $\Lambda^r T_x^*(X)$ with the $\mathbb{R}$-vector space of r-linear alternating forms on $T_x(X)$. This is possible because $T_x^*(X)$ is dual to $T_x(X)$. By an r-linear alternating form on $T_x(X)$ we mean an r-linear map

$$\omega : \underbrace{T_x(X) \times \cdots \times T_x(X)}_{r \text{ times}} \longrightarrow \mathbb{R}$$

such that, for any $\nu_1, \ldots, \nu_r \in T_x(X)$ and any permutation σ of $\{1, \ldots, r\}$,

$$\omega(\nu_{\sigma(1)}, \ldots, \nu_{\sigma(r)}) = \operatorname{sign}(\sigma)\omega(\nu_1, \ldots, \nu_r),$$

where $\operatorname{sign}(\sigma)$ is the signature of the permutation σ. In particular, $\omega(\nu_1, \ldots, \nu_r) = 0$ if $\nu_i = \nu_j$ for some pair (i, j) such that $i \neq j$.

Remark. We have $\Lambda^1 T_x^*(X) = T_x^*(X)$ and $\Lambda^r T_x^*(X) = \{0\}$ for any $r > \dim T_x^*(X) = \dim X$.

The *exterior algebra* of $T_x^*(X)$ is the sum $\Lambda^\bullet T_x^*(X) = \bigoplus_{r\geqslant 0} \Lambda^r T_x^*(X)$.

A. Exterior product

The *exterior product* of an r-form ω in $\Lambda^r T_x^*(X)$ and an s-form η in $\Lambda^s T_x^*(X)$ is an $(r+s)$-form denoted by $\omega \wedge \eta$ and defined by

$$\omega \wedge \eta(\nu_1, \ldots, \nu_{r+s}) = \frac{1}{r!s!} \sum_{\sigma} \operatorname{sign}(\sigma)\omega(\nu_{\sigma(1)}, \ldots, \nu_{\sigma(r)})\eta(\nu_{\sigma(r+1)}, \ldots, \nu_{\sigma(r+s)}),$$

where the sum is taken over all permutations σ of the set $\{1, \dots, r+s\}$. If r or s is zero – if $r = 0$, for example – then ω is a real number and we set $\omega \wedge \eta = \omega\eta$.

The above equation, together with the distributivity law for addition, allows us to define the exterior product of any two elements in $\Lambda^\bullet T_x^*(X)$. This produces an internal composition law on $\Lambda^\bullet T_x^*(X)$ which we denote by $\wedge$. It is easy to check that $\wedge$ is associative but not commutative. However, we do have

$$\omega \wedge \eta = (-1)^{rs}\eta \wedge \omega \quad \text{if } \omega \in \Lambda^r T_x^*(X) \text{ and } \eta \in \Lambda^s T_x^*(X).$$

Let (U, h) be a chart of X in a neighbourhood of x and let $(x_1, \dots, x_n)$ be the associated local coordinates. For any $r \in \{1, \dots, n\}$ the family of elements of the form

$$\{(dx_{j_1})_x \wedge \cdots \wedge (dx_{j_r})_x, 1 \leqslant j_1 < \cdots < j_r \leqslant n\}$$

is a basis of $\Lambda^r T_x^*(X)$. In particular, $\dim \Lambda^r T_x^*(X) = \binom{n}{r}$ and any r-form ω in $\Lambda^r T_x^*(X)$ can be written in the form

$$\omega = \sum_J a_J (dx_J)_x,$$

where the sum is taken over all strictly increasing r-tuplets $(j_1, \dots, j_r)$ of $\{1, \dots, n\}^r$. Here, we set

$$(dx_J)_x = (dx_{j_1})_x \wedge \cdots \wedge (dx_{j_r})_x \quad \text{if } J = (j_1, \dots, j_r).$$

We note that the coordinates a_J of ω in this basis are given by

$$a_J = \omega\Big(\big(\frac{\partial}{\partial x_{j_1}}\big)_x, \dots, \big(\frac{\partial}{\partial x_{j_r}}\big)_x\Big) \quad \text{where } J = (j_1, \dots, j_r).$$

Definition 5.1. Let X be a $\mathcal{C}^q$ differentiable manifold for some $q \geqslant 1$ and let A be an open set in X. A differential *r-form* or a *differential form of degree* r on A is a map $\omega : A \to \Lambda^\bullet T^*(X) = \bigcup_{x \in X} \Lambda^\bullet T_x^*(X)$ such that $p \circ \omega = \mathrm{Id}$, where p is the natural projection from $\Lambda^\bullet T^*(X)$ to X.

Let (U, h) be a chart of X and let $(x_1, \dots, x_n)$ be the associated local coordinates. If ω is a differential r-form on A then for any point $x \in A \cap U$

$$\omega\big|_{A \cap U}(x) = \sum_{\substack{J=(j_1,\dots,j_r)\\ j_1 < \cdots < j_r}} a_J(x)(dx_J)_x.$$

Let dx_J be the differential form of degree r defined by $x \mapsto (dx_J)_x$. Then

$$\omega\big|_{A \cap U} = \sum_{\substack{J=(j_1,\dots,j_r)\\ j_1 < \cdots < j_r}} a_J dx_J,$$

where the functions a_J are defined on $A \cap U$. The differential form ω is said to be $\mathcal{C}^\ell$ for some $\ell < q$ if and only if the functions a_J are $\mathcal{C}^\ell$ on $A \cap U$. Let $\mathcal{C}^\ell_r(A)$ be the vector space of differential forms of degree r which are $\mathcal{C}^\ell$ on A.

If $\omega \in \mathcal{C}^\ell_r(A)$ and $\eta \in \mathcal{C}^\ell_s(A)$ then the differential form $\omega \wedge \eta$ defined by $\omega \wedge \eta(x) = \omega(x) \wedge \eta(x)$ for any $x \in A$ is contained in $\mathcal{C}^\ell_{r+s}(A)$. The differential form dx_J, where $J = (j_1, \dots, j_r)$, is the exterior product $dx_{j_1} \wedge \cdots \wedge dx_{j_r}$.

Consider the vector space $\bigoplus_{r \geqslant 0} \mathcal{C}^\ell_r(A)$. The exterior product $\wedge$ defined above is an internal composition law on $\bigoplus_{r \geqslant 0} \mathcal{C}^\ell_r(A)$ and the space $\bigoplus_{r \geqslant 0} \mathcal{C}^\ell_r(A)$ is therefore an algebra called the algebra of $\mathcal{C}^\ell$ differential forms on A. This algebra is denoted by $\mathcal{C}^\ell_\bullet(A)$.

B. Exterior derivative

As seen in example 3.5, the differential df of a $\mathcal{C}^1$ function f on an open set A in X defines a continuous 1-form on A; there is therefore a map $d : \mathcal{C}^1(A) \to \mathcal{C}^0_1(A)$ which satisfies the Leibniz rule

$$d(fg) = g df + f dg.$$

(This can be easily checked using the definition.)

We want to extend d to the whole of the algebra $\mathcal{C}^\ell_\bullet(A)$ for any $\ell \geqslant 1$.

Theorem 5.2. *Let X be a $\mathcal{C}^q$ differentiable manifold for some $q \geqslant 2$ and let ℓ be an integer such that $0 \leqslant \ell < q$. There is then a unique linear map $d : \mathcal{C}^1_\bullet(X) \to \mathcal{C}^0_\bullet(X)$ such that*

i) *df is the differential of f for any $f \in \mathcal{C}^1(X)$*
ii) *if $1 \leqslant \ell < q$ and $r \geqslant 0$ then $d\omega \in \mathcal{C}^{\ell-1}_{r+1}(X)$ for any $\omega \in \mathcal{C}^\ell_r(X)$,*
iii) *if $f \in \mathcal{C}^\ell(X)$ and $2 \leqslant \ell \leqslant q$ then $d(df) = 0$,*
iv) *if $\omega_1 \in \mathcal{C}^1_r(X)$ and $\omega_2 \in \mathcal{C}^1_s(X)$ then*

$$d(\omega_1 \wedge \omega_2) = d\omega_1 \wedge \omega_2 + (-1)^r \omega_1 \wedge d\omega_2.$$

The map d thus defined is called the exterior derivative *on X.*

Proof. We note first that if d exists then it is a local operator. In other words, if ω_1 and ω_2 are two differential forms of degree r on X which are equal on an open set U in X then $d\omega_1$ and $d\omega_2$ are equal on U.We prove this by showing that if $\omega \in \mathcal{C}^1_r(X)$ vanishes on U then $d\omega = 0$ on U. Consider a point $x \in U$ and a function $f \in \mathcal{C}^1(X)$ such that $f(x) = 0$ and $f \equiv 1$ on some neighbourhood of $X \setminus U$: if $\omega \in \mathcal{C}^1_r(X)$ vanishes on U then we can write $\omega = f\omega$. By property 4) of d,

$$d\omega = d(f\omega) = df \wedge \omega + f d\omega$$

and since $f(x) = 0$ and $\omega(x) = 0$, we get $d\omega(x) = 0$.

To prove the uniqueness of any operator d satisfying 1), 2), 3) and 4) it is therefore enough to consider the case where X is a chart domain U of X. If (U, h) is a chart of X and $(x_1, \ldots, x_n)$ are the associated local coordinates then $\omega \in \mathcal{C}^1_r(U)$ can be written in the form

$$\omega = \sum_J a_J dx_J,$$

where for any $J = (j_1, \ldots, j_r) \in \mathbb{N}^r$ such that $j_1 < \cdots < j_r$, $dx_J = dx_{j_1} \wedge \cdots \wedge dx_{j_r}$ and $a_J \in \mathcal{C}^1(U)$.

Since d is linear and $d(dx_j) = 0$ on applying 4) we get

$$d\omega = \sum_J da_J \wedge dx_J$$

and d is therefore entirely determined by its value on functions.

As the operator d, if it exists, is local and unique, it is enough to show that it exists on any chart domain U. Moreover, by linearity it is enough to define d on $\mathcal{C}^1_r(U)$. Let (U, h) be a chart of X and let $(x_1, \ldots, x_n)$ be the associated local coordinates. If $\omega \in \mathcal{C}^1_r(U)$,

$$\omega = \sum_{\substack{J=(j_1,\ldots,j_r)\\ j_1<\cdots<j_r}} a_J dx_J \quad \text{on } U$$

then we set

$$d\omega = \sum_{\substack{J=(j_1,\ldots,j_r)\\ j_1<\cdots<j_r}} da_J \wedge dx_J = \sum_{\substack{J=(j_1,\ldots,j_r)\\ j_1<\cdots<j_r}} \sum_{j=1}^n \frac{\partial a_J}{\partial x_j} dx_j \wedge dx_J \quad \text{on } U.$$

It is easy to check that the map d satisfies 1), 2), 3) and 4). □

Corollary 5.3. *If* $2 \leqslant \ell \leqslant q$ *then* $d(d\omega) = 0$ *for any* $\omega \in \mathcal{C}^\ell_\bullet(X)$.

Proof. It will be enough to prove the result in a chart domain. Let (U, h) be a chart and let $(x_1, \ldots, x_n)$ be the associated local coordinates. If $\omega \in \mathcal{C}^\ell_r(X)$ then

$$\omega\big|_U = \sum_{\substack{J=(j_1,\ldots,j_r)\\ j_1<\cdots<j_r}} a_J dx_J$$

and

$$d\omega\big|_U = \sum_{\substack{J=(j_1,\ldots,j_r)\\ j_1<\cdots<j_r}} da_J \wedge dx_J.$$

Then

$$d(d\omega\big|_U) = \sum_{\substack{J=(j_1,\ldots,j_r)\\ j_1<\cdots<j_r}} (d(da_J) \wedge dx_J - da_J \wedge d(dx_J))$$

using 4) and the linearity of d. By property 3) of d, $d(da_J) = 0$ and as $d(dx_J) = d(dx_{j_1} \wedge \cdots \wedge dx_{j_r})$ it follows from 4) and 3) that $d(dx_J) = 0$. We finally get $d(d\omega\big|_U) = 0$. □

C. Pullback

Let X and Y be two differentiable manifolds and let $\mu : X \to Y$ be a $\mathcal{C}^1$ map. For any $x \in X$ the map μ induces a map $d\mu_x$ from T_xX to $T_{\mu(x)}Y$, defined as follows: $d\mu_x(\nu)$ is the class of the curve $\mu \circ \alpha$, where α represents $\nu \in T_xX$. Transposing, we get a map $\mu_x^* : \Lambda^r T^*_{\mu(x)}Y \to \Lambda^r T^*_x X$, defined by

$$\mu_x^*\omega(\nu_1, \dots, \nu_r) = \omega\big(d\mu_x(\nu_1), \dots, d\mu_x(\nu_r)\big), \quad \text{for any } \nu_1, \dots, \nu_r \in T_xX.$$

If ω is a differential form on an open set W in Y then we define a differential form $\mu^*\omega$ on $\mu^{-1}(W)$ by setting $(\mu^*\omega)(x) = \mu_x^*(\omega(\mu(x)))$ for any x in $\mu^{-1}(W)$. This is the *pullback* of ω under μ. We check that $\mu^* f = f \circ \mu$ for any function $f \in \mathcal{C}^1(Y)$ and, since the operator d is just the usual derivative in this case,

$$d(\mu^* f)(x) = d(f \circ \mu)(x) = df(\mu(x)) \circ d\mu(x) = \mu^*(df)(x).$$

Let (U, h) and (V, k) be two charts of X and Y respectively defining local coordinates $(x_1, \dots, x_n)$ and $(y_1, \dots, y_m)$. We assume that $\mu(U) \subset V$. Let ω be a differential form on an open set W in Y whose restriction to $V \cap W$ can be written in the form

$$\omega\big|_{V\cap W} = \sum_{\substack{J=(j_1,\dots,j_r)\\ j_1<\cdots<j_r}} b_J dy_{j_1} \wedge \cdots \wedge dy_{j_r}.$$

Then

$$\mu^*\omega = \sum_{\substack{J=(j_1,\dots,j_r)\\ j_1<\cdots<j_r}} b_J \circ \mu \; d(y_{j_1} \circ \mu) \wedge \cdots \wedge d(y_{j_r} \circ \mu).$$

Proposition 5.4. *Let X and Y be two $\mathcal{C}^q$ differentiable manifolds and let μ be a $\mathcal{C}^q$ map from X to Y.*

i) *The pullback μ^* is a homomorphism of algebras from $\mathcal{C}^0_\bullet(Y)$ to $\mathcal{C}^0_\bullet(X)$ such that*

$$\mu^*(\mathcal{C}^\ell_r(Y)) \subset \mathcal{C}^\ell_r(X) \quad \text{for any } 0 \leqslant \ell < q \text{ and } r \geqslant 0.$$

ii) *μ^* commutes with the exterior derivatives d_X and d_Y on X and Y. More precisely, if ω is a $\mathcal{C}^\ell$ differential form on Y for some $1 \leqslant \ell < q$ then*

$$d_X(\mu^*\omega) = \mu^*(d_Y\omega).$$

iii) *If Z is a $\mathcal{C}^q$ differentiable manifold and $\lambda : Y \to Z$ is a $\mathcal{C}^q$ map then*

$$(\lambda \circ \mu)^* = \mu^* \circ \lambda^*.$$

Proof. i) follows from the explicit expression in local coordinates.

We get ii) by taking a chart domain V on Y. Writing $\omega \in \mathcal{C}^\ell_r(Y)$ in the form

$$\omega = \sum_{\substack{J=(j_1,\dots,j_r)\\ j_1<\cdots<j_r}} b_J dy_J \quad \text{on } V$$

we see that

$$\begin{aligned}
\mu^*\omega &= \sum_{\substack{J=(j_1,\ldots,j_r)\\ j_1<\cdots<j_r}} b_J \circ \mu \, d(y_{j_1} \circ \mu) \wedge \cdots \wedge d(y_{j_r} \circ \mu) \\
d_X(\mu^*\omega) &= \sum_{\substack{J=(j_1,\ldots,j_r)\\ j_1<\cdots<j_r}} d(b_J \circ \mu) \wedge d(y_{j_1} \circ \mu) \wedge \cdots \wedge d(y_{j_r} \circ \mu) \\
&= \sum_{\substack{J=(j_1,\ldots,j_r)\\ j_1<\cdots<j_r}} \mu^*(db_J) \wedge \mu^*(dy_{j_1}) \wedge \cdots \wedge \mu^*(dy_{j_r}) \\
&= \mu^*\Big(\sum_{\substack{J=(j_1,\ldots,j_r)\\ j_1<\cdots<j_r}} db_J \wedge dy_J \Big) \\
&= \mu^*(d_Y\omega)
\end{aligned}$$

since μ^* is an algebra homomorphism and ii) holds for functions.

We prove iii) by simply applying the definition of the pullback map. □

6 Integration of differential forms

In this section we will show how to integrate differential n-forms on an oriented differentiable manifold of dimension n.

A. Orientable manifolds

Definition 6.1. A $\mathcal{C}^q$ manifold X of dimension n is said to be *orientable* if there is a continuous differential form Ω of degree n on X which does not vanish on X. Two differential forms Ω_1 and Ω_2 define the same orientation on X if there is a continuous positive function f on X such that $\Omega_1 = f\Omega_2$. An orientable manifold on which we have chosen an orientation is said to be *oriented.*

Remark. The vector space $\Lambda^n T_x^*(X)$ has dimension 1 for any $x \in X$, so any continuous non-vanishing n-forms Ω_1 and Ω_2 on X differ by a continuous non-vanishing function on X. This function therefore has constant sign on any connected component of X. If X is orientable and connected there are therefore exactly two possible orientations on X.

If X is oriented by Ω then the system of local coordinates $(x_1, \ldots, x_n)$ associated to the chart (U, h) is said to be positively oriented if the differential form $dx_1 \wedge \cdots \wedge dx_n$ defines the same orientation as Ω,

Let $\mu : (X_1, \Omega_1) \to (X_2, \Omega_2)$ be a $\mathcal{C}^1$ diffeomorphism between two oriented differentiable manifolds X_1 and X_2. We say that μ is orientation preserving if $\mu^*\Omega_2 = f\Omega_1$ for some positive function f on X_1.

Examples.

1) If $(t_1, \ldots, t_n)$ are the usual coordinates on $\mathbb{R}^n$ then the differential n-form $dt_1 \wedge \cdots \wedge dt_n$ defines an orientation on $\mathbb{R}^n$.
2) A non-empty open set D in $\mathbb{R}^n$ is an oriented manifold with the orientation induced by the orientation on $\mathbb{R}^n$.
3) If $D \subset \mathbb{R}^n$ is an open set with $\mathcal{C}^k$ boundary for some $k \geqslant 1$ then the usual orientation on D induces an orientation on bD in the following way. Consider a point $x \in bD$ and let r be a $\mathcal{C}^k$ defining function for D in a neighbourhood U of x. In other words,

$$U \cap D = \{y \in U \mid r(y) < 0\} \quad \text{and} \quad dr(y) \neq 0 \text{ for any } y \in U \cap bD.$$

Restricting U, we can assume there is a positively oriented system of coordinates defined by a chart (U, h) such that $h = (r, x_2, \ldots, x_n)$. The $(n-1)$ last coordinates then form a system of local coordinates on $bD \cap U$. We orient $bD \cap U$ using the $(n-1)$-form $i^*(dx_2 \wedge \cdots \wedge dx_n)$, where i is the inclusion of $bD \cap U$ in U. This orientation is independent of the choice of defining function r on D and can be extended to the whole of bD using a partition of unity (cf. Chap. II, § 8).

If X is a $\mathcal{C}^q$ manifold of dimension n for some $q \geqslant 1$ and if (U_i, h_i) and (U_j, h_j) are two charts in X then

$$h_i \circ h_j^{-1} : h_j(U_i \cap U_j) \longrightarrow \mathbb{R}^n$$

is a $\mathcal{C}^q$ map. We set

$$d_{ij}(x) = \det[J(h_i \circ h_j^{-1})(h_j(x))] \quad \text{for any } x \in U_i \cap U_j,$$

where J is the Jacobian matrix.

Proposition 6.2. *The manifold X is orientable if and only if X has an atlas $\{(U_i, h_i)\}_{i \in I}$ such that $d_{ij}(x) > 0$ for every $i, j \in I$ and any $x \in U_i \cap U_j$.*

Proof. We can assume that X is connected. Let Ω be a continuous differential form of degree n on X which does not vanish on X. If (U_a, h_a) is a chart whose associated local coordinates are $(x_1, \ldots, x_n)$ then we set $\Omega_a = dx_1 \wedge \cdots \wedge dx_n$. This is a continuous n-form without zeros on U_a. For any $a \in X$ there is a chart (U_a, h_a) such that $a \in U_a$ and $\Omega = g_a \Omega_a$, where g_a is a continuous strictly positive function on U_a. (We simply replace $h_a(x) = (x_1, \ldots, x_n)$ by $(x_1, \ldots, x_{n-1}, -x_n)$ if necessary.) Moreover, $\Omega_a = d_{ab} \Omega_b$ on $U_a \cap U_b$, and it follows that $d_{ab} = \frac{g_b}{g_a} > 0$ on $U_a \cap U_b$.

Conversely, suppose that $\{(U_i, h_i)\}_{i \in I}$ is an atlas of X such that $d_{ij} > 0$ on $U_i \cap U_j$ for any $i, j \in I$. As above, we set $\Omega_i = dx_1 \wedge \cdots \wedge dx_n$, where the $(x_1, \ldots, x_n)$ are the local coordinates associated to the chart (U_i, h_i). Let $(\chi_i)_{i \in I}$ be a $\mathcal{C}^q$ partition of unity subordinate to the open cover $(U_i)_{i \in I}$ and set $\Omega = \sum_{i \in I} \chi_i \Omega_i$. The differential form Ω is $\mathcal{C}^{k-1}$ and has degree n on X. Moreover, for any $a \in X$, if I_a denotes the set of indices $i \in I$ such that $a \in \operatorname{supp} \chi_i$, then

$$\Omega(a) = \sum_{i \in I_a} \chi_i(a) \Omega_i(a) = \Big(\sum_{i \in I_a} \chi_i(a) d_{ii_0}(a) \Big) \Omega_{i_0}(a)$$

for any $i_0 \in I_a$. Since $\sum_{i \in I_a} \chi_i(a) = 1$, $d_{ii_0}(a) > 0$ for any $i \in I$ and $\chi_i \geqslant 0$, we have $\Omega(a) \neq 0$. □

B. Integration of differential forms

We start by defining integration of differential forms on open sets in $\mathbb{R}^n$ and then extend it to differentiable manifolds.

1) *Open sets in* $\mathbb{R}^n$. Let U be an open set in $\mathbb{R}^n$ and let $\eta \in \mathcal{C}^0_n(U)$ be a continuous differential form of degree n with compact support in U. There is then a unique continuous function f with compact support on U such that

$$\eta = f dx_1 \wedge \cdots \wedge dx_n.$$

We set

$$\int_U \eta = \int_U f dx_1 \cdots dx_n,$$

where the right-hand side is the integral of f on U with respect to Lebesgue measure on $\mathbb{R}^n$.

If W is an open set in $\mathbb{R}^n$ and $F : W \to F(W) = U$ is a $\mathcal{C}^1$ diffeomorphism then the change of variable formula for the Lebesgue integral on $\mathbb{R}^n$ says that

$$\int_{F(W)} f(x) dx_1 \cdots dx_n = \int_W f(F(t)) |\det(dF(t))| dt_1 \cdots dt_n.$$

If F is orientation-preserving then

$$\det(dF(t)) > 0 \quad \text{for all } t \in W$$

and therefore

$$\begin{aligned}\int_W f(F(t)) |\det(dF(t))| dt_1 \cdots dt_n &= \int_W f \circ F dF_1 \wedge \cdots \wedge dF_n \\ &= \int_W F^*(f dx_1 \wedge \cdots \wedge dx_n)\end{aligned}$$

and hence $\int_{F(W)} \eta = \int_W F^*(\eta)$.

2) *Manifolds.* Let X be an oriented $\mathcal{C}^q$ differentiable manifold of dimension n and let (U, h) be a chart on X whose local system of coordinates $(x_1, \ldots, x_n)$ is positively oriented. If $\omega \in \mathcal{C}^0_n(U)$ is a continuous differential form of degree n with compact support on U then we set

$$\int_U \omega = \int_{h(U)} (h^{-1})^* \omega,$$

where the right-hand side is the integral of the form $(h^{-1})^*\omega$ on the open set $h(U)$ in $\mathbb{R}^n$ defined in 1).

We now check that this definition is independent of the choice of chart (U, h). Let (U, k) be another chart whose local system of coordinates is positively oriented. The map $F = h \circ k^{-1} : k(U) \to h(U)$ is then an orientation-preserving $\mathcal{C}^1$ diffeomorphism and by 1)

$$\int_{h(U)} (h^{-1})^*\omega = \int_{k(U)} F^*(h^{-1})^*\omega = \int_{k(U)} (h^{-1} \circ F)^*\omega = \int_{k(U)} (k^{-1})^*\omega$$

which proves that our definition of the integral is independent of the choice of chart.

We now consider the problem of integrating differential forms whose support is not contained in a chart domain.

Let X be an orientable $\mathcal{C}^q$ differentiable manifold of dimension n. By Proposition 6.2 there is then an atlas $\mathcal{U}$ of X whose chart domains are connected and have the property that for any two charts (U, k) and (U', h') in $\mathcal{U}$ we have $\det J(h' \circ h^{-1})(y) > 0$ whenever $y = h(x)$ for some $x \in U \cap U'$, where J denotes the Jacobian matrix. Suppose that X is connected and oriented and consider an atlas $\mathcal{U} = (U_i, h_i)_{i \in I}$ corresponding to the orientation on X. Let $(\chi_i)_{i\in I}$ be a partition of unity subordinate to the atlas $(U_i)_{i \in I}$. If ω is a continuous differential form of degree n with compact support in X then we set

$$\int_X \omega = \sum_{i \in I} \int_{U_i} \chi_i \omega_i.$$

It is easy to show that this definition is independent of the choice of the partition of unity and the atlas. If X is not connected then we set $\int_X \omega = \sum_{i\in I} \int_{X_i} \omega$, where the sets X_i are the connected components of X.

The expression $\int_X \omega$ defined above is called the *integral of the differential n-form ω on the oriented manifold X*. We note that if we change the orientation on X then the integral is multiplied by -1.

Remark. We have only defined the integral of continuous differential forms with compact support: it is clear that this definition can be extended to other classes of forms, such as forms with L^1 coefficients, as in the case of $\mathbb{R}^n$.

7 Stokes' theorem

We will not prove the most general version of Stokes' theorem in this section, but only the special case used in this book.

Theorem 7.1. *Let X be an oriented $\mathcal{C}^q$ differentiable manifold of dimension n for some $q \geqslant 2$, and let $D \Subset X$ be a relatively compact open set in X with $\mathcal{C}^1$ boundary. If $\omega \in \mathcal{C}^1_{n-1}(\overline{D})$ then*

$$\int_{bD} \omega = \int_D d\omega.$$

Remarks.

1) The regularity hypothesis on ω means that ω is defined and continuous on $\overline{D}$ and the coefficients of ω in the local system of coordinates associated to a chart (U, h) are $\mathcal{C}^1$ on $U \cap \overline{D}$. In other words, the partial derivatives of the coefficients of ω defined on $U \cap D$ can be extended continuously to $U \cap \overline{D}$.
2) The orientation of bD is assumed to be the orientation induced by the orientation of D. We set

$$\int_{bD} \omega = \int_{bD} i^*\omega,$$

where i is the inclusion of bD in X.

Proof (of Theorem 7.1). As the set $\overline{D}$ is compact, we can find a finite number of charts $(U_i, h_i)_{1 \leqslant i \leqslant \ell}$ of X such that $\overline{D} \subset \bigcup_{i=1}^{\ell} U_i$ and if $U_i \cap bD \neq \varnothing$ for some i then $h_i = (r, h_i')$, where $h_i' : U_i \to \mathbb{R}^{n-1}$ is defined by restricting the local coordinates on bD to $U_i \cap bD$ and $D \cap U_i = \{x \in U_i \mid -1 < r(x) < 0\}$. Let (χ_i) be a $\mathcal{C}^q$ partition of unity subordinate to the open cover $(U_i)_{1 \leqslant i \leqslant \ell}$. By linearity it is enough to prove that

$$\int_{bD \cap U_i} \chi_i \omega = \int_{D \cap U_i} d(\chi_i \omega) \quad \text{for any } 1 \leqslant i \leqslant \ell.$$

We start by considering the case where $bD \cap U_i \neq \varnothing$. The chart $(U_i \cap bD, \widetilde{h}_i = h_i'|_{bD \cap U_i})$ is then a positively oriented chart of bD. For ease of notation we write U instead of U_i, h instead of h_i and χ instead of χ_i.

We set $(h^{-1})^*(\chi\omega) = \sum_{j=1}^n g_j(t) dt_1 \wedge \cdots \wedge dt_{j-1} \wedge dt_{j+1} \wedge \cdots \wedge dt_n$, where the functions $g_j \in \mathcal{C}^1(U \cap \overline{D})$ have compact support. Then

$$(\widetilde{h}^{-1})^* i^*(\chi\omega) = g_1(0, t_2, \ldots, t_n) dt_2 \wedge \cdots \wedge dt_n$$

and $$(h^{-1})^* d(\chi\omega) = d((h^{-1})^* \chi\omega) = \sum_{j=1}^n (-1)^{j-1} \frac{\partial g_j}{\partial t_j} dt_1 \wedge \cdots \wedge dt_n.$$

As $h(D \cap U) \subset \{t \in \mathbb{R}^n \mid -1 < t_1 < 0\}$, we have

$$\int_{D\cap U} d(\chi\omega) = \sum_{j=1}^{n}(-1)^{j-1}\int_{\{t\in\mathbb{R}^n\mid -1<t_1<0\}} \frac{\partial g_j}{dt_j} dt_1 \cdots dt_n.$$

As the functions g_j have compact support, we have

$$\int_{\mathbb{R}} \frac{\partial g_j}{\partial t_j} dt_j = 0, \quad j = 2, \ldots, n \quad \text{and} \quad \int_{-1}^{0} \frac{\partial g_1}{\partial t_1} dt_1 = g(0, t_2, \ldots, t_n),$$

from which it follows that

$$\int_{D\cap U} d(\chi\omega) = \int_{\mathbb{R}^{n-1}} g_1(0, t_2, \ldots, t_n) dt_2 \cdots dt_n$$

$$= \int_{\widetilde{h}(bD\cap U)} (\widetilde{h}^{-1})^* i^*(\chi\omega) = \int_{bD\cap U} \chi\omega.$$

Assume now that $U_i \cap bD = \varnothing$. We can assume that $U_i \subset D$ and since $\int_{bD\cap U_i} \chi_i\omega = 0$ it is enough to prove that $\int_{U_i} d(\chi_i\omega) = 0$. Repeating the above calculations, we are led to integrate differentials of $\mathcal{C}^1$ functions with compact support in $\mathbb{R}^n$, where the index $j = 1$ now has the same behaviour as the others. It follows that $\int_{U_i} d(\chi_i\omega) = 0$. □

Remarks 7.2.

1) If X is a compact manifold and $X = D$ then $bD = \varnothing$ and $\int_X d\omega = 0$ for any form $\omega \in \mathcal{C}^1_{n-1}(X)$.
2) It is easy to extend Stokes' theorem to domains with piecewise $\mathcal{C}^1$ boundaries in the following way. There is a finite cover $\{U_1, \ldots, U_\ell\}$ of bD by open sets in X and there are functions $r_i \in \mathcal{C}^1(U_i)$ such that

$$D \cap \Big(\bigcup_{i=1}^{\ell} U_i\Big) = \Big\{x \in \bigcup_{i=1}^{\ell} U_i \mid r_i(x) < 0, \forall\, x \in U_i, \forall\, i = 1, \ldots, \ell\Big\}$$

and $dr_{i_1} \wedge \cdots \wedge dr_{i_\nu} \neq 0$ on $U_{i_1} \wedge \cdots \wedge U_{i_\nu}$ for every subset $\{i_1, \ldots, i_\nu\}$ in $\{1, \ldots, \ell\}$. We set $\Sigma_i = \{x \in U_i \mid r_i(x) = 0\}$ and $S_i = \Sigma_i \cap bD$: the sets $\overset{\circ}{S}_i$ are then manifolds and $bD = \bigcup_{i=1}^{\ell} S_i$. If we set $\int_{bD} \omega = \sum_{i=1}^{\ell} \int_{\overset{\circ}{S}_i} \omega$ then Stokes' formula still holds.

B

Sheaf theory

This appendix contains the sheaf theoretic results needed to understand the proof of the Dolbeault isomorphism in Chapter V.

Definition 1. Let X be a topological space. A *presheaf* $\mathcal{F}$ on X is given by the following data:

i) a non-empty set $\mathcal{F}(U)$ for every open set U in X,
ii) a restriction map

$$\rho_{UV} : \mathcal{F}(V) \longrightarrow \mathcal{F}(U)$$

for every pair (U, V) of open sets in X such that $U \subset V$ which has the properties that

a) for any open set U in X $\rho_{UU} = I$,
b) whenever $U \subset V \subset W$ we have $\rho_{UW} = \rho_{UV} \circ \rho_{VW}$.

If $s \in \mathcal{F}(V)$ we will often simply write $s|_V$ for $\rho_{UV} s$.

If the sets $\mathcal{F}(U)$ are abelian groups (resp. rings) and the maps ρ_{UV} are group (resp. ring) homomorphisms, then the presheaf $\mathcal{F}$ is a *presheaf of abelian groups* (resp. of rings). In this case $\mathcal{F}(\varnothing) = \{0\}$.

If $\mathcal{F}$ and $\mathcal{G}$ are presheaves of abelian groups (or rings) then a *morphism of presheaves* $\varphi : \mathcal{F} \to \mathcal{G}$ is a collection of homomorphisms $\varphi_U : \mathcal{F}(U) \to \mathcal{G}(U)$ which commute with restriction maps, i.e. which have the property that $\rho^{\mathcal{G}}_{UV} \circ \varphi_V = \varphi_U \circ \rho^{\mathcal{F}}_{UV}$.

Definition 2. A presheaf $\mathcal{F}$ is a *sheaf* if and only if it satisfies the following gluing axioms.

(R_1) Consider elements $s_1, s_2 \in \mathcal{F}(U)$. If the set U can be written in the form $U = \bigcup_{i \in I} U_i$ and $\rho_{U_i U} s_1 = \rho_{U_i U} s_2$ for any $i \in I$ then $s_1 = s_2$.

(R_2) If $U = \bigcup_{i \in I} U_i$ and for any $i \in I$ there is an $s_i \in \mathcal{F}(U_i)$ such that the compatibility conditions

$$\rho_{(U_i \cap U_j) U_i} s_i = \rho_{(U_i \cap U_j) U_j} s_j$$

hold then there is an $s \in \mathcal{F}(U)$ such that $\rho_{U_i U} s = s_i$.

Examples. If X is a complex analytic manifold then $\mathcal{O}$, $\mathcal{C}^{\alpha}_{p,q}$, $0 \leqslant \alpha \leqslant \infty$ and $\mathcal{D}'_{p,q}$ are all sheaves on X.

A. Čech cohomology of a sheaf $\mathcal{F}$

Let X be a topological space, let $\mathcal{F}$ be a sheaf of abelian groups on X and let $\mathcal{U} = (U_i)_{i \in I}$ be an open cover of X. For any $p \in \mathbb{N}$ we denote elements of I^{p+1} by $\alpha = (\alpha_0, \ldots, \alpha_p)$ and for any such α we set $U_\alpha = U_{\alpha_0} \cap \cdots \cap U_{\alpha_p}$. A *p-cochain* c of the open cover $\mathcal{U}$ with values in $\mathcal{F}$ is a map associating an element $c_\alpha \in \mathcal{F}(U_\alpha)$ to any $\alpha \in I^{p+1}$ which is an alternating function of α. The set $C^p(\mathcal{U}, \mathcal{F})$ of p-cochains of $\mathcal{U}$ with values in $\mathcal{F}$ is equipped with an abelian group structure induced by the abelian group structure on $\mathcal{F}$.

We define a *coboundary* operation $\delta^p : C^p(\mathcal{U}, \mathcal{F}) \to C^{p+1}(\mathcal{U}, \mathcal{F})$ by

$$(\delta^p c)_\alpha = \sum_{j=0}^{p+1} (-1)^j c_{\alpha_0 \cdots \widehat{\alpha}_j \cdots \alpha_{p+1}}\big|_{U_\alpha},$$

where the notation $\widehat{\alpha}_j$ means the index α_j has been suppressed. We set $C^p(\mathcal{U}, \mathcal{F}) = 0$ and $\delta^p = 0$ whenever $p < 0$. In degrees 0 and 1 for example,

$$p = 0,\ c = (c_\alpha),\ (\delta^0 c)_{\alpha\beta} = (c_\beta - c_\alpha)\big|_{U_{\alpha\beta}}$$
$$p = 1,\ c = (c_{\alpha_\beta}),\ (\delta^1 c)_{\alpha\beta\gamma} = (c_{\beta\gamma} - c_{\alpha\gamma} + c_{\alpha\beta})\big|_{U_{\alpha\beta\gamma}}.$$

It is an easy exercise for the reader that $\delta^{p+1} \circ \delta^p = 0$.

Let $Z^p(\mathcal{U}, \mathcal{F}) = \{c \in C^p(\mathcal{U}, \mathcal{F}) \mid \delta^p c = 0\}$ be the group of *p-cocycles* of $\mathcal{F}$ and let

$$E^p(\mathcal{U}, \mathcal{F}) = \{\delta^{p-1} c \mid c \in C^{p-1}(\mathcal{U}, \mathcal{F})\}$$

be the group of *p-coboundaries* of $\mathcal{F}$. $E^p(\mathcal{U}, \mathcal{F})$ is then a subgroup of $Z^p(\mathcal{U}, \mathcal{F})$. We introduce the quotient group

$$H^p(\mathcal{U}, \mathcal{F}) = Z^p(\mathcal{U}, \mathcal{F}) / E^p(\mathcal{U}, \mathcal{F})$$

which we call the pth *Čech cohomology group* of $\mathcal{F}$ with respect to $\mathcal{U}$. We note that if c is a 0-cocycle then $c_\alpha - c_\beta = 0$ in $U_\alpha \cap U_\beta$ for any α and β. The family $(c_\alpha)_{\alpha \in I}$ then defines an element $s \in \mathcal{F}(X)$ such that $s\big|_{U_\alpha} = c_\alpha$ and hence

$$H^0(\mathcal{U}, \mathcal{F}) = \mathcal{F}(X).$$

Let $\mathcal{V} = (V_j)_{j \in J}$ now be another cover of X which is a refinement of $\mathcal{U}$ – or in other words, such that there is a map $\rho : J \to I$ such that $V_j \subset U_{\rho(j)}$ for every $j \in J$. We can then define a map $\rho^\bullet : C^\bullet(\mathcal{U}, \mathcal{F}) \to C^\bullet(\mathcal{V}, \mathcal{F})$ by setting

$$(\rho^p c)_{\alpha_0 \cdots \alpha_p} = c_{\rho(\alpha_0) \cdots \rho(\alpha_p)}\big|_{V_{\alpha_0 \cdots \alpha_p}}.$$

It is clear that this map commutes with δ and therefore defines a map $\rho^* : H^p(\mathcal{U}, \mathcal{F}) \to H^p(\mathcal{V}, \mathcal{F})$. Note that the map ρ^* is independent of the choice of ρ.

Indeed, if ρ' is another map with the same properties as ρ then the maps $\rho^\bullet$ and $\rho'^\bullet$ are homotopic. We define a map $h^p : C^p(\mathcal{U},\mathcal{F}) \to C^{p-1}(\mathcal{V},\mathcal{F})$ by

$$(h^p c)_{\alpha_0\cdots\alpha_{p-1}} = \sum_{0\leqslant j\leqslant p-1} (-1)^j c_{\rho(\alpha_0)\cdots\rho(\alpha_j)\rho'(\alpha_j)\cdots\rho'(\alpha_{p-1})}\big|_{V_{\alpha_0\cdots\alpha_{p-1}}}.$$

A straightforward calculation gives us the homotopy formula

$$\delta^{p-1}\circ h^p + h^{p+1}\delta^p = \rho'^p - \rho^p.$$

It follows that if c is a cocycle then $\rho'^\bullet c - \rho^\bullet c = \delta h^\bullet c$ is a coboundary and hence ρ^* is equal to ρ'^*.

We can think of an open cover of X as being a subset of $\mathcal{P}(X)$. The family of all open covers of X is then a set and we can consider the inductive limit of the groups $H^p(\mathcal{U},\mathcal{F})$ with respect to the maps ρ^*.

The pth *Čech cohomology group* of $\mathcal{F}$ on X, $H^p(X,\mathcal{F})$, is the inductive limit

$$H^p(X,\mathcal{F}) = \varinjlim_{\mathcal{U}} H^p(\mathcal{U},\mathcal{F}),$$

where $\mathcal{U}$ runs over the set of open covers of X. More precisely, this means that the elements of $H^p(X,\mathcal{F})$ are equivalence classes in the disjoint union of the groups $H^p(\mathcal{U},\mathcal{F})$, where an element in $H^p(\mathcal{U},\mathcal{F})$ and an element in $H^p(\mathcal{V},\mathcal{F})$ are identified if their images are equal in some $H^p(\mathcal{W},\mathcal{F})$, where $\mathcal{W}$ is an open cover of X which is a refinement of both $\mathcal{U}$ and $\mathcal{V}$.

Proposition 3. *Let X be a $\mathcal{C}^\infty$ differentiable manifold and let $\mathcal{E}$ be the sheaf of $\mathcal{C}^\infty$ germs of functions on X. If $\mathcal{F}$ is a sheaf of $\mathcal{E}$-modules on X then $H^p(\mathcal{U},\mathcal{F}) = 0$ for any $p > 0$ and any open cover $\mathcal{U}$ on X. In particular, $H^p(X,\mathcal{F}) = 0$ for any $p > 0$.*

Proof. Consider a cocycle $c \in Z^p(\mathcal{U},\mathcal{F})$ and let $(\varphi_i)_{i\in I}$ be a $\mathcal{C}^\infty$ partition of unity subordinate to $\mathcal{U}$. For any $\alpha \in I^p$ we set

$$c'_\alpha = \sum_{i\in I} \varphi_i c_{i\alpha}.$$

It is clear that $c' \in C^{p-1}(\mathcal{U},\mathcal{F})$ because $\mathcal{F}$ is a sheaf of $\mathcal{E}$-modules, and, for any $\alpha \in I^{p+1}$,

$$\begin{aligned}(\delta^{p-1}c')_\alpha &= \sum_{i\in I}\varphi_i \sum_{j=0}^{p}(-1)^j c_{i\alpha_0\cdots\widehat{\alpha}_j\cdots\alpha_p} = \sum_{i\in I}\varphi_i\big(c_\alpha - (\delta^p c)_{i\alpha_0\cdots\alpha_p}\big)\\ &= \sum_{i\in I}\varphi_i c_\alpha = c_\alpha\end{aligned}$$

because c is a cocycle. This completes the proof of the proposition. □

B. The long exact cohomology sequence

Let $\mathcal{F}, \mathcal{G}$ and $\mathcal{H}$ be three sheaves of abelian groups on X and let φ and ψ be sheaf morphisms such that the sequence

$$0 \longrightarrow \mathcal{F} \longrightarrow \mathcal{G} \longrightarrow \mathcal{H} \longrightarrow 0$$

is exact. In other words, φ is injective, ψ is surjective and the image of φ is equal to the kernel of ψ.

This exact sequence defines an exact sequence

$$0 \longrightarrow C^p(\mathcal{U}, \mathcal{F}) \longrightarrow C^p(\mathcal{U}, \mathcal{G}) \longrightarrow C^p(\mathcal{U}, \mathcal{H})$$

but the last map is not surjective in general. Let $C^p_{\mathcal{G}}(\mathcal{U}, \mathcal{H})$ be the image of $C^p(\mathcal{U}, \mathcal{G})$ in $C^p(\mathcal{U}, \mathcal{H})$. It is easy to see that δ sends $C^p_{\mathcal{G}}(\mathcal{U}, \mathcal{H})$ to $C^{p+1}_{\mathcal{G}}(U, \mathcal{H})$. We have therefore defined a complex of liftable cochains and we let $H^p_{\mathcal{G}}(\mathcal{U}, \mathcal{H})$ be the associated cohomology groups. We therefore have an exact sequence

$$0 \longrightarrow C^p(\mathcal{U}, \mathcal{F}) \longrightarrow C^p(\mathcal{U}, \mathcal{G}) \longrightarrow C^p_{\mathcal{G}}(U, \mathcal{H}) \longrightarrow 0.$$

The following theorem then follows immediately from the snake lemma, which is presented at the end of this appendix.

Theorem 4. *There is a connecting morphism*

$$\delta^* : H^p_{\mathcal{G}}(\mathcal{U}, \mathcal{H}) \longrightarrow H^{p+1}(\mathcal{U}, \mathcal{F})$$

and a long exact sequence

$$0 \longrightarrow H^0(\mathcal{U}, \mathcal{F}) \xrightarrow{\varphi^*} H^0(\mathcal{U}, \mathcal{G}) \xrightarrow{\psi^*} H^0_{\mathcal{G}}(\mathcal{U}, \mathcal{H}) \xrightarrow{\delta^*} H^1(\mathcal{U}, \mathcal{F})$$
$$\xrightarrow{\varphi^*} H^1(\mathcal{U}, \mathcal{G}) \xrightarrow{\psi^*} H^1_{\mathcal{G}}(\mathcal{U}, \mathcal{H}) \xrightarrow{\delta^*} H^2(\mathcal{U}, \mathcal{F}) \longrightarrow \cdots,$$

where φ^* and ψ^* are the natural maps induced by φ and ψ.

If $\mathcal{V} = (V_j)_{j \in J}$ is a refinement of $\mathcal{U}$ and $\rho : J \to I$ is a map such that $V_j \subset U_{\rho(j)}$ for any $j \in J$ then we can define as above a map

$$\rho^* : H^p_{\mathcal{G}}(\mathcal{U}, \mathcal{H}) \longrightarrow H^p_{\mathcal{G}}(\mathcal{V}, \mathcal{H})$$

which is independent of the choice of ρ.

Theorem 5. *If X is paracompact (i.e. if X is Hausdorff and any open cover of X has a locally finite refinement) and*

$$0 \longrightarrow \mathcal{F} \longrightarrow \mathcal{G} \longrightarrow \mathcal{H} \longrightarrow 0$$

is an exact sequence of sheaves then there is a long exact sequence

$$0 \longrightarrow H^0(X, \mathcal{F}) \longrightarrow H^0(X, \mathcal{G}) \longrightarrow H^0(X, \mathcal{H}) \longrightarrow H^1(X, \mathcal{F}) \longrightarrow$$
$$\longrightarrow H^1(X, \mathcal{G}) \longrightarrow H^1(X, \mathcal{H}) \longrightarrow H^2(X, \mathcal{F}) \longrightarrow \cdots$$

which is the inductive limit over open covers $\mathcal{U}$ of X of the exact sequences of Theorem 4.

Proof. It is enough to prove that the natural map

$$\varinjlim_{\mathcal{U}} H^p_{\mathcal{G}}(\mathcal{U}, \mathcal{H}) \longrightarrow \varinjlim_{\mathcal{U}} H^p(\mathcal{U}, \mathcal{H})$$

is an isomorphism. To do this we will show that any cochain of $\mathcal{H}$ can be lifted to $\mathcal{G}$ upon passing to a refinement.

Lemma 6. *For any $c \in C^p(\mathcal{U}, \mathcal{H})$ there is an open cover $\mathcal{V} = (V_j)_{j \in J}$ which is a refinement of $\mathcal{U}$ and a map $\rho : J \to I$ such that $\rho^p c \in C^p_{\mathcal{G}}(\mathcal{V}, \mathcal{H})$.*

Proof. Since X is paracompact we can assume that $\mathcal{U}$ is locally finite. We can choose an open cover $(W_i)_{i \in I}$ such that $\overline{W}_i \subset U_i$ for any i. For any $x \in X$ we choose an open neighbourhood V_x of x such that

i) $V_x \subset W_i$ for any $x \in W_i$,
ii) if $x \in U_i$ or $V_x \cap W_i \neq \varnothing$ then $V_x \subset U_i$,
iii) if $x \in U_\alpha$ for some $\alpha \in I^{p+1}$ then $c_\alpha \in C^p(U_\alpha, \mathcal{H})$ can be lifted to $\mathcal{G}(V_x)$.

Such a neighbourhood V_x exists since by definition of sheaf morphisms $\mathcal{H}$ can be lifted locally to a section of $\mathcal{G}$ and since x is only contained in a finite number of the sets W_i (resp. U_i) we only need to lift a finite number of sections and hence i) and ii) only represent a finite number of conditions.

Now choose $\rho : X \to I$ such that $x \in W_{\rho(x)}$ for any x. Condition i) then implies that $V_x \subset W_{\rho(x)}$ so the open cover $\mathcal{V} = (V_x)_{x \in X}$ is a refinement of $\mathcal{U}$. If $V_{x_0 \cdots x_p} \neq \varnothing$ then

$$V_{x_0} \cap W_{\rho(x_j)} \supset V_{x_0} \cap V_{x_j} \neq \varnothing \quad \text{for any } 0 \leqslant j \leqslant p$$

and hence $V_{x_0} \subset U_{\rho(x_0) \cdots \rho(x_p)}$ by ii). Condition iii) implies that the section $c_{\rho(x_0) \cdots \rho(x_p)}$ can be lifted to $\mathcal{G}(V_{x_0})$ and in particular it can be lifted to $\mathcal{G}(V_{x_0 \cdots x_p})$. It follows that $\rho^p c$ can be lifted to $\mathcal{G}$. □

C. The snake lemma

A complex of abelian groups $(K^\bullet, d)$ is a sequence

$$K^0 \xrightarrow{d^0} K^1 \xrightarrow{d^1} K^2 \longrightarrow \cdots \longrightarrow K^q \xrightarrow{d^q} K^{q+1} \longrightarrow \cdots,$$

where the terms K^q are abelian groups and the mappings d^q are group homomorphisms such that $d^{q+1} \circ d^q = 0$. The cohomology groups associated to the complex $(K^\bullet, d)$ are defined by $H^q(K^\bullet) = \ker d^q / \operatorname{Im} d^{q-1}$.

A morphism φ from the complex $(K^\bullet, d)$ to the complex $(L^\bullet, \delta)$ is a sequence $(\varphi_q)_{q \in \mathbb{N}}$ of group homomorphisms $\varphi^q : K^q \to L^q$ satisfying the commutation relations

$$\varphi^{q+1} \circ d^q = \delta^q \circ \varphi^q.$$

It follows that $\varphi^q(\ker d^q) \subset \ker \delta^q$ and $\varphi^q(\operatorname{Im} d^{q-1}) \subset \operatorname{Im} \delta^{q-1}$, and hence φ induces a homomorphism $\widetilde{\varphi}^q : H^q(K^\bullet) \to H^q(L^\bullet)$ for every q.

Lemma 7. *Let*

$$0 \longrightarrow K^\bullet \xrightarrow{\varphi} L^\bullet \xrightarrow{\psi} M^\bullet \longrightarrow 0$$

be a short exact sequence of complexes of abelian groups. There is then a connecting homomorphism

$$\vartheta^q : H^q(M^\bullet) \longrightarrow H^{q+1}(K^\bullet)$$

such that the long cohomology sequence

$$0 \longrightarrow H^0(K^\bullet) \longrightarrow H^0(L^\bullet) \longrightarrow H^0(M^\bullet) \xrightarrow{\vartheta^0} H^1(K^\bullet) \longrightarrow$$
$$\longrightarrow H^1(L^\bullet) \longrightarrow H^1(M^\bullet) \xrightarrow{\vartheta^1} H^2(K^\bullet) \longrightarrow \cdots$$

is exact.

Moreover, for any commutative diagram of short exact sequences of complexes of abelian groups

$$\begin{array}{ccccccccc} 0 & \longrightarrow & K^\bullet & \longrightarrow & L^\bullet & \longrightarrow & M^\bullet & \longrightarrow & 0 \\ & & \downarrow & & \downarrow & & \downarrow & & \\ 0 & \longrightarrow & \widehat{K}^\bullet & \longrightarrow & \widehat{L}^\bullet & \longrightarrow & \widehat{M}^\bullet & \longrightarrow & 0, \end{array}$$

the associated diagram of long exact cohomology sequences

$$\begin{array}{ccccccccc} \longrightarrow & H^q(K^\bullet) & \longrightarrow & H^q(L^\bullet) & \longrightarrow & H^q(M^\bullet) & \xrightarrow{\vartheta^q} & H^{q+1}(K^\bullet) & \longrightarrow \\ & \downarrow & & \downarrow & & \downarrow & & \downarrow & \\ \longrightarrow & H^q(\widehat{K}^\bullet) & \longrightarrow & H^q(\widehat{L}^\bullet) & \longrightarrow & H^q(\widehat{M}^\bullet) & \xrightarrow{\vartheta^q} & H^{q+1}(\widehat{K}^\bullet) & \longrightarrow \end{array}$$

is commutative.

Proof. We start by constructing the connecting homomorphism ϑ. Consider the following commutative diagram whose lines are exact:

$$\begin{array}{ccccccccc} & & \downarrow & & \downarrow & & \downarrow & & \\ 0 & \longrightarrow & K^q & \xrightarrow{\varphi^q} & L^q & \xrightarrow{\psi^q} & M^q & \longrightarrow & 0 \\ & & \downarrow{\scriptstyle d^q} & & \downarrow{\scriptstyle \delta^q} & & \downarrow{\scriptstyle \gamma^q} & & \\ 0 & \longrightarrow & K^{q+1} & \xrightarrow{\varphi^{q+1}} & L^{q+1} & \xrightarrow{\psi^{q+1}} & M^{q+1} & \longrightarrow & 0 \\ & & \downarrow{\scriptstyle d^{q+1}} & & \downarrow{\scriptstyle \delta^{q+1}} & & \downarrow{\scriptstyle \gamma^{q+1}} & & \\ 0 & \longrightarrow & K^{q+2} & \xrightarrow{\varphi^{q+2}} & L^{q+2} & \xrightarrow{\psi^{q+2}} & M^{q+2} & \longrightarrow & 0 \\ & & \downarrow & & \downarrow & & \downarrow & & \end{array}$$

If $m \in \ker \gamma^q$ represents the element $\{m\}$ in $H^q(M^\bullet)$ then $\vartheta\{m\} = \{k\} \in H^{q+1}(K)$ is the class obtained by the following construction:

$$\begin{array}{ccccc} & & \ell \in L^q & \xrightarrow{\psi^q} & m \in M^q \\ & & \downarrow \delta^q & & \downarrow \gamma^q \\ k \in K^{q+1} & \xrightarrow{\varphi^{q+1}} & \delta^q \ell \in L^{q+1} & \xrightarrow{\psi^{q+1}} & 0 \in M^{q+1} \end{array}$$

The element ℓ is chosen such that $\psi^q(\ell) = m$, which is possible since ψ^q is surjective. As $\psi^{q+1}(\delta^q \ell) = \gamma^q(m) = 0$ there is a unique element $k \in K^{q+1}$ such that $\varphi^{q+1}(k) = \delta^q \ell$ by the exactness of the line $q+1$. In fact, k is contained in $\ker d^{q+1}$ because φ^{q+2} is injective. Indeed,

$$\varphi^{q+2}(d^{q+1}k) = \delta^{q+1}(\varphi^{q+1}k) = \delta^{q+1}(\delta^q \ell) = 0 \Longrightarrow d^{q+1}k = 0.$$

To prove that ϑ^q is well defined it will be enough to show that the cohomology class of $\{k\}$ only depends on $\{m\}$ and not on the chosen representative m. Let $\widetilde{m}$ be another representative of m. Then $\widetilde{m} = m + \gamma^{q-1}\mu$. By the surjectivity of ψ^{q-1}, there is a $\lambda \in L^{q-1}$ such that $\psi^{q-1}(\lambda) = \mu$. Consider an element $\widetilde{\ell} \in L^q$ such that $\psi^q(\widetilde{\ell}) = m + d\mu = \psi^q(\ell + \delta^{q-1}\lambda)$. Since the line q is exact we have $\widetilde{\ell} = \ell + \delta^{q-1}\lambda + \varphi^q(\kappa)$ for some $\kappa \in K^q$ and hence $\delta^q\widetilde{\ell} = \delta^q \ell + \delta^q \varphi^q(\kappa) = \varphi^{q+1}(\widetilde{k})$, where $\widetilde{k} = k + d^q\kappa$ has the same cohomology class as k.

Let us now prove that this long cohomology sequence is exact. We prove first that $\ker \vartheta^q = \operatorname{Im} \widetilde{\psi}^q$. If $\{m\} \in \operatorname{Im} \widetilde{\psi}^q$ then we can choose m such that $m = \psi^q(\ell)$ and $\delta^q \ell = 0$. It then follows from the definition of ϑ^q that $\vartheta^q\{m\} = 0$. Conversely, if $\vartheta^q\{m\} = \{k\} = 0$ then $k = d^q\kappa$ and hence $\delta^q \ell = \varphi^{q+1}(k) = \varphi^{q+1}(d^q\kappa) = \delta^q(\varphi^q(\kappa))$ if $m = \psi^q(\ell)$. It follows that $\ell - \varphi^q(\kappa) \in \ker \delta^q$ and $m = \psi^q(\ell - \varphi^q(\kappa))$ and hence $\{m\} \in \operatorname{Im} \widetilde{\psi}^q$.

Let us now prove that $\operatorname{Im} \vartheta^q = \ker \widetilde{\varphi}^{q+1}$. Let $\{k\}$ be an element of $\ker \widetilde{\varphi}^{q+1}$. Then $\varphi^{q+1}(k) \in \operatorname{Im} \delta^q$ and there is therefore an ℓ such that $\varphi^{q+1}(k) = \delta^q \ell$. If $m = \psi^q(\ell)$, then $\{k\} = \vartheta^q\{m\}$ by definition of ϑ^q. The opposite inclusion is obvious by definition of ϑ^q.

The proof of the fact that $\operatorname{Im} \widetilde{\varphi}^q = \ker \widetilde{\psi}^q$ and the commutativity of the last diagram are left to the reader. □

C

Functional analysis

Throughout this section, E and F will be Banach spaces over $\mathbb{C}$.

Definition 1. A linear map T from E to F is a *compact operator* if for any bounded subset U in E the set $T(U)$ is relatively compact in F.

Remark. It follows immediately from the definition that any compact linear map is continuous.

We recall the classical properties of compact operators.

Theorem 2. *Let E be a Banach space and let T be a compact linear map from E to itself. Then*

i) $\dim \ker(I+T) < +\infty$.
ii) $\operatorname{Im}(I+T)$ *is a closed subspace of* E.
iii) $\dim \ker(I+T) = \dim(E/\operatorname{Im}(I+T))$.

The interested reader will find the proofs of these results in [Ru].

Definition 3. An *operator* from E to F is a linear map T defined on a subspace $\mathcal{D}(T)$ of E with image in F. The space $\mathcal{D}(T)$ is the *domain of definition* of T.

The *graph* $\mathcal{G}(T)$ of the operator T is the subspace of $E \times F$ consisting of pairs $(x, T(x))$, where x runs over the elements of $\mathcal{D}(T)$.

An operator T from E to F is said to be *closed* if the graph $\mathcal{G}(T)$ of T is a closed subspace of $E \times F$.

Remark. By the closed graph theorem an operator T from E to F is a continuous linear map from E to F if and only if $\mathcal{D}(T) = E$ and T is closed.

The following propositions are proved in Appendix 2 of [He/Le1]. We reprove them here for the reader's convenience.

Proposition 4. *Let T be a closed operator from E to F whose domain of definition is $\mathcal{D}(T)$. If $T(\mathcal{D}(T))$ has finite codimension in F then $T(\mathcal{D}(T))$ is a closed subspace of F.*

Proof. Let n be the codimension of the subspace $T(\mathcal{D}(T))$ in F. Choose a linear map $S : \mathbb{C}^n \to F$ such that $\operatorname{Im} S + T(\mathcal{D}(T)) = F$. Let T' be the operator from $E \oplus \mathbb{C}^n$ to F whose domain of definition is $\mathcal{D}(T') = \mathcal{D}(T) \oplus \mathbb{C}^n$ and which is defined by $T'(x) = T(x)$ for any $x \in \mathcal{D}(T) + \{0\}$ and $T'(x) = S(x)$ for any $x \in \{0\} \oplus \mathbb{C}^n$. The operator T' is closed and $T'(\mathcal{D}(T')) = F$.

We equip $\mathcal{D}(T')$ with the graph norm associated to T', i.e. for any $x \in \mathcal{D}(T')$ we set $\|x\|_{\Gamma_{T'}} = \|x\| + \|T'(x)\|$. The space $\mathcal{D}(T')$ is then a Banach space. Indeed, if $(x_n)_{n\in\mathbb{N}}$ is a Cauchy sequence in $(\mathcal{D}(T'), \| \ \|_{\Gamma_{T'}})$ then $(x_n)_{n\in\mathbb{N}}$ and $(T'(x_n))_{n\in\mathbb{N}}$ are Cauchy sequences in the Banach spaces $E \oplus \mathbb{C}^n$ and F. There are therefore points $x \in E \oplus \mathbb{C}^n$ and $y \in F$ such that $\lim_{n\to\infty} x_n = x$ and $\lim_{n\to\infty} T'(x_n) = y$. As the operator T' is closed $x \in \mathcal{D}(T')$ and $y = T(x)$ and it follows that x_n tends to x in $(\mathcal{D}(T'), \| \ \|_{\Gamma_{T'}})$.

Set $\ker T' = \{x \in \mathcal{D}(T') \mid T'(x) = 0\}$: this is a closed subspace of $(\mathcal{D}(T'), \| \ \|_{\Gamma_{T'}})$ because T' is a closed operator. The vector space $\mathcal{D}(T')/\ker T'$ is therefore a Banach space. Let $\widehat{T}'$ be the operator from $\mathcal{D}(T')/\ker T'$ to F induced by T': this is a bijection from $\mathcal{D}(T')/\ker T'$ to F which is continuous because $\widehat{T}'$ is a closed operator. The open mapping theorem then implies that its inverse $(\widehat{T}')^{-1}$ is a continuous linear operator from F to $\mathcal{D}'(T')/\ker T'$. Then

$$T(\mathcal{D}(T)) = \left((\widehat{T}')^{-1}\right)^{-1}(\mathcal{D}(T) + \{0\}/\ker T')$$

and $\mathcal{D}(T) + \{0\}/\ker T'$ is a closed subspace of $\mathcal{D}(T')/\ker T'$; the continuity of $(\widehat{T}')^{-1}$ then implies that $T(\mathcal{D}(T))$ is a closed subspace of F. □

Proposition 5. *Let T be a closed operator from E to F with domain of definition $\mathcal{D}(T)$. Assume that*

i) *$T(\mathcal{D}(T)) = F$.*
ii) *There is a continuous linear map S from F to E such that $\operatorname{Im} S \subset \mathcal{D}(T)$ and $I - TS$ is a compact linear operator from F to itself.*

There is then a continuous linear map $\widetilde{T}$ from F to E such that $\operatorname{Im} \widetilde{T} \subset \mathcal{D}(T)$ and $T\widetilde{T} = I$.

Proof. Set $K = TS - I$: by hypothesis K is a compact linear map from F to F and $TS = I + K$. By Theorem 2, i), $\ker TS$ is finite dimensional, so it has a topological complement G in F. In other words, G is a closed subspace of F and $F = G \oplus \ker TS$. By Theorem 2 ii) and iii), $\operatorname{Im} TS$ is a closed subspace of finite codimension in F, so it also has a topological complement H in F, where H is a closed vector space in F and $F = H \oplus \operatorname{Im} TS$. Moreover, $\dim H = \dim \ker TS$.

Since $T(\mathcal{D}(T)) = F$, we can construct a linear map $R : \ker TS \to \mathcal{D}(T)$ such that $TR(\ker TS) = H$ and $TR(x) \neq 0$ for any $x \neq 0$ in $\ker TS$. (If

$(e_1, \ldots, e_k)$ is a basis for $\ker TS$, $(h_1, \ldots, h_k)$ is a basis for H and ℓ_i, $i = 1, \ldots, k$, is an element of $\mathcal{D}(T)$ such that $T(\ell_i) = h_i$ then we simply set $R(e_i) = \ell_i$.) We then define a continuous linear map $\widetilde{S}$ from F to E by setting

$$\widetilde{S}(x) = S(x) \quad \text{if } x \in G$$
$$\widetilde{S}(x) = R(x) \quad \text{if } x \in \ker TS.$$

The map $T\widetilde{S}$ is then invertible and $\widetilde{T} = \widetilde{S}(T\widetilde{S})^{-1}$ is the map we seek. □

References

[Bo] S. BOCHNER – "Analytic and meromorphic continuation by means of Green's formula", *Ann. of Math. (2)* **44** (1943), p. 652–673.

[Br1] H.J. BREMERMANN – "Über die Äquivalenz der pseudokonvexen Gebiete und der Holomorphiegebiete im Raum von n komplexen Veränderlichen", *Math. Ann.* **128** (1954), p. 63–91.

[Br2] ———, "Complex convexity", *Trans. Amer. Math. Soc.* **82** (1956), p. 17–51.

[Ca/Th] H. CARTAN & P. THULLEN – "Zur Theorie der Singularitäten der Funktionen mehrerer komplexen Veränderlichen", *Math. Ann.* **106** (1932), no. 1, p. 617–647.

[Ci] E.M. ČIRKA – "Analytic representation of CR-functions", *Mat. Sb. (N.S.)* **98 (140)** (1975), no. 4(12), p. 591–623.

[Ci/St] E.M. ČIRKA & E.L. STOUT – "Removable singularities in the boundary", in *Contributions to complex analysis and analytic geometry*, Aspects Math., vol. E26, Vieweg, Braunschweig, 1994, p. 43–104.

[Do1] P. DOLBEAULT – "Formes différentielles et cohomologie sur une variété analytique complexe. I", *Ann. of Math. (2)* **64** (1956), p. 83–130.

[Do2] ———, "Formes différentielles et cohomologie sur une variété analytique complexe. II", *Ann. of Math. (2)* **65** (1957), p. 282–330.

[Eh] L. EHRENPREIS – "A new proof and an extension of Hartogs' theorem", *Bull. Amer. Math. Soc.* **67** (1961), p. 507–509.

[Gr] H. GRAUERT – "On Levi's problem and the imbedding of real-analytic manifolds", *Ann. of Math. (2)* **68** (1958), p. 460–472.

[Gr/Li] H. GRAUERT & I. LIEB – "Das Ramirezsche Integral und die Lösung der Gleichung $\overline{\partial} f = \alpha$ im Bereich der beschränkten Formen", *Rice Univ. Studies* **56** (1970), no. 2, p. 29–50.

[Gu] R.C. GUNNING – *Introduction to holomorphic functions of several variables*, Wadsworth & Brooks/Cole Advanced Books & Software, Pacific Grove, CA, 1990.

[Har] F. HARTOGS – "On boundaries of complex analytic varieties. I", *Münch. Ber.* **36** (1906), p. 223–242.

[Ha/La] F.R. HARVEY & H.B. LAWSON, JR. – "On boundaries of complex analytic varieties. I", *Ann. of Math. (2)* **102** (1975), no. 2, p. 223–290.

[He1] G.M. HENKIN – "Integral representation of functions which are holomorphic in strictly pseudoconvex regions, and some applications", *Mat. Sb. (N.S.)* **78 (120)** (1969), p. 611–632.

[He2] ———, "Integral representation of functions in strongly pseudoconvex regions, and applications to the $\overline{\partial}$-problem", *Mat. Sb. (N.S.)* **82 (124)** (1970), p. 300–308.

[He/Le1] G.M. HENKIN & J. LEITERER – *Theory of functions on complex manifolds*, Monographs in Mathematics, vol. 79, Birkhäuser Verlag, Basel, 1984.

[He/Le2] ———, *Andreotti–Grauert theory by integral formulas*, Progress in Mathematics, vol. 74, Birkhäuser Boston Inc., Boston, MA, 1988.

[Ho1] L. HÖRMANDER – "L^2 estimates and existence theorems for the $\overline{\partial}$ operator", *Acta Math.* **113** (1965), p. 89–152.

[Ho2] ———, *An introduction to complex analysis in several variables*, revised ed., North-Holland Publishing Co., Amsterdam, 1973.

[Ko/Ni] J.J. KOHN & L. NIRENBERG – "A pseudo-convex domain not admitting a holomorphic support function", *Math. Ann.* **201** (1973), p. 265–268.

[Ko] W. KOPPELMAN – "The Cauchy integral for differential forms", *Bull. Amer. Math. Soc.* **73** (1967), p. 554–556.

[Kr] S.G. KRANTZ – *Function theory of several complex variables*, John Wiley & Sons Inc., New York, 1982.

[Ky] A.M. KYTMANOV – *The Bochner-Martinelli integral and its applications*, Birkhäuser Verlag, Basel, 1995.

[Lan] S. LANG – *Algebra*, second ed., Addison-Wesley Publishing Company, Reading, MA, 1984.

[Lau] H.B. LAUFER – "On the infinite dimensionality of the Dolbeault cohomology groups", *Proc. Amer. Math. Soc.* **52** (1975), p. 293–296.

[L-T1] C. LAURENT-THIÉBAUT – "Produits de courants et formule des résidus", *Bull. Sci. Math. (2)* **105** (1981), no. 2, p. 113–158.

[L-T2] ———, "Sur l'extension des fonctions CR dans une variété de Stein", *Ann. Mat. Pura Appl. (4)* **150** (1988), p. 141–151.

[L-T/Le] C. LAURENT-THIÉBAUT & J. LEITERER – *Andreotti–Grauert theory on real hypersurfaces*, Quaderni, Scuola Normale Superiore di Pisa, 1995.

[Lel1] P. LELONG – "Les fonctions plurisousharmoniques", *Ann. Sci. École Norm. Sup. (3)* **62** (1945), p. 301–338.

[Lel2] ———, "La convexité et les fonctions analytiques de plusieurs variables complexes", *J. Math. Pures Appl. (9)* **31** (1952), p. 191–219.

[Lev] E.E. LEVI – "Studii siu punti singolari essenziali delle funzioni analitiche di due o piu variabili complesse", *Annali di Mat. (3)* **17** (1910), p. 61–87.

[Lu1] G. LUPACCIOLU – "Some global results on extension of CR-objects in complex manifolds", *Trans. Amer. Math. Soc.* **321** (1990), no. 2, p. 761–774.

[Lu2] ———, "Characterization of removable sets in strongly pseudoconvex boundaries", *Ark. Mat.* **32** (1994), no. 2, p. 455–473.

[Lu/To] G. LUPACCIOLU & G. TOMASSINI – "Un teorema di estensione per le CR-funzioni", *Ann. Mat. Pura Appl., IV. Ser.* **137** (1984), p. 257–263.

[Ma1] E. MARTINELLI – "Alcuni teoremi integrali per le funzioni analitiche di piu variabili complesse", *Mem. Accad. Ital.* **9** (1938), p. 269–283.

[Ma2] ———, "Sopra una dimostrazione di R. Fueter per un teorema di Hartogs", *Comment. Math. Helv.* **15** (1943), p. 340–349.

[Mi] J. MILNOR – *Morse theory*, Annals of Mathematics Studies, vol. 51, Princeton University Press, Princeton, N.J., 1963.

[Na1] R. NARASIMHAN – *Several complex variables*, Chicago Lectures in Mathematics, The University of Chicago Press, Chicago, Ill.–London, 1971.

[Na2] ______, *Analysis on real and complex manifolds*, second ed., Advanced Studies in Pure Mathematics, Masson & Cie, Éditeur, Paris, 1973.

[No] F. NORGUET – "Sur les domaines d'holomorphie des fonctions uniformes de plusieurs variables complexes. (Passage du local au global.)", *Bull. Soc. Math. France* **82** (1954), p. 137–159.

[Ok] K. OKA – *Collected papers*, Springer-Verlag, Berlin, 1984, Translated from the French by R. Narasimhan, With commentaries by H. Cartan, Edited by R. Remmert.

[Ram] E. RAMÍREZ DE ARELLANO – "Ein Divisionsproblem und Randintegraldarstellungen in der komplexen Analysis", *Math. Ann.* **184** (1970), p. 172–187.

[Ra] R.M. RANGE – *Holomorphic functions and integral representations in several complex variables*, Graduate Texts in Mathematics, vol. 108, Springer-Verlag, New York, 1986.

[Rh] G. DE RHAM – *Variétés différentiables. Formes, courants, formes harmoniques*, Actualités Scientifiques et Industrielles, vol. 1222b, Hermann, Paris, 1973.

[Ros] H. ROSSI – "The local maximum modulus principle", *Ann. of Math. (2)* **72** (1960), p. 1–11.

[Ru] W. RUDIN – *Functional analysis*, second ed., International Series in Pure and Applied Mathematics, McGraw-Hill Inc., New York, 1991.

[Sc] L. SCHWARTZ – *Théorie des distributions*, Hermann, Paris, 1966.

[Sl] Z. SŁODKOWSKI – "Analytic set-valued functions and spectra", *Math. Ann.* **256** (1981), no. 3, p. 363–386.

[Sto] G. STOLZENBERG – "Uniform approximation on smooth curves", *Acta Math.* **115** (1966), p. 185–198.

[St] E.L. STOUT – *The theory of uniform algebras*, Bogden & Quigley, Inc., Tarrytown-on-Hudson, N. Y., 1971.

[We] B.M. WEINSTOCK – "Continuous boundary values of analytic functions of several complex variables", *Proc. Amer. Math. Soc.* **21** (1969), p. 463–466.

Index of notation

Spaces of functions or differential forms

$H^{p,q}_{\Theta}(X)$	Dolbeault cohomology group with support in the family Θ	II.7
$H^{p,q}_{0,1/2}(\overline{D})$	Dolbeault cohomology group $Z^0_{p,q}(\overline{D})/E^{1/2}_{p,q}(D)$	VII.3
$H^q(X, F')$	Čech cohomology group of the sheaf F	V.4
$\Lambda^\alpha(D)$	space of Hölder continuous functions of order α on D	III.2
$\mathcal{O}(D)$	space of holomorphic functions on D	I.1
$\mathcal{O}(D, \mathbb{C}^m)$	space of holomorphic functions from D to $\mathbb{C}^m$	I.5
Ω^p	sheaf of germs of holomorphic p-forms	V.4
$\mathrm{PSH}(D)$	space of plurisubharmonic functions on D	VI.2
$Z^\alpha_{p,q}(X)$	space of $\overline{\partial}$-closed differential (p,q) forms with coefficients in $\mathcal{C}^\alpha(\lambda)$	V.4
$Z^{-\infty}_{p,q}(X)$	space of $\overline{\partial}$-closed currents on X	V.4
$Z^{p,q}(X)$	space of $\mathcal{C}^\infty$ differential (p,q)-forms which are $\overline{\partial}$-closed on X	II.7
$Z^{p,q}_{\Theta}(X)$	space of $\mathcal{C}^\infty$ differential (p,q)-forms which are $\overline{\partial}$-closed and whose support is contained in the family Θ	II.7

Sets

$B(\xi, r)$	ball of centre ξ and radius r in $\mathbb{C}^n$	III.1
$\widehat{K}_\Omega$	holomorphically convex hull of K with respect to Ω	VI.1
$\widehat{K}^\bullet_\Omega$	psh-convex hull of K with respect to Ω	VI.3
$P(\xi, r)$	polydisc of centre ξ and radius r	I.2
$\partial_0 P$	special boundary of the polydisc P	I.2
$SS(T)$	singular support of the current T	II.3
$T_x X$	tangent space to X at x	A.4
$T^*_x X$	cotangent space to X at x	A.3
$T^{1,0}_x(X)$	holomorphic tangent space to X at x	II.5
$T^{0,1}_x(X)$	antiholomorphic tangent space to X at x	II.5

Norms

$\Vert f\Vert_{k,D}$	the $\mathcal{C}^k$ norm on D	I.1
$\Vert f\Vert_D$	the uniform norm on D	I.1
$\vert f\vert_{\alpha,D}$	the Hölder norm of order α on D	III.2
$\vert z\vert$	the norm on $\mathbb{C}^n$	I.1

Other symbols

$\mathcal{K}(T, S)$	the Kronecker index of the currents T and S	II.3
$[Y]$	the integration current on Y	II.1

Index